THIRD WORLD URBANIZATION

THIRD WORLD URBANIZATION

EDITED BY

JANET ABU-LUGHOD
RICHARD HAY, JR.

METHUEN
NEW YORK LONDON
TORONTO

Library of Congress Catalog Card Number: 76-53367
ISBN: 0-416-60141-3
Manufactured in the United States of America
Designed by First Impression

First published by Maaroufa Press, Inc., 1977
Reprinted by Methuen, Inc., 1979, 1986

Published in U.S.A. by
Methuen, Inc.
29 West 35th Street
New York, N.Y. 10001

In Canada by
Methuen Publications
150 Laird Drive,
Toronto,
Ontario M4G 3V7

In the rest of the world by
Methuen & Co. Ltd.
11 New Fetter Lane,
London EC4P 4EE

List of Authors

Charles Abrams
Samir Amin
Giovanni Arrighi
Ronald H. Chilcote
Wayne A. Cornelius, Jr.
Friedrich Engels
Frank Furedi
Jacques Gernet
Richard Hay, Jr.
Mary Racelis Hollnsteiner
Lea Jellinek
Ibn Khaldun
Anthony Leeds
V. I. Lenin
Paul M. Lubeck
T. G. McGee
Niccolo Machiavelli
Karl Marx
Enzo Mingione
Alejandro Portes
Diego Robles Rivas
Bryan R. Roberts
Donald Francis Roy
D. Slater
Edward W. Soja
Richard J. Tobin
United Nations
Barbara Ward (Lady Jackson)

CONTENTS

PREFACE

Despite the growing significance of the Third World and the critical nature of its urbanization, there are few synthetic books covering more than one region of the Third World which can be used either by scholars seeking an overview of the process of world urbanization or by students in the growing number of courses now being offered in the field of comparative urbanism. This lack became frustratingly apparent two years ago when we introduced a new program in Comparative Urban Studies at Northwestern University and began to set up courses in the departments of sociology, anthropology, geography and political science to examine urban developments in Third World countries. Some earlier anthologies were out of print; others were dated, their information long superseded. But the most distressing problem was that the field of urbanization, particularly with reference to developing countries, seemed to us to have stagnated at theoretically-sterile conceptualizations or, even worse, had deteriorated into fragmented empirical-descriptive reports, whether observing with sympathy or "noting with alarm" the rapidly declining condition of individual cities.

This book attempts to rectify this deficiency. As we searched beyond the standard works on urbanization we found exciting literature reporting new types of theoretical approaches and analytical research being undertaken in all parts of the developing world—in Asia, Africa, and Latin America. We found a growing band of scholars, whether known to one another or not, who seemed to be reaching similar evaluations of very different situations in these three regions of the Third World. It is this new literature, often appearing in relatively obscure publications known only to area specialists, that we have sought to bring together in this anthology. We have ruthlessly edited these selections in order to create a volume which could be read with profit even by persons who were not fully familiar with the region under study and, by our editing, we have sought to create a coherent "line of reasoning" implicit in the ordering of the selections and the major "lessons" to be derived from each piece. In this, we have had to oversimplify the often subtle and complex thinking of individual authors. To them we apologize, and to our readers we commend the fuller original pieces, especially as they move beyond the "general" logic of this volume to more particular analyses of individual variations.

The editors have worked together for more than two years in the Program of Comparative Urban Studies at Northwestern. It is therefore difficult to determine where the work of one ends and the contribution of another begins. In sharing responsibility for the Program, and in participating together in the faculty and graduate student seminars and in an interdisciplinary undergraduate course on Cities of the Third World, we came to share a set of ideas and enthusiasms. As the book took shape, we found ourselves hunting for new answers to questions which had not even been in our minds when we began. We thoroughly enjoyed the ferment and excitement of the process and hope that the book conveys the spirit of this continuing search. It is certainly not intended to stand as a definitive conclusion to it.

The intellectual excitement we felt in preparing this book has been contagiously present throughout the two years of Northwestern's Comparative Urban Studies

Program. Funded initially by a seed-grant from the U.S. Office of Education, International Programs Office, in 1974–75, and again in 1975–76, the Program engaged the interest of over forty faculty members drawn from a variety of academic disciplines. It benefited from original research papers presented to a faculty-graduate student colloquium by some 12 speakers during 1974–75, from the enthusiastic research and controversy generated in several graduate level seminars (some for credit, some extra-curricular), and from numerous guest speakers who came from as far away as Australia to share their ideas with us.

The book in particular benefited enormously from the thirty students who served as willing guinea pigs the first time we offered our course on Cities of the Third World. The students read poorly-reproduced copies of unpublished papers, trudged to the reference room of the library to study sources that could not be circulated, spent long hours in the reserve reading room to share a single copy of some obscure source. They told us frankly what made sense to them, what sources their independent research supported and which ones it did not, and they kept after us with questions which forced us into more study and research. While we recognize that it is too late for them to benefit from this book, it is really a belated gift to them. It is the book from which we think we could have given a first-rate course; we hope that it will benefit other teachers and future students.

A preface is always a welcome chance to thank sponsors and colleagues—while also absolving them of our sins of commission and omission. This we do in a *non-perfunctory* way. We acknowledge our intellectual indebtedness to John Friedmann, School of Architecture and Urban Planning at the University of California, whose earlier program in Comparative Urban Studies and Planning helped inspire our own and whose publications introduced us to many of the authors included in this volume. John Walton, a friend and colleague from the Department of Sociology at Northwestern, shared with us both his knowledge of Latin American urbanism and his interests in urban theory. He introduced us to Professor Alejandro Portes of the Department of Sociology at Duke University and to Anthony Leeds, Professor of Anthropology at Boston University. The work of both these scholars has immeasurably aided us. Professor T. G. McGee of the National University of Australia also visited our program and thereby widened our horizons to Asia. He recommended a number of the readings we have included in this volume and explored with us the similarities and differences between the Asian case and urbanization elsewhere in the Third World.

The bibliography with which this volume concludes is a contracted version of two detailed annotated guides to the literature prepared by the Comparative Urban Studies Program and available from the Center for Urban Affairs. The Latin American materials were compiled and annotated by Priscilla Helm Walton and the African materials were assembled by Hugh Mullenbach. These regional references have been supplemented for this book by sources on the Asian city.

Selection of materials included in this volume was made jointly by myself as Director of the Program, Richard Hay, Teaching Assistant in the Program, and Hugh Mullenbach, Administrative Assistant in the Program. I bear primary responsibility for the introductions which precede the selections and for most of the condensations. The statistical chapter, including the generation of original tables on world urbanization, is the responsibility of Richard Hay, while Hugh Mullenbach bears primary responsibility for the bibliography. Although ill health forced Hugh's withdrawal from the project, we gratefully acknowledge his contributions.

We express appreciation to the Center for Urban Affairs of Northwestern University, with which our program has been associated, and especially to its director, Professor Louis Masotti, who was always supportive of our efforts. Both the Center and the Department of Sociology made available to us the typists who worked long, fast and accurately on the materials for this book. We extend a special thanks to Sue Vehlow, Donna Buckingham, Diane Eastman and Alice Murray for their help.

I would also like to thank the publisher of Maaroufa Press, Mr. T. Tieken, for catching me in a manic moment and thereby committing me to turn idle thoughts into action. His encouragement throughout helped us to sustain the delusion that others might find this book as fascinating as we do. And his assignment of Ms. Evie Righter as editor for the Press introduced us to a delightful person as well as a creative user of blue pencil. We thank her for help and her company.

Janet Abu-Lughod
Evanston, Illinois
December 2, 1976

The settlers' town is a strongly built town, all made of stone and steel. It is a brightly lit town; the streets are covered with asphalt, and the garbage cans swallow all the leavings, unseen, unknown and hardly thought about. The settler's feet are never visible, except perhaps in the sea; but there you're never close enough to see them. His feet are protected by strong shoes although the streets of his town are clean and even, with no holes or stones. The settler's town is a well-fed town, an easygoing town; its belly is always full of good things. The settlers' town is a town of white people, of foreigners.

The town belonging to the colonized people, or at least the native town, the Negro village, the medina, the reservation, is a place of ill fame, peopled by men of evil repute. They are born there, it matters little where or how; they die there, it matters not where, nor how. It is a world without spaciousness; men live there on top of each other, and their huts are built one on top of the other. The native town is a hungry town, starved of bread, of meat, of shoes, of coal, of light. The native town is a crouching village, a town on its knees, a town wallowing in the mire.

Reprinted from Frantz Fanon, *The Wretched of the Earth* (New York: Grove Press, Inc., 1965), p. 39, by permission of the publisher.

INTRODUCTION

There has been a revolution in the world within the past quarter of a century—a revolution no less important because it erupted only sporadically and at scattered points, each event seemingly a unique occurrence with its own set of causes and goals. The Third World has emerged from the shadowy position it occupied in the nineteenth century, when it was either the domain of or irrelevant to the western world of advanced capitalist development, and has become a force which increasingly challenges the balance of power "worked out" in the Northern hemisphere between the First (capitalist) and Second (socialist) Worlds. Demographically, it now represents the largest majority of mankind and is growing the fastest; economically, it contains the bulk of growing labor power and the largest future markets; and politically, it has won victory after victory in the United Nations where it has demanded a "New Economic Order," i.e., a redivision of the wealth. The Third World can no longer be ignored. It must be taken into account by scholars, politicians and citizens on this shrinking planet.

Gone forever are the days when the entire "nonwestern" world was viewed simplistically as an undifferentiated mass of societies which had "not yet learned to be western." In those halcyon days of western hegemony, the standard against which other societies were judged (and so often found wanting!) was the First World, that transatlantic core of Europe, the United States and Canada which rose to world power after the sixteenth century and which, by the nineteenth, controlled its destiny. The first World War—its very name an appropriate ethnocentric symbol since it involved only a small portion of the world—marked the beginning of the end. By the 1930s it was clear that universal involvement was underway, a fact confirmed in the 1940s by a war which did, indeed, involve most of the world. It was in the wake of that conflict that the term "nonwestern" was born. Under this new label, western scholars began to address the rest of mankind as a complex and changing agglomeration of newly independent nations who were to be helped across the difficult threshold of modernization. A congruence of interests between the developed and the so-called underdeveloped worlds was assumed, perhaps too naively.

The most recent phase of this changing relationship is now upon us. The assistance offered by the west is viewed with suspicion by many Third World countries who fear that this aid is intended to reestablish in economic terms the hegemony the west had forfeited in political terms through decolonization. Third World scholars and political leaders are increasingly critical of earlier western writings on "development" for too often locating the causes of underdevelopment in the culture of the Third World itself (blaming the victim) while ignoring historic and international factors which create external obstacles to development. Initiatives in defining the factors that account for underdevelopment are now being taken by the Third World itself. From Latin American and African scholars have come some important reformulations: the theory of "dependency" which suggests that the disparities between rich and poor, city and countryside, powerful and powerless within Third World countries have been fostered and sustained by the very process of their development in subordination to First World economic power; the theory of the "development of underdevelopment," which posits that underdevelopment is not so much a starting point *from* which countries evolve as it is the intended outcome of a process whereby they are integrated as exploited peons in the international economy (Amin, 1974; Frank, 1967). From a Caribbean scholar has come a parallel theory of the psychology of oppression, whereby the colonized victim is not only blamed for his own troubles but taught to accept and internalize this view of himself as incompetent (Fanon, 1963, 1965).

These reformulations often shock Americans into defensive postures. But the shock has forced a reconsideration of many of the accepted truisms about economic and social development, about the relationship between our own system and that of the rest of the world, and about the applicability of our own experience in development to countries elsewhere. Within the context of this reconsideration, we have begun to reformulate our understanding of the process of urbanization in relationship to development, of the nature of cities and their role in that process, of the very nature of urban life itself. Out of this reformulation can come both a deeper understanding of the situation as it exists and is perceived within the Third World and a deeper understanding of urban processes at work in our own society.

The purpose of this book is therefore two-fold. First and foremost, it is an introduction to urbanization in the Third World within a unified theoretical framework which we believe is powerful in illuminating many of the phenomena consistently found in the cities of Asia, Africa and Latin America. Second, it is an attempt to separate ourselves somewhat from the specific problems which beset American cities, in the hope that through an attempt to understand the pains, travails and protests of others, we can gain new insights into our own cities. We have attempted in this book to incorporate the views, evaluations and aspirations of the Third World and to present not only the multiple problems which face them in their growing cities but some of their attempts to deal with them.

By any set of standards, the topic of Third World urbanization is intrinsically significant. If indeed we wish to understand urban ways, we cannot afford to ignore or dismiss as "deviant" more than half of the world's largest cities, at least half of the world's urbanites. Of the 1,968 cities today with at least 100,000 inhabitants, slightly in excess of half are located in Asia, Africa and Latin America. Of the 1.5 billion residents living in big cities in the world, at least half live in cities of the Third World.* This situation will grow even more marked in the near future. Projections by the United Nations indicate that, by the year 2000, some 62 percent of the world's *urban* population will be in areas we call "developing" while only 38 percent will be in cities of the so-called developed world (U.N., 1969).

This situation has come into being only recently. In 1920, when the field of urban sociology was born, almost three-fourths of the world's urbanites lived in western cities, while only a little more than a fourth lived in the less-developed regions. At that time, perhaps understandably, urbanists focused upon the industrialized west; today this is no longer possible. The subject of this book, therefore, is central to the study

*Our calculations from 1975 U.N. data.

of urbanism and urbanization, since the kinds of cities described and analyzed in this volume are now increasingly *typical.*

Some have tried to defend the legitimacy of studying Third World cities in terms of the problems and potential dangers they represent. Thus, U.N. official Maurice Strong expressed a not atypical position when he singled out "fast-growing cities in the developing world as the places where a [world-wide epidemic] disaster was most likely to occur, . . . [a] situation . . . fraught with peril for *our* world" (*New York Times,* 4 June 1975, italics added). We do not share his alarm nor do we feel that our concern with such cities need be defended on the grounds of enlightened self-interest. However, it is important to acknowledge that the field of urban studies has always received an added impetus to attention whenever cities have been defined as "posing a danger."

During two periods in recent scholarly history we have witnessed a resurgence of interest in the structure and functions of cities and their attendant ways of life. Each of these periods was a time of special problems. The first occurred in the United States in the 1920s (i.e., the Chicago School of urban sociology) and seems, in retrospect, to have been related to the culminating (and cut-off) point of massive foreign immigration to United States cities—an immigration which had filled American cities with "strangers" whose integration was considered problematic, whose presence was considered "threatening." The second occurred in the 1950s and 1960s in the United States and seems to have been related to a similar culmination of a massive movement of Blacks from the rural south to the urban north and with similarly anxious responses. Some of the present concern with Third World cities seems to derive from these same fears and preoccupations. Whatever the motive, however, the net result of these three separate impulses to reexamine "the city" has been to decrease the degree of provincialism with which cities and their varied ways of life have been viewed. We are gradually overcoming some of the deficiencies which characterized the field of urban studies at earlier times, deficiencies which might be summarized as follows:

1. Studies were too often restricted in time and space. In general, the field focused almost exclusively upon a limited number of twentieth century American cities and attempted to formulate generalizations about "urban life" on the basis of too small and too narrow a sample. Because American cities had a relatively short history, time was understandably a neglected variable. Comparative study of the evolution of urban types and urban "cultures," which could have helped identify the peculiarities of contemporary American cities, was not central to the field, nor were cultural variations or the wider issues of international political and economic power brought into urban studies.
2. Studies too often made the faulty assumption that the city together with its immediate hinterland was a relatively self-contained object with independent internal dynamics of growth, organization and change. (See the critique of Castells, 1976.) This set of assumptions could not have been entertained had Third World cities been included among the objects of study.

3. Studies were too often conducted under the dominant paradigm of Adam Smith's invisible hand. The causes of urban growth, of particular urban patterns, of the distribution of land uses and population groups *within* cities, of the processes of change and transformation, were postulated as occurring at a subliminal "subconsensual" level of ecological competition for space and economic "niches." The variables of economic and political power, of legal codes, of the capitalist system of wage labor and capital investment in the built environment—these tended to be underplayed and under-researched. The operation of these variables is particularly clear in Third World cities. A study of them can help to correct previous tendencies to take these variables as constants.

We believe, therefore, that this book can serve both students interested in cities of the Third World and those whose immediate concerns are with American cities. After stretching their focus beyond the United States, the latter may return to American cities with a new understanding of the basic and underlying processes of urban life.

The readings selected for inclusion in this anthology have been organized as suggested answers to a set of logically ordered questions. The first section asks whether the issues now being raised concerning cities in the Third World are really new ones, without precedent in earlier times, or whether they are merely today's particular manifestations of recurrent issues addressed by earlier thinkers whose analyses may help our quest for understanding. We present some brief excerpts from the writings of a fourteenth century Tunisian (Ibn Khaldun), a fifteenth century Florentine (Niccolo Machiavelli), a nineteenth century Berliner (Karl Marx), and an early twentieth century Russian (V. I. Lenin). These readings have a remarkably contemporary ring. Most of the major inequities which command our attention today in studying urbanization in the Third World—the inequality between city and countryside, the asymmetry between colonialist and colonized, the gap between the bourgeoisie and the under- and unemployed "proto" proletariat, the shocking contrast between rich and poor—these have been issues of deep concern for hundreds of years. Especially remarkable is that the ideas of these authors are clear precursors of the "new" theories discussed in Section III.

The second section of the book poses a more concrete issue. While some of the contrasts and cleavages discussed in the introductory section appear to be universal and to predate the modern world, what is striking when one examines the historical record is that, over human history, the relative positions of the major protagonists have been reversed. Man's earliest urban empires were located not in the west but in those very parts of the world now considered "backward": the Indus Valley of Pakistan, the Yellow River Valley of China, and the Nile and Mesopotamian valleys of Egypt and Iraq. It was not until the sixteenth century, in fact, that the cities of northern Europe began to assume any importance and that the rest of the world began to occupy its subordinate position as "periphery" to the "core" centered in western Europe (Wallerstein, 1974).

A striking illustration of this point is found in Gernet's description of the Chinese city of Hangchow which in the thirteenth century was probably the largest in the world. But while Hangchow was particularly impressive, it was hardly unique in the Third World. One could easily have chosen similarly advanced urban centers in Africa and the Orient to demonstrate that the present contrast between developed and underdeveloped regions was also to be found in the Middle Ages, but then it was the west which occupied the position of underdeveloped.

How this reversal took place is, of course, the subject of many scholarly and detailed histories which lie beyond the scope of this book. But in the introduction to her new book, *The Home of Man,* Barbara Ward (Lady Jackson) succinctly and sharply summarizes some of the major features. She argues that European advances since the sixteenth century and the present hegemony of the west were made possible not only by *internal* developments in science and technology but also because Europe could draw freely, albeit by force, upon the resources and labor power of the peoples it subjugated; indeed, the draining of resources from the Third World and the export of European products to a docile and controlled market in the Third World were indispensable impetuses to European economic growth. Thus, to suggest that the Third World can "catch up" to the First through exclusively internal transformations is to miss the entire point of history. Yet the Third World cannot follow the same exploitative methods in its development that the west was able to utilize—simply because there is no one left to exploit.

Ward's piece describes the general status reversals that resulted from Europe's post-sixteenth century expansion. But that is too generalized a picture. Certainly, within Europe there were vast differences among countries in terms of the benefits they derived from colonies, and within the Third World colonialism had quite different impacts in different regions. In Asia the impact was chiefly economic and, except in Australia and New Zealand, culturally and demographically superficial. In Africa, there were enormous variations, since in some areas there were venerable civilizations and high levels of urbanization, whereas in others European settlements were imposed upon a virtual *tabula rasa.* The case in Latin America was quite different. There, a foreign model of urbanization, guided by uniform laws, was imposed by the Spanish upon a continent with little urban history (with the important exceptions of Mayan, Incan and Aztec culture areas). Cities newly founded along the receding frontier were critical in subduing the indigenous inhabitants and in expanding the zones of Spanish settlement and control. Some effects on contemporary cities in Latin America of these historical precedents are explored by Portes. These reveal why, in many ways, the cities of Latin America diverge from others in the Third World.

The section then turns to statistical and geographic data to show the present distribution of population and urbanization in the world today, utilizing the most recently compiled estimates of 1976 and including projections concerning the future. From these data it is quite clear that a new reversal of positions is already upon us. While as recently as the nineteenth century most large cities in the world were located in the developed world and most urbanites were westerners (Weber, 1965), at present one can already say that the majority of cities are in the Third World and the most "typical" (in the statistical sense) urbanite in the world lives in the kinds of cities discussed in this book. That this trend will continue and become intensified in the foreseeable future is also clearly evident from U.N. projections.

Earlier it was believed that such high rates of urbanization were healthy, since they indicated economic growth and progress. It was assumed that industrialization and urbanization went together. The fallacies of this perspective have now been amply exposed. For while the gap between levels of urbanization (i.e., the percentage of total population in a country living in urban-size places) of the developed and developing worlds has been closing rapidly, the gap between their levels of economic prosperity (as measured, for example, by per capita national income or gross national product) has grown progressively wider at the very same time. This, more than any other anomaly in the field, has forced a reconsideration of the theory of economic development (or modernization, as it was sometimes called) and the role of cities in the process of industrialization and economic growth.

The third section of the book, therefore, examines the changing theoretical perspective, asking what new theories exist which might better account for past conditions, explain the present relationship between the developed capitalist west and the Third World, and prescribe the kinds of policies needed to redress the current situation of inequity. That the Third World now occupies an unenviable position cannot be doubted. Its poverty, relative to the developed world, is clear. It remains in a subordinate economic position, still producing primary products for an international market which buys them cheaply, and still subsidizing the production costs of goods manufactured by western firms through the provision of cheap labor, either in the form of migrants (the so-called "guest workers" of Europe who do the dirtiest jobs and receive below union wages) or increasingly in the form of local workers for branches of multinational firms. In return, it must import expensive capital goods from the west—capital goods which are based upon a technology that maximizes the use of its scarcest resource (capital) while minimizing the use of its most plentiful resource (labor). Local elites in the capital cities of the Third World, exposed to the wants and consumerism of the west, also import expensive manufactured luxury goods disproportionate to the capabilities of their economies and wasteful of precious foreign exchange funds. Furthermore, many of the strategies and policies designed to redress inequities have actually served to increase them by tying Third World economies even more closely to an international economic order in which their bargaining position, OPEC notwithstanding, remains weak. To some extent this weakness in the economic sphere derives from the Third World's subordinate political position—a subordination which has remained, even after the earlier goals of decolonization and political independence were achieved.

These facts do not fit well with previous sociological theories which tended to attribute the "backwardness" of the Third World (and consequently its poverty and its urban problems) either to a "lag" in the process of their inevitable evolution to the level of "advanced" societies or to a lack of will and capacity on their part. Alejandro Portes in his paper on national development takes a hard look at these earlier and widely accepted theories and advances some pointed criticisms concerning their internal logic, the value positions they hide, and their lack of relevance to concrete historical and contemporary cases. He suggests that the newer theory of dependency—now gaining increased acceptance both as an explanation for past failures in development and as a prescription for policies needed to achieve development in the Third World—may offer more promising leads for researchers interested in urbanization in the Third World. The article by Chilcote reviews the literature on dependency theory that has been generated chiefly by Latin American social scientists, showing the issues it has addressed and the anomalies it has tried to account for. Interestingly enough, African scholars were simultaneously and somewhat independently devising similar theories to explain different conditions in their part of the world. One of the most influential African theorists is Samir Amin who, because his works have been available until recently only in French, is less well known to American social scientists than the Latin American authors. We include an excerpt from Amin's analysis of the history of Africa to illustrate the usefulness of his framework for explaining in cogent fashion changing conditions on that continent.

What relevance do these new theoretical perspectives have for understanding the *urban* problems of the Third World? This is the question we address in the fourth and most important section of the book. We assume that urban problems are merely manifestations in particular geographic sites of problems endemic to the entire society, but we also assume that urban areas offer a particularly vivid and concentrated arena within which to see the operation of these larger forces. Thus, while it may not be possible to study cities in isolation from their context, it is valuable to focus upon them *sui generis* to evaluate the special role they play in national development (or underdevelopment) and the special way they highlight the socio-economic and political conflicts within society.

Regardless of theoretical perspective, urban scholars agree on *which* issues in Third World urbanization are particularly pressing. The problems that have been singled out for special study have revolved chiefly around four closely interrelated trends, each of which is addressed in a subsection of Section IV.

1. Regional Imbalance:

In many parts of the Third World there is not only a discontinuity between urban and rural areas but also a sharp discrepancy between the capital city or chief port (either of colonial origin or of renewed importance when the country became linked to international trade) and the remaining cities and towns of the urban hierarchy. Often, this has led to a condition of excessive "primacy" or, as some have termed it, "macrocephaly," in which one city so dominates the remainder of the country that it prevents other parts from developing. How this situation came into being and what some of the consequences of it are is addressed in subsection A.

2. Migration:

Regional imbalances have been intensified by the propensity of persons born and raised in rural parts of the country to migrate, primarily to the largest cities, in search of jobs and a better future for their children. This, coupled with natural increase which is often higher in cities than in the countryside, leads to very rapid growth of the largest cities which are called upon to absorb vast numbers of new urbanites; the resulting strain on urban resources often conflicts with the need to invest these resources in agricultural and industrial enterprises. Subsection B examines the issue of migration and its effects upon both the city and the prospects for economic development.

3. Jobs:

The rapidly increasing urban population requires a commensurate increase in the number of jobs in urban areas, if this population is not to starve or be reduced to total dependence. In most Third World countries, expansion in the industrialized sector of the economy has lagged behind population growth—especially where the industrial sector has favored imported capital-intensive technologies. The result has been an inflation of the often marginally productive tertiary or service sector of the urban economy. The causes and consequences of this process of tertiarization are explored in subsection C.

4. Housing:

Shelter and public utilities are also overstrained by the large increases in population. Densities build up in the existing housing stock, causing rapid deterioration, and there is a virtually insatiable need for additional residential facilities on the part of groups which can certainly not afford new housing built to western specifications (often favored by planners and required by legal codes). The result has been the proliferation of self-built housing, often on lands which have had to be occupied surreptitiously by squatting. These issues are taken up in subsection D.

Both "classical" urbanization theory and the newer "dependency" approach define these issues as problems—but there the resemblance ends. The explanations diverge concerning the true *causes* of the problems and indeed which aspects are defined as problems. The analysts disagree on the evaluation of the *mechanisms* available to solve the problems. And, flowing from these differences, scholars of each persuasion advocate markedly contrasting *policies* to cope with the problems. This book can only scratch the surface of these important issues. Each subsection includes a general statement of the issue, a summary of the positions which might be taken on it, and some examples drawn from various parts of the Third World to demonstrate the scope and research dimensions of the problem. Our treatment is far from exhaustive, but even the small sample of cases should sensitize the reader to the complex debates over the far from simple dilemmas that face Third World cities.

The final section of the book focuses more directly upon the goals for urbanization in the Third World and examines some alternative national urban development policies that are being followed in a variety of countries to achieve such goals. For it is only in the context of goals that one can identify and specify problems, and it is only in relation to goals that one can evaluate present conditions and policies for amelioration. These evaluations are, in turn, dependent upon how one has conceptualized the causes, because policies are in a fundamental sense prescriptions for modifying and changing the underlying "causes."

The goals for urban *and* rural settlements were specified by the international community in "The Declaration of Principles" which was passed by an overwhelming majority of member states of the United Nations at the World Conference on Human

Settlements (HABITAT, 1976) held in Vancouver, Canada in June of 1976. This declaration embodies the hopes of the Third World to redress the inequities that currently condemn so many urbanites to lives of privation and hardship. (It is sad to note that most western nations, including the United States, failed to ratify this declaration.) Section V begins with this declaration.

Alternative policies and programs for national urban development are being followed by various countries in the world, indicating that there is no agreement on how these goals are to be reached. Some of these alternative approaches are summarized in the report prepared by the U.N. Secretariat as a background paper for the HABITAT 1976 conference.

Of all the innovations in policy that have been devised to tackle the recurrent issues of distributive justice between urban and rural areas and within urban areas themselves, perhaps none has been followed with as close attention—both by policy makers in the Third World and urban scholars, regardless of nationality—as the Chinese experiment. Because the approach is so different and the scale so vast, and because China remains relatively unknown to us, it has been difficult to evaluate the results of their attempts or to determine whether their approach offers a viable model for other parts of the Third World. According to the article by Mingione which concludes the volume, the Chinese experiment with agro-industrial settlements is an attempt to eliminate, once and for all, many of the inequities discussed in the opening section of our book: the gap between city and countryside, the inequity between rulers and ruled, the distance between rich and poor. These are goals to which the Third World has given verbal support. Whether they can be achieved in China, or anywhere else for that matter, remains the open question on which the book ends but does not conclude.

A topic as vast and complicated as urbanization in the Third World clearly cannot be covered in sufficient depth and specificity within so short a volume. We have contended throughout that each region of the world faces different challenges as it struggles for economic growth, for higher standards of living and fairer distributions of social benefits, for unified nations in which cities are not only good to live in but creative contributors to welfare and development. Each region draws upon its own cultural strengths, depends upon its own set of natural and man-transformed resources, and is constrained or assisted by its own location in the world. To understand the basic issues of Third World urbanization in general is not equivalent therefore to understanding the situation of cities in specific regions nor to understanding the problems and potentials of any given city. For this, much more is needed. We have included a carefully selected bibliography to guide additional reading. But, in the final analysis, many of the studies required to yield a deep understanding of the changes taking place in Third World cities have not yet been made, and many of the experiments with techniques for creating more viable cities, not only in the Third World but in the developed world as well, have not yet been tried. The reading, research and policy agendas remain open.

SECTION I

CITIES AND INEQUALITY

INTRODUCTION

Despite all the discussions of "one world," of the rural-urban continuum and the classless (or middle class) society, there is something very persistent about the basic oppositions which have preoccupied human thought. Underlying these oppositions is the issue of inequality. Three types of oppositions, each relating to a basic gap in the distribution of power, status and wealth, can be identified. First, there is the gap between the conqueror and the conquered, which in undisguised form is represented by colonialism and in hidden form by neocolonialism and economic dependency. The second antithesis is clearly between the city and the countryside which, throughout much of human history, has been a relationship of controller to controlled, of exploiter to exploited. The third opposition, impressively parallel in its effects, is the transposition within the city itself of the class discrepancies of the wider context, whether these are racially, politically, or economically conditioned.

Contrast or "opposition," in all these instances, is merely a euphemism for gross inequality. In this section the readings attempt to identify the major dimensions along which these inequalities are arrayed and to demonstrate that they have been objects of study and concern for quite some time. Our intent is to show that some of the *causes* of these inequities have been located by these early writers in particularly contemporary types of factors. It is remarkable that the new theories of "dependency" presented in a later section of the book are hinted at in these selections written long before the emergence of the Third World as a concept or power group.

The contrast between urban and non-urban life, described so sharply in the fourteenth century by a philosopher of history, Ibn Khaldun, can still be observed in many parts of the Third World, even though it seems to be increasingly blurred in the developed world. Now, just as then, this contrast must be attributed to the economic division of labor between the city and the countryside. The role of changes in the "mode of production," however, in transforming and indeed intensifying this distinction between urban and rural is explored in greater depth by Marx who wrote from a vantage point *beyond* the industrial revolution. And the tendency for the international system of capitalism to project *onto a world scale* this same distinction between core and periphery through an international division of labor between the advanced capitalist nations (chiefly in the west) and those of primary extraction (chiefly in the Third World) is foresightedly recognized by Lenin in an essay written long before the term "multinationals" had been invented.

The major mechanism whereby the international division of labor was established in the initial period was, of course, colonialism. It is therefore surprising (and sobering) to find in the writings of Niccolo Machiavelli, a fifteenth century Italian who lived *before* the "Age of Discovery," a cogent and exhaustive summary of the alternative methods a conqueror might employ *vis à vis* a subject nation: he might either remove or destroy its wealth; he might implant a settler regime to extract the surplus; or he might coopt a local elite for the purpose of indirect rule. It is difficult to find a fourth method in the long history of colonialism which came after Machiavelli. In addition, Machiavelli noted, together with Ibn Khaldun, the tendency for conquered peoples to internalize the values of their conquerors and to denigrate their own culture, blaming it for their defeat. Frantz Fanon (1963) has recently shown how the first step in decolonization must be the rejection of this tendency to imitate the conqueror, if authentic independence is to be achieved.

In Section III we return to a number of these themes. The reader may wish to reread Section I after studying the selections in Section III in order to trace how the earlier ideas have been incorporated into current approaches.

CHAPTER 1

An Arab Philosophy of History: Selections from the Prolegomena of Ibn Khaldun of Tunis (1332–1406)

IMITATION OF THE CONQUERORS BY THE VANQUISHED

The vanquished always seek to imitate their victors in their dress, insignia, belief, and other customs and usages. This is because men are always inclined to attribute perfection to those who have defeated and subjugated them. Men do this either because the reverence they feel for their conquerors makes them see perfection in them or because they refuse to admit that their defeat could have been brought about by ordinary causes, and hence they suppose that it is due to the perfection of the conquerors. Should this belief persist long, it will change into a profound conviction and will lead to the adoption of all their characteristics. This imitation may come about either unconsciously or because of a mistaken belief that the victory of the conquerors was due not to their superior solidarity and strength but to [inferiority of] the customs and beliefs of the conquered. Hence, arises the further belief that such an imitation will remove the causes of defeat.

Therefore we see the defeated always imitating the victors in their way of dressing, of carrying their arms, in their equipment and in all their mode of living. . . .

COUNTRYMEN AND TOWNSMEN

Countrymen are morally superior to townsmen. This is because the soul is, by its nature, prepared to receive any impressions of good or evil that may be stamped on it. As the Prophet (Mohammad), peace be upon him, said,

Reprinted from Charles Issawi, trans., *An Arab Philosophy of History: Selections from the Prolegomena of Ibn Khaldun of Tunis (1332–1406)* (New York: Grove Press, Inc., 1958), pp. 53, 66, 68, 80–81, 92–93, 117–18, by permission of Grove Press, Inc. and the English publisher, John Murray Ltd. First edition June 1950, reprinted July 1955, printed in Great Britain by Butler & Tanner Ltd., Frome and London and published by John Murray (Publishers) Ltd. This material has been edited for this publication.

"All children are born with the same unformed natures. It is their parents who make of them Jews, Christians, or Zoroastrians."

And the greater the number of impressions of one kind which the soul has received, the further it moves away from the opposite kind and the more difficult it becomes for it to acquire that other kind. Thus the moral man, who has been trained in good habits and has formed inclinations towards virtue, finds it difficult to tread the path of evil; and conversely for the evil man with vicious habits.

Now townsmen are so immersed in luxury, pleasure-seeking, and worldliness, and so accustomed to indulge their desires, that their souls are smeared with vice and stray far from the path of virtue. . . .

Countrymen, though also worldly minded, are forced to confine themselves to bare necessities; they do not seek to indulge their desire for luxury and pleasures. Their habits and actions are relatively simple, hence they are less subject than townsmen to reproach on the grounds of vice and evil doing. In a word, they are nearer to the natural state than are townsmen; their souls have received fewer evil imprints derived from vicious habits than have those of townsmen and, therefore, respond more readily to treatment. . . .

Countrymen are more courageous than townsmen. The reason for this is that townsmen, being accustomed to a peaceful and tranquil life of pleasure-seeking, delegate the task of defending their persons and properties to their governors and garrisons. Surrounded by thick walls, protected by defences, they live in security and forget the use of arms. Successive generations of such a way of life breed a people accustomed, like women and children, to look to others for protection; and with time this habit of dependence becomes a second nature.

Countrymen and nomads, on the other hand, live a more isolated life, far from large towns and garrisons, undefended by walls or defences. Hence they look for protection to themselves alone, not trusting others. Always armed and watchful, except for brief moments during camp meetings, sleeping on their mounts, they are ever on the lookout for any sign of danger. Hence they do not fear to wander unaccompanied in the countryside or the wilderness, being full of confidence in their own courage and power. For courage has become one of their deepest qualities and audacity a second nature to them, emerging whenever occasion calls. And townsmen, however much they may mix with nomads in the steppe or accompany them on their travels, will always be dependent on them and unable to do anything for themselves, as any one can see. . . . And the reason for this, as we have said before, is that man is the creature of his habits and customs, not of his inborn nature and temperament; for that to which men are accustomed soon becomes to them a second nature or deep-rooted inclination, replacing their original nature and impulses. . . .

STAGES OF ECONOMIC DEVELOPMENT

Know, then, that the differences between peoples arise principally from the differences in their occupations; for their very union springs out of the need for co-operation in the securing of a livelihood.

And first, before comforts and luxuries, come those occupations which deal with the bare necessities of life. Hence some men devote themselves to agriculture, sowing and planting, and some tend animals such as sheep, cows, goats, bees, and silkworms, with a view to using their produce. And those who devote themselves to agriculture and animal husbandry are compelled by necessity to go out into the open country, which has the space, which in towns is lacking, for fields, pastures, plantations, and so on. Such people must therefore necessarily pursue a nomadic life and for that reason they will unite, co-operate in economic matters, and have food, dwelling, and shelter only to the extent which answers the bare necessities of life, without any of the superfluities.

Should their standard of living, however, rise, so that they begin to enjoy more than the bare necessities, the effect will be to breed in them a desire for repose and tranquillity. They will therefore co-operate to secure superfluities; their food and clothing will increase in quantity and refinement; they will enlarge their houses and plan their towns for defence. A further improvement in their conditions will lead to habits of luxury, resulting in extreme refinement in cooking and the preparation of food; in choosing rich clothing of the finest silk; in raising lofty mansions and castles and furnishing them luxuriously, and so on. At this stage the crafts develop and reach their height. Lofty castles and mansions are built and decorated sumptuously, water is drawn to them and a great diversity takes place in the way of dress, furniture, vessels, and household equipment.

Such are the townsmen, who earn their living in industry or trade. Their gains are greater than those working in agriculture or animal husbandry and their standard of living higher, being in line with their wealth. We have shown, then, that both the nomadic and the urban stages are natural and necessary. . . .

Our contention that country life precedes town life, and is its origin, is confirmed by the fact that investigation into the ancestry of the inhabitants of any town will reveal that most of them originated in the countryside adjoining that town, to which their ancestors came when they had improved their condition. . . .

SURPLUS CREATION

[A] single individual is incapable of satisfying his needs by himself, but must co-operate with other members of society. The product of such co-operative labour will exceed by far the needs of the group. Thus, in the production of wheat, for example, we do not see each individual providing for his own needs; rather, we see six or ten persons co-operating: a blacksmith, a carpenter to repair the tools; an ox-tender, a man to plough the soil, and another to reap the grain; and so forth for the different kinds of agricultural work, each man specializing in one operation.

The result of such co-operative labour is to produce a quantity of food which is sufficient for many times the number of persons engaged in the work; co-operative labour more than satisfies the needs of those engaged in it. Consequently, when the inhabitants of a district or town devote their efforts to providing necessities, they find out that they need only part of their labour for that purpose; the rest of their labour is available for the production of luxuries, or goods required by the inhabitants of other districts and exchanged with them for goods of equal value imported from these other districts—all of which leads to riches. . . .

Consequently, the income of such a community will necessarily rise, and prosperity will soon lead to luxury and refinement in matters of housing, household equipment, dress, servants, mounts, etc. Now the demand for such things attracts men skilled in their production; this leads to prosperity in such crafts and services, higher incomes for those engaged in them, and a rise in the income and expenditure of the whole community.

This increase in prosperity leads to a further increase in economic activity which leads to a rise in incomes and increasing luxury, the new wants so created will lead to the creation of new industries and services, with consequent increases in income and prosperity. And this process can go on two or three times, because all the new activities minister to luxury, unlike the original activities which ministered to necessities. . . .

Thus the inhabitants of a more populous city are more prosperous than their counterparts in a less populous one: the judge in the former being better off than the judge in the latter; the trader, than the trader; the craftsman, than the craftsman; the man in the street, than the man in the street; the prince, than the prince; and the policeman, than the policeman. . . .

NATURAL AGES OF THE STATE

[G]enerally speaking, it is rare that the age of the state should exceed three generations, a generation being the average age of an individual, that is forty years or the time necessary for full growth and development. . . .

We said that the age of the state rarely exceeds three generations because the first generation still retains its nomadic roughness and savagery, and such nomadic characteristics as a hard life, courage, predatoriness, and the desire to share glory. All this means that the strength of the solidarity uniting the people is still firm, which makes that people feared and powerful and able to dominate others.

The second generation, however, have already passed from the nomadic to the sedentary way of life, owing to the power they wield and the luxury they enjoy. They have abandoned their rough life for an easy and luxurious one. Instead of all sharing in the power and glory of the state, one wields it alone, the rest being too indolent to claim their part. Instead of aggressiveness and the desire for conquest we see in them contentment with what they have. All this relaxes the ties of solidarity, to a certain extent, and humility and submissiveness begin to appear in them; yet they still retain much of their pristine spirit because of what they have seen and remembered of the previous generation, with its self-confidence, pursuit of glory, and power to defend and protect itself. They cannot entirely give up all these characteristics, even though they have abandoned some of them. They still hope to regain the conditions prevailing in the previous generation, or even have the illusion that these virtues are still to be found in them.

As for the third generation, they have completely forgotten the nomadic and rough stage, as though it had never existed. They have also lost their love of power and their social solidarity through having been accustomed to being ruled. Luxury corrupts them, because of the pleasant and easy way of living in which they have been brought up. As a result, they become a liability on the state, like women and children who need to be protected. Solidarity is completely relaxed and the arts of defending oneself and of attacking the enemy are forgotten.

They deceive people by their insignia, dress, horse-riding and culture; yet all the while they are more cowardly than women. If then a claimant or aggressor appear, they are incapable of pushing him back. Consequently, the head of the state is compelled to rely on others for defence, making extensive use of clients and mercenaries, who may to some extent replace the original free warriors. . . .

CHAPTER 2

The Prince and the Discourses

NICCOLO MACHIAVELLI

THE WAY TO GOVERN CITIES OR DOMINIONS THAT, PREVIOUS TO BEING OCCUPIED, LIVED UNDER THEIR OWN LAWS

When those states which have been acquired are accustomed to live at liberty under their own laws, there are three ways of holding them. The first is to despoil them; the second is to go and live there in person; the third is to allow them to live under their own laws, taking tribute of them, and creating within the country a government composed of a few who will keep it friendly to you. Because this government, being created by the prince, knows that it cannot exist without his friendship and protection, and will do all it can to keep them. What is more, a city used to liberty can be more easily held by means of its citizens than in any other way, if you wish to preserve it. . . .

[A]ll cities are founded either by natives of the country or by strangers. The little security which the natives found in living dispersed; the impossibility for each to resist isolated, either because of the situation or because of their small number, the attacks of any enemy that might present himself; the difficulty of uniting in time for defence at his approach, and the necessity of abandoning the greater number of their retreats, which quickly became a prize to the assailant,—such were the motives that caused the first inhabitants of a country to build cities for the purpose of escaping these dangers. They resolved, of their own accord, or by the advice of someone who had most authority amongst them, to live together in some place of their selection that might offer them greater conveniences and greater facility of defence. Thus, amongst many others were Athens and Venice; the first was built under the authority of Theseus, who had gathered the dispersed inhabitants; and the second owed its origin to the fact that several tribes had taken refuge on the little islands situated at the head of the Adriatic Sea, to escape from war, and from the Barbarians who after

---

Reprinted from Niccolo Machiavelli, *The Prince and the Discourses,* trans. Luigi Ricci, revised by E. R. P. Vincent (London: Oxford University Press, 1968), p. 18 and pp. 105–109, by permission of the publisher.

the fall of the Roman Empire had overrun Italy. These refugees of themselves, and without any prince to govern them, began to live under such laws as seemed to them best suited to maintain their new state. In this they succeeded, happily favored by the long peace, for which they were indebted to their situation upon a sea without issue, where the people that ravaged Italy could not harass them, being without any ships. Thus from that small beginning they attained that degree of power in which we see them now.

The second case is when a city is built by strangers; these may be either freemen, or subjects of a republic or of a prince, who, to relieve their states from an excessive population, or to defend a newly acquired territory which they wish to preserve without expense, send colonies there. The Romans founded many cities in this way within their empire. Sometimes cities are built by a prince, not for the purpose of living there, but merely as monuments to his glory; such was Alexandria, built by Alexander the Great. But as all these cities are at their very origin deprived of liberty, they rarely succeed in making great progress, or in being counted amongst the great powers. Such was the origin of Florence; for it was built either by the soldiers of Sylla, or perhaps by the inhabitants of Mount Fiesole, who, trusting to the long peace that prevailed in the reign of Octavian, were attracted to the plains along the Arno. Florence, thus built under the Roman Empire, could in the beginning have no growth except what depended on the will of its master.

The founders of cities are independent when they are people who, under the leadership of some prince, or by themselves, have been obliged to fly from pestilence, war, or famine, that was desolating their native country, and are seeking a new home. These either inhabit the cities of the country of which they take possession, as Moses did; or they build new ones, as was done by Aeneas. In such case we are able to appreciate the talents of the founder and the success of his work, which is more or less remarkable according as he, in founding the city, displays more or less wisdom and skill. Both the one and the other are recognized by the selection of the place where he has located the city, and by the nature of the laws which he establishes in it. And as men work either from necessity or from choice, and as it has been observed that virtue has more sway where labor is the result of necessity rather than of choice, it is a matter of consideration whether it might not be better to select for the establishment of a city a sterile region, where the people, compelled by necessity to be industrious, and therefore less given to idleness, would be more united, and less exposed by the poverty of the country to occasions for discord; as was the case with Ragusa, and several other cities that were built upon an ungrateful soil. Such a selection of site would doubtless be more useful and wise if men were content with what they possess, and did not desire to exercise command over others.

Now, as people cannot make themselves secure except by being powerful, it is necessary in the founding of a city to avoid a sterile country. On the contrary, a city should be placed rather in a region where the fertility of the soil affords the means of becoming great, and of acquiring strength to repel all who might attempt to attack it, or oppose the development of its power. As to the idleness which the fertility of a country tends to encourage, the laws should compel men to labor where the sterility of the soil does not do it; as was done by those skilful and sagacious legislators who have inhabited very agreeable and fertile countries, such as are apt to make men idle and unfit for the exercise of valor. These by way of an offset to the pleasures and softness of the climate, imposed upon their soldiers the rigors of a strict discipline and severe exercises, so that they became better warriors than what nature produces in the harshest climates and most sterile countries. . . .

I say, then, that for the establishment of a city it is wisest to select the most fertile spot, especially as the laws can prevent the ill effects that would otherwise result from that very fertility. . . .

CHAPTER 3

> The foundation of every division of labor that is well developed, and brought about by exchange in commodities, is the separation between town and country. It may be said, that the whole economic history of the society is summed up in the movement of this antithesis.
>
> Karl Marx, *Capital,* Vol. I, p. 352

The City, The Division of Labor, and the Emergence of Capitalism

KARL MARX
FRIEDRICH ENGELS

The greatest division of material and mental labour is the separation between town and country. The antagonism between town and country begins with the transition from barbarism to civilization, from tribe to State, from locality to nation, and runs through the whole history of civilization to the present day (the Anti-Corn Law League). The existence of the town implies, at the same time, the necessity of administration, police, taxes, etc., in short, of the municipality, and thus of politics in general. Here first became manifest the division of the population into two great classes, which is directly based on the division of labour and on the instruments of production. The town already is in actual fact the concentration of the population, of the instruments of production, of capital, of pleasures, of needs, while the country demonstrates just the opposite fact, their isolation and separation. The antagonism of town and country can only exist as a result of private property. It is the most crass expression of the subjection of the individual under the division of labour, under a definite activity forced upon him—a subjection which makes one man into a restricted town-animal, the other into a restricted country-animal, and daily creates anew the conflict between their interests. Labour is here again the chief thing, power *over* individuals, and as long as the latter exists, private

Reprinted from Karl Marx and Friedrich Engels, *The German Ideology,* ed. R. Pascal (New York: International Publishers, 1939), pp. 43–57, by permission of the publisher. This material has been edited for this publication.

property must exist. The abolition of the antagonism between town and country is one of the first conditions of communal life, a condition which again depends on a mass of material premises and which cannot be fulfilled by the mere will, as anyone can see at first glance. (These conditions have still to be enumerated.) The separation of town and country can also be understood as the separation of capital and landed property, as the beginning of the existence and development of capital independent of landed property—the beginning of property having its basis only in labour and exchange. . . .

The flight of the serfs into the towns went on without interruption right through the Middle Ages. These serfs, persecuted by their lords in the country, came separately into the towns, where they found an organized community, against which they were powerless, in which they had to subject themselves to the station assigned to them by the demand for their labour and the interest of their organized urban competitors. These workers, entering separately, were never able to attain to any power, since if their labour was of the guild type which had to be learned, the guildmasters bent them to their will and organized them according to their interest; or if their labour was not such as had to be learned, and therefore not of the guild type, they became day-labourers and never managed to organize, remaining an unorganized rabble. The need for day-labourers in the towns created the rabble. These towns were true "associations," called forth by the direct need of providing for the protection of property, and multiplying the means of production and defence of the separate members. The rabble of these towns was devoid of any power, composed as it was of individuals strange to one another who had entered separately, and who stood unorganized over against an organized power, armed for war, and jealously watching over them. The journeymen and apprentices were organized in each craft as it best suited the interests of the masters. The filial relationship in which they stood to their masters gave the latter a double power—on the one hand because of their influence on the whole life of the journeymen, and on the other because, for the journeymen who worked with the same master, it was a real bond, which held them together against the journeymen of other masters and separated them from these. And finally, the journeymen were bound to the existing order by their simple interest in becoming masters themselves. . . .

In the towns, the division of labour between the individual guilds was as yet quite natural, and, in the guilds themselves, not at all developed between the individual workers. Every workman had to be versed in a whole round of tasks, had to be able to make everything that was to be made with his tools. The limited commerce and the scanty communication between the individual towns, the lack of population and the narrow needs did not allow of a higher division of labour, and therefore every man who wished to become a master had to be proficient in the whole of his craft. Thus there is found with medieval craftsmen an interest in their special work and in proficiency in it, which was capable of rising to a narrow artistic sense. For this very reason, however, every medieval craftsman was completely absorbed in his work, to which he had a contented, slavish relationship, and to which he was subject to a far greater extent than the modern worker, whose work is a matter of indifference to him.

Capital in these towns was a natural capital, consisting of a house, the tools of the craft, and the natural hereditary customers; and not being realizable, on account of the backwardness of commerce and the lack of circulation, it descended from father to son. Unlike modern capital, which can be assessed in money and which may be indifferently invested in this thing or that, this capital was directly connected with the particular work of the owner, inseparable from it and to this extent "estate" capital.

The next extension of the division of labour was the separation of production and commerce, the formation of a special class of merchants; a separation which, in the towns bequeathed by a former period, had been handed down (among other things with the Jews) and which very soon appeared in the newly formed ones. With this there was given the possibility of commercial communications transcending the immediate neighbourhood, a possibility, the realization of which depended on the existing means of communication, the state of public safety in the countryside, which was determined by political conditions (during the whole of the Middle Ages, as is well known, the merchants travelled in armed caravans), and on the cruder or more advanced needs (determined by the stage of culture attained) of the region accessible to intercourse. With commerce the prerogative of a particular class, with the extension of trade through the merchants beyond the immediate surroundings of the town, there immediately appears a reciprocal action between production and commerce. The towns enter into relations *with one another*, new tools are brought from one town into the other, and the separation between production and commerce soon calls forth a new division of production between the individual towns, each of which is soon exploiting a predominant branch of industry. The local restrictions of earlier times begin gradually to be broken down.

In the Middle Ages the citizens in each town were compelled to unite against the landed nobility to save their skins. The extension of trade, the establishment of communications, led the separate towns to get to know other towns, which had asserted the same interests in the struggle with the same antagonist. Out of the many local corporations of burghers there arose only gradually the burgher *class*. The conditions of life of the individual burghers became, on account of their antagonism to the existing relationships and of the mode of labour determined by these, conditions which were common to them all and independent of each individual. The burghers had created the conditions in so far as they had torn themselves free from feudal ties, and were created by them in so far as they were determined by their antagonism to the feudal system which they found in existence. When the individual towns began to enter into associations, these common conditions developed into class conditions. The same conditions, the same antagonism, the same interests necessarily called forth on the whole similar customs everywhere. The bourgeoisie itself, with its conditions, develops only gradually, splits according to the division of labour into various fractions and finally absorbs all earlier possessing classes (while it develops the majority of the earlier non-possessing, and a part of the earlier possessing, class into a new class, the proletariat) in the measure to which all earlier property is transformed into industrial or commercial capital. The separate individuals form a class only in so far as they have to carry on a common battle against another class; otherwise they are on hostile terms with each other as competitors. On the other hand, the class in its turn achieves an independent existence over against the individuals, so that the latter find their conditions of existence predestined, and hence have their position in life and their personal development assigned to them by their class, become subsumed under it. . . .

The immediate consequence of the division of labour between the various towns was the rise of manufactures, branches of production which had outgrown the guild-system. Manufactures first flourished, in Italy and later in Flanders, under the historical premise of commerce with foreign nations. In other countries, England and France for example, manufactures were at first confined to the home market. Besides the premises already mentioned manufactures depend on yet another: an already advanced concentration of population, particularly in the countryside, and of capital, which began to accumulate in the hands of individuals, partly in the guilds in spite of the guild regulations, partly among the merchants.

That labour which from the first presupposed a machine, even of the crudest sort, soon showed itself the most capable of development. Weaving, earlier carried on in the country by the peasants as a secondary occupation to procure their clothing, was the first labour to receive an impetus and a further development through the extension of commerce. Weaving was the first and remained the principal manufacture. The rising demand for clothing materials, consequent on the growth of population, the growing accumulation and mobilization of natural capital through accelerated circulation, the demand for luxuries called forth by the latter and favoured generally by the gradual extension of commerce, gave weaving a quantitative and qualitative stimulus, which wrenched it out of the form of production hitherto existing. Alongside the peasants weaving for their own use, who continued with this sort of work, there emerged a new class of weavers in the towns, whose fabrics were destined for the whole home market and usually for foreign markets too. Weaving, an occupation demanding in most cases little skill and soon splitting up into countless branches, by its whole nature resisted the trammels of the guild. Weaving was therefore carried on mostly in villages and market-centres without guild organization, which gradually became towns, and indeed the most flourishing towns in each land. With guild-free manufacture, property relations also quickly changed. The first advance beyond natural, estate-capital was provided by the rise of merchants whose capital was from the beginning movable, capital in the modern sense as far as one can speak of it, given the circumstances of those times. The second advance came with manufacture, which again made mobile a mass of natural capital, and altogether increased the mass of movable capital as against that of natural capital. At the same time, manufacture became a refuge of the peasants from the guilds which excluded them or paid them badly, just as earlier the guild-towns had served as a refuge for the peasants from the oppressive landed nobility. . . .

With manufacture was given simultaneously a changed relationship between worker and employer. In the guilds the patriarchal relationship between journeyman and master maintained itself; in manufacture its place was taken by the monetary relation between worker and capitalist—a relationship which in the countryside and in small towns retained a patriarchal tinge, but in the larger, the real manufacturing towns, quite early lost almost all patriarchal complexion.

Manufacture and the movement of production in general received an enormous impetus through the extension of commerce which came with the discovery of America and the sea-route to the East Indies. The new products imported thence, particularly the masses of gold and silver which came into circulation and totally changed the position of the classes towards one another, dealing a hard blow to feudal landed property and to the workers; the expeditions of adventurers, colonization; and above all the extension of markets into a world-market, which had now become possible and was daily becoming more and more a fact, called forth a new phase of historical development, into which in general we cannot here enter further. Through the colonization of the newly discovered countries the commercial struggle of the nations amongst one another was given new fuel and accordingly greater extension and animosity.

The expansion of trade and manufacture accelerated the accumulation of movable capital, while in the guilds, which were not stimulated to extend their production natural capital remained stationary or even declined. Trade and manufacture created the big bourgeoisie: in the guilds was concentrated the petty bourgeoisie, which no longer was dominant in the towns as formerly, but had to bow to the might of the great merchants and manufacturers. Hence the decline of the guilds, as soon as they came into contact with manufacture.

The material, commercial relations of nations took on, in the epoch of which we have been speaking, two different forms. At first the small quantity of gold and silver in circulation involved the ban on the export of these metals; and industry, for the most part imported from abroad and made necessary by the need for employing the growing urban population, could not do without those privileges which could be granted not only, of course, against home competition, but chiefly against foreign. The local guild privilege was in these prohibitions extended over the whole nation. Customs duties originated from the tributes exacted by the feudal lords from merchants passing through their territories, tributes later imposed likewise by the towns, and which, with the rise of the modern states, were the treasury's most obvious means of raising money. . . .

The second period began in the middle of the seventeenth century and lasted almost to the end of the eighteenth. Commerce and navigation had expanded more rapidly than manufacture, which played a secondary role; the colonies were becoming considerable consumers; and after long struggles the separate nations shared out the opening world-market among themselves. This period begins with the Navigation Laws and colonial monopolies. The competition of the nations among themselves was excluded as far as possible by tariffs, prohibitions, and treaties; and in the last resort the competitive struggle was carried on and decided by wars (especially naval wars). The mightiest maritime nation, the English, retained preponderance in trade and manufacture. Here, already, we find concentration on one country. Manufacture was all the time sheltered by protective duties in the home market, by monopolies in the colonial market, and abroad as much as possible by differential duties. The working-up of home-produced material was encouraged (wool and linen in England, silk in France), the export of home-produced raw material forbidden (wool in England), and that of imported material neglected or suppressed (cotton in England). The nation dominant in sea-trade and colonial power naturally secured for itself also the greatest quantitative and qualitative expansion of manufacture. Manufacture could not be carried on without protection, since, if the slightest change takes place in other countries, it can lose its market and be ruined; under reasonably favourable conditions it may easily be introduced into a country, but for this very reason can easily be destroyed. At the same time through the mode in which it is carried on, particularly in the eighteenth century, in the countryside, it is so interwoven with the vital relationships of a great mass of individuals, that no country dare jeopardize their existence by permitting free competition. In so far as it manages to export, it therefore depends entirely on the extension or restriction of commerce, and exercises a relatively very small reaction on the latter. Hence its secondary importance and the influence of the merchants in the eighteenth century. It was especially the merchants and shippers who more than anybody else pressed for State protection and monopolies; the manufacturers demanded and indeed received protection, but all the time were inferior in political importance to the merchants. The commercial towns, particularly the maritime towns, won to some extent the civilized outlook of the big bourgeoisie, but in the factory towns an extreme petty-bourgeois outlook persisted. . . . The eighteenth century was the century of trade. Pinto says this expressly: "Le commerce fait la marotte du siècle," ("Commerce is the rage of the century"); and, "depuis quelque temps il n'est plus question que de commerce, de navigation et de marine" ("for some time now people have been talking only about commerce, navigation, and the navy"). . . .[1]

The concentration of trade and manufacture in one country, England, developing irresistibly in the seventeenth century, gradually created for this country a relative world-market, and thus a demand for the manufactured products of this country, which could no longer be met by the industrial productive forces hitherto existing. This demand, outgrowing the productive forces, was the motive power which, by producing big industry—the application of elemental forces to industrial ends, machinery and the most complex division of labour—called into existence the third period of private ownership since the Middle Ages. . . .

Big industry universalized competition in spite of these protective measures (it is practical free trade; the duty is only a palliative, a barrier *within* free trade), established means of communication and the modern world-market, subordinated trade to itself, transformed all capital into industrial capital, and thus produced the rapid circulation (the financial system is perfected) and the centralization of the various forms of capital. By universal competition it forced all individuals to strain their energy to the utmost. It destroyed as far as possible ideology, religion, morality, etc., and where it could not do this, made them into a palpable lie. It produced world-history for the first time, in so far as it made all civilized nations and every individual member of them dependent for the satisfaction of their wants on the whole world, thus destroying the former natural exclusiveness of separate nations. It made natural science subservient to capital and took from the division of labour the last semblance of its natural character. It destroyed natural growth in general, as far as this is possible while labour exists, and resolved all natural relationships into money relationships. In the place of natural towns it created the modern, large industrial cities which have sprung up over-night. Wherever it penetrated, it destroyed the crafts and all earlier stages of industry. It completed the victory of the commercial town over the countryside.

1. The movement of capital, although considerably accelerated, still remained, however, relatively slow. The splitting-up of the world-market into separate parts, each of which was exploited by a particular nation, the exclusion of competition among themselves on the part of the nations, the clumsiness of production itself and the fact that finance was only evolving from its early stages, greatly impeded circulation. The consequence of this was a haggling, mean and niggardly spirit which still clung to all merchants and to the whole mode of carrying on trade. Compared with the manufacturers, and above all with the craftsmen, they were certainly big bourgeois; compared with the merchants and industrialists of the next period they remain petty bourgeois, cf. Adam Smith.

CHAPTER 4

Imperialism, The Highest Stage of Capitalism

V. I. LENIN

DIVISION OF THE WORLD AMONG THE GREAT POWERS

In his book, on "the territorial development of the European colonies," A. Supan [1906:254], the geographer, gives the following brief summary of this development at the end of the nineteenth century:

Percentage of Territory Belonging to the European Colonial Powers (Including the United States)

	1876	1900	Increase or decrease
Africa	10.8	90.4	+79.6
Polynesia	56.8	98.9	+42.1
Asia	51.5	56.6	+ 5.1
Australia	100.0	100.0	—
America	27.5	27.2	— 0.3

"The characteristic feature of this period," he concludes, "is, therefore, the division of Africa and Polynesia." As there are no unoccupied territories—that is, territories that do not belong to any state—in Asia and America, it is necessary to amplify Supan's conclusion and say that the characteristic feature of the period under review is the final partitioning of the globe—final, not in the sense that *repartition* is impossible; on the contrary, repartitions are possible and inevitable—but in the sense that the colonial policy of the capitalist countries has *completed* the seizure of the unoccupied territories on our planet. For the first time the world is completely divided up, so that in the future *only* redivision is possible, i.e., territories can only pass from one "owner" to another, instead of passing

Reprinted from V. I. Lenin, "Imperialism, The Highest Stage of Capitalism," *Lenin Selected Works,* Vol. 1 (Moscow: Progress Publishers, 1970), pp. 726–65. This material has been edited for this publication.

as ownerless territory to an "owner."

Hence, we are living in a peculiar epoch of world colonial policy, which is most closely connected with the "latest stage in the development of capitalism," with finance capital. For this reason, it is essential first of all to deal in greater detail with the facts, in order to ascertain as exactly as possible what distinguishes this epoch from those preceding it, and what the present situation is. In the first place, two questions of fact arise here: is an intensification of colonial policy, a sharpening of the struggle for colonies, observed precisely in the epoch of finance capital? And how, in this respect, is the world divided at the present time? . . .

For Great Britain, the period of the enormous expansion of colonial conquests was that between 1860 and 1880, and it was also very considerable in the last twenty years of the nineteenth century. For France and Germany this period falls precisely in these twenty years. The development of pre-monopoly capitalism, of capitalism in which free competition was predominant, reached its limits in the 1860s and 1870s. We now see that it is *precisely after that period* that the tremendous "boom" in colonial conquests begins, and that the struggle for the territorial division of the world becomes extraordinarily sharp. It is beyond doubt, therefore, that capitalism's transition to the stage of monopoly capitalism, to finance capital, *is connected* with the intensification of the struggle for the partitioning of the world.

Hobson, in his work on imperialism, marks the years 1884–1900 as the epoch of intensified "expansion" of the chief European states. According to his estimate, Great Britain during these years acquired 3,700,000 square miles of territory with 57,000,000 inhabitants; France, 3,600,000 square miles with 36,500,000; Germany, 1,000,000 square miles with 14,700,000; Belgium, 900,000 square miles with 30,000,000; Portugal, 800,000 square miles with 9,000,000 inhabitants. The scramble for colonies by all of the capitalist states at the end of the nineteenth century and particularly since the 1880s is a commonly known fact in the history of diplomacy and foreign policy. . . .

To present as precise a picture as possible of the territorial division of the world and of the changes that have occurred during the last decades in this respect, I shall utilise the data furnished by Supan in the work already quoted on the colonial possessions of all the powers of the world. Supan takes the years 1876 and 1900; I shall take the year 1876—a year very aptly selected, for it is precisely by that time that the pre-monopolist stage of development of West-European capitalism can be said to have been, in the main, completed—and the year 1914. . . .

We clearly see . . . how "complete" was the partition of the world at the turn of the twentieth century. After 1876 colonial possessions increased to enormous dimensions. . . . In 1876 three powers had no colonies, and a fourth, France, had scarcely any. By 1914 these four powers had acquired colonies with an area of 14,000,000 square kilometers, i.e, about half as much again as the area of Europe, with a population of nearly 100,000,000. The unevenness in the rate of expansion of colonial possessions is very great. If, for instance, we compare France, Germany and Japan, which do not differ very much in area and population, we see that the first has acquired almost three times as much colonial territory as the other two combined. In regard to finance capital, France, at the beginning of the period we are considering, was also, perhaps, several times richer than Germany and Japan put together. In addition to . . . purely economic conditions, geographical and other conditions also affect the dimensions of colonial possessions. However strong the process of levelling the world, of levelling the economic and living conditions in different countries, may have been in the past decades as a result of the pressure of large-scale industry, exchange and finance capital, considerable differences still remain; and among the six countries mentioned we see, firstly, young capitalist countries (America, Germany, Japan) whose progress has been extraordinarily rapid; secondly, countries with an old capitalist development (France and Great Britain), whose progress lately has been much slower than that of the previously mentioned countries, and thirdly, a country most backward economically (Russia), where modern capitalist imperialism is enmeshed, so to speak, in a particularly close network of pre-capitalist relations. . . .

Colonial policy and imperialism existed before the latest stage of capitalism, and even before capitalism. . . .

The principal feature of the latest stage of capitalism is the domination of monopolist associations of big employers. These monopolies are most firmly established when *all* the sources of raw materials are captured by one group, and we have seen with what zeal the international capitalist associations exert every effort to deprive their rivals of all opportunity of competing, to buy up, for example, ironfields, oilfields, etc. Colonial possession alone gives the monopolies complete guarantee against all contingencies in the struggle against competitors, including the case of an adversary wanting to be protected by a law establishing a state monopoly. The more capitalism is developed, the more strongly the shortage of raw materials is felt, the more intense the competition and the hunt for sources of raw materials throughout the world, the more desperate the struggle for . . . colonies. . . .

Finance capital is interested not only in the already discovered sources of raw materials but also in potential sources, because present-day technical development is extremely rapid, and land which is useless today may be improved tomorrow if new methods are devised (to this end a big bank can equip a special expedition of engineers, agricultural experts, etc.), and if large amounts of capital are invested. This also applies to prospecting for minerals, to new methods of processing up and utilising raw materials, etc., etc. Hence, the inevitable striving of finance capital to enlarge its spheres of influence and even its actual territory. In the same way that the trusts capitalise their property at two or three times its value, taking into account its "potential" (and not actual) profits and the further results of monopoly, so finance capital in general strives to seize the largest possible amount of land in all kinds of places, and by every means, taking into account potential sources of raw materials and fearing to be left behind in the fierce struggle for the last remnants of independent territory. . . .

The interests pursued in exporting capital also give an impetus to the conquest of colonies, for in the colonial market it is easier to employ monopoly methods . . . to eliminate competition, to ensure supplies, to secure the necessary "connections," etc.

The non-economic superstructure which grows up on the basis of finance capital, its politics and its ideology, stimulates the striving for colonial conquest. "Finance capital does not want liberty, it wants domination," as Hilferding very truly says. . . .

Since we are speaking of colonial policy in the epoch of capitalist imperialism, it must be observed that finance capital and its foreign policy, which is the struggle of the great powers for the economic and political division of the world, give rise to a number of *transitional* forms of state dependence. Not only are there two main groups of countries, those owning colonies, and the colonies themselves, but also the diverse forms of dependent countries which, politically, are formally independent, but in fact, are enmeshed in the net of financial and diplomatic dependence, typical of this epoch. We have already referred to one form of dependence —the semi-colony. An example of another is provided by Argentina.

"South America, and especially Argentina," writes Schulze-Gaevernitz [1906] in his work on British imperialism, "is so dependent financially on London that it ought to be described as almost a British commercial colony. . . ."

IMPERIALISM, AS A SPECIAL STAGE OF CAPITALISM

We must now try to sum up, to draw together the threads of what has been said above on the subject of imperialism. Imperialism emerged as the development and direct continuation of the fundamental characteristics of capitalism in general. But capitalism only became capitalist imperialism at a definite and very high stage in its development, when certain of its fundamental characteristics began to change into their opposites, when the features of the epoch of transition from capitalism to a higher social and economic system had taken shape and revealed themselves in all spheres. Economically, the main thing in this process is the displacement of capitalist free competition by capitalist monopoly. Free competition is the basic feature of capitalism, and of commodity production generally;

monopoly is the exact opposite of free competition, but we have seen the latter being transformed into monopoly before our eyes, creating large-scale industry and forcing out small industry, replacing large-scale by still larger-scale industry, and carrying concentration of production and capital to the point where out of it has grown and is growing monopoly: cartels, syndicates and trusts, and merging with them, the capital of a dozen or so banks, which manipulate thousands of millions. At the same time the monopolies, which have grown out of free competition, do not eliminate the latter, but exist above it and alongside it, and thereby give rise to a number of very acute, intense antagonisms, frictions and conflicts. Monopoly is the transition from capitalism to a higher system.

If it were necessary to give the briefest possible definition of imperialism we should have to say that imperialism is the monopoly stage of capitalism. Such a definition would include what is most important, for, on the one hand, finance capital is the bank capital of a few very big monopolist banks, merged with the capital of the monopolist associations of industrialists; and, on the other hand, the division of the world is the transition from a colonial policy which has extended without hindrance to territories unseized by any capitalist power, to a colonial policy of monopolist possession of the territory of the world, which has been completely divided up....

... And so, without forgetting the conditional and relative value of all definitions in general, ... we must give a definition of imperialism that will include the following five of its basic features:

(1) The concentration of production and capital has developed to such a high stage that it has created monopolies which play a decisive role in economic life; (2) the merging of bank capital with industrial capital, and the creation, on the basis of this "finance capital," of a financial oligarchy; (3) the export of capital as distinguished from the export of commodities acquires exceptional importance; (4) the formation of international monopolist capitalist associations which share the world among powers (5) the territorial division of the whole world among the biggest capitalist powers is completed. Imperialism is capitalism at that stage of development at which the dominance of monopolies and finance capital is established; in which the export of capital has acquired pronounced importance; in which the division of the world among the international trusts has begun, in which the division of all territories of the globe among the biggest capitalist powers has been completed....

Finance capital and the trusts do not diminish but increase the differences in the rate of growth of the various parts of the world economy....

Further, imperialism is an immense accumulation of money capital in a few countries, amounting, as we have seen, to 100,000–150,000 million francs in securities. Hence the extraordinary growth of a class, or rather, a stratum of rentiers, i.e, people who live by "clipping coupons," who take no part in any enterprise whatever, whose profession is idleness. The export of capital, one of the most essential economic bases of imperialism, still more completely isolates the rentiers from production and sets the seal of parasitism on the whole country that lives by exploiting the labour of several overseas countries and colonies. . . .

For that reason the term "rentier state" . . . or usurer state, is coming into common use in the economic literature that deals with imperialism. The world has become divided into a handful of usurer states and a vast majority of debtor states. "At the top of the list of foreign investments," says Schulze-Gaevernitz, "are those placed in politically dependent or allied countries: Great Britain grants loans to Egypt, Japan, China and South America. Her navy plays here the part of the bailiff in case of necessity. Great Britain's political power protects her from the indignation of her debtors. . . ."

"Great Britain," says Schulze-Gaevernitz [1906], "is gradually becoming transformed from an industrial into a creditor state. Notwithstanding the absolute increase in industrial output and the export of manufactured goods, there is an increase in the relative importance of income from interest and dividends, issues of securities, commissions and speculation in the whole of the national economy. In my opinion it is precisely this that forms the economic basis of imperialist ascendancy. The creditor is more firmly attached to the debtor than the seller is to the buyer. . . ."

The description of "British imperialism" in Schulze-Gaevernitz's book reveals the same parasitical traits. The national income of Great Britain approximately doubled from 1865 to 1898, while the income "from abroad" increased *ninefold* in the same period. . . .

THE PLACE OF IMPERIALISM IN HISTORY

We have seen that in its economic essence imperialism is monopoly capitalism. This in itself determines its place in history, for monopoly that grows out . . . of free competition, is the transition from the capitalist system to a higher socio-economic order. We must take special note of the four principal . . . manifestations of monopoly capitalism, which are characteristic of the epoch we are examining.

Firstly, monopoly arose out of the concentration of production at a very high stage. This refers to the monopolist capitalist associations, cartels, syndicates and trusts. We have seen the important part these play in present-day economic life. At the beginning of the twentieth century, monopolies had acquired complete supremacy in the advanced countries, and although the first steps towards the formation of the cartels were taken by countries enjoying the protection of high tariffs (Germany, America), Great Britain, with her system of free trade, revealed the same basic phenomenon, only a little later, namely, the birth of monopoly out of the concentration of production.

Secondly, monopolies have stimulated the seizure of the most important sources of raw materials, especially for the basic and most highly cartelised industries in capitalist society: the coal and iron industries. The monopoly of the most important sources of raw materials has enormously increased the power of big capital, and has sharpened the antagonism between cartelised and non-cartelised industry.

Thirdly, monopoly has sprung from the banks. The banks have developed from modest middleman enterprises into the monopolists of finance capital. Some three to five of the biggest banks in each of the foremost capitalist countries have achieved the "personal link-up" between industrial and bank capital, and have concentrated in their hands the control of thousands upon thousands of millions which form the greater part of the capital and income of entire countries. A financial oligarchy, which throws a close network of dependence relationships over all the economic and political institutions of present-day bourgeois society without exception—such is the most striking manifestation of this monopoly.

Fourthly, monopoly has grown out of colonial policy. To the numerous "old" motives of colonial policy, finance capital has added the struggle for the sources of raw materials, for the export of capital, for spheres of influence, i.e., for spheres for profitable deals, concessions, monopoly profits and so on, economic territory in general. When the colonies of the European powers, for instance, comprised only one-tenth of the territory of Africa (as was the case in 1876), colonial policy was able to develop by methods other than those of monopoly—by the "free grabbing" of territories, so to speak. But when nine-tenths of Africa had been seized (by 1900), when the whole world had been divided up, there was inevitably ushered in the era of monopoly possession of colonies and, consequently, of particularly intense struggle for the division and the redivision of the world.

The extent to which monopolist capital has intensified all the contradictions of capitalism is generally known. It is sufficient to mention the high cost of living and the tyranny of the cartels. This intensification of contradictions constitutes the most powerful driving force of the transitional period of history, which began from the time of the final victory of world finance capital.

Monopolies, oligarchy, the striving for domination and not for freedom, the exploitation of an increasing number of small or weak nations by a handful of the richest or most powerful nations—all these have given birth to those distinctive characteristics of imperialism which compel us to define it as parasitic or decaying capitalism. More and more prominently there emerges, as one of the tendencies of imperialism, the creation of the "rentier state," the usurer state, in which the bourgeoisie to an ever-increasing degree lives on the proceeds of capital exports and by "clipping coupons." It would be a mistake to believe that this tendency to decay precludes the rapid growth of capitalism. It does not. In the epoch of imperialism, certain branches of industry, certain strata of the bourgeoisie, and certain countries betray, to a greater or lesser degree, now one and now another of these tendencies. On the whole, capitalism is growing far more rapidly than before; but this growth is not only becoming more and more uneven in general, its unevenness also manifests itself, in particular, in the decay of the countries which are richest in capital (Britain). . . .

SECTION II

URBANIZATION: PAST TO PRESENT

INTRODUCTION

The cleavages and contrasts highlighted in Section I are given specificity and application in this section where concrete historical cases and larger overall trends are examined. At the end of Section II, statistical data are presented which compare levels of urbanization within various world regions, trace how these have changed over the past few hundred years, and show that, despite a reduction in the "urbanization gap" between the developed and developing worlds, the wealth gap remains great and even appears to be widening.

Statistics, however, are merely a short-hand for a reality which is far more complex and lies imbedded in concrete historical circumstances which, were we to do justice to them, would require volumes. Certainly this is not possible. Instead, we have included short extracts from several sources to illustrate a few basic propositions which should be clearly grasped before proceeding to a study of contemporary Third World cities. Put baldly, these propositions are:

1. Urbanization is indigenous to the Third World and indeed originated there. Urbanization is not, therefore, coterminous with westernization. *Necessary* similarities should not be expected.
2. Only since the sixteenth century have European cities gained prominence. Before that time, the major world centers were in the east and some of these cities reached remarkable levels of size, organization and elegance.
3. The major spurt in western urbanization occurred between the sixteenth century when colonial conquests first began and the end of the nineteenth century when industrialization *and* empire-building were the dual motors of development.
4. During the period of their most rapid growth, the nineteenth century, European cities suffered from problems whose nature and magnitude were similar to those currently observed in Third World cities.
5. However, despite these parallels, the situations are not comparable. The Third World cannot use the same techniques for economic development which the western world employed since these techniques involved expropriating labor power and natural resources from the underdeveloped world and controlling international markets. On the contrary, the Third World begins with a set of handicaps inherited from the colonial era which are apparently being perpetuated, despite the formal disappearance of political empires.
6. Finally, each subregion of the Third World presents special conditions which derive from its past as well as from its present efforts to shape its future. In studying Third World urbanization, therefore, one should be cautious of overgeneralization or the tendency to deduce the specific circumstances of any one region from one's knowledge of another region.

The first four readings in Section II are designed to give substance to these bald statements. The description of the Chinese capital of Hangchow in the late thirteenth century addresses the first two propositions. The second selection, taken from Engels' *The Condition of the Working-Class in England,* is intended to illustrate the fourth point, while the extract from Barbara Ward's new book, *The Home of Man,* addresses the third and fifth propositions. Finally, Alejandro Portes' analysis of the effects of early Spanish colonial settlement upon the contemporary organization of cities throughout Latin America illustrates the sixth proposition concerning historic specificity. Brief as these selections are, they help capture some of the concrete and complex realities which underly the cold statistics with which this section concludes.

One hardly needs to be reminded that urbanism is not a modern phenomenon but indeed goes back some five thousand years. Every school child knows how cities were cradled in the great river valleys of Mesopotamia, Egypt, the Indian subcontinent and China. We recall these facts but seldom draw from them the explicit lesson that no culture nor geographic region has had a monopoly over the talent of city-building.

It has only been within the past few hundred years that the positions of east and west have been completely reversed. When Rome's light was extinguished by the European nomadic tribes, remember that Constantinople's continued to glow. When Europe was in its "Dark Ages," recall how brightly the light shown on the Arab World on the eastern and southern shores of the Mediterranean. The crusaders in the eleventh century returned to their dismal castle keeps with fabulous tales of the magnificent cities in Egypt and Greater Syria. Marco Polo came back with even more wondrous descriptions of the cities of China (including Hangchow). Even the Spanish "discoverers" of the New World were awed by Mayan, Incan and Aztec ceremonial centers, and the British explorers were surprised to find cities in "Darkest Africa."

Third World cities were once impressive entities, as the city of Hangchow so clearly illustrates. The city had a population in 1275 of over a million inhabitants, a figure which was not reached on the European continent until 1810 (London). One is impressed by the physical size and amenities of the city, by the competence of its bureaucratic apparatus which evidently took a complete census every three years, by the elaborate networks of trade and transport which supplied the city with every necessity and not a few luxuries.

But on another level, the description of Hangchow can be read to indicate how persistent are the themes of conquest, domination, luxury for the rich, misery for the poor, in the long history of urban life. For Hangchow was built by one set of conquerors and was soon to fall into the hands of another. Its ability to command rice supplies from a thousand miles away clearly evidenced imperial control over a vast hinterland. And finally, class stratification was rampant, for while the wealthy lolled about in those pretty boats on the beautiful lake, the poor were huddled together in tall, densely-packed tenements, and when the rich went promenading, six to a carriage, along the wide paved ceremonial way, it was the poor who carried them on their shoulders.

Such are the persisting ambiguities of the urban achievement. Much the same can be said for the great achievements of modern economic development. Whether we look at the rococco cathedrals of Spain encrusted with Incan gold, the elegant splendor of Versailles, the stolid prosperity of Edwardian London, or the skyscrapers and consumer glut of a capital of today; whether we quake before the brute power of noisy factories, its transmuted understatement in the bank president's glass-steel-carpet office, or its barren wastes strewn along fifty-eight miles of Pentagon corridors —hidden within these achievements are the costs, usually borne by the ignored and the invisible.

In the great thrust toward modernization in the west, the costs were borne by the poor within and by the colonized without. We turn first to the poor within, using as our example the newly urbanized proletariat of London. The extract we have selected focuses on such problems as inadequate housing, over-population, insufficient employment and the proliferation of make-shift jobs in the "tertiary" sector—peddling, jobbing, begging. These are problems we associate more with Kuala Lumpur than with London. But while it is important to recognize the parallels and to note the existence in nineteenth century western cities of problems similar to those that face today's Third World city, it is also important to acknowledge some of the important differences between the cases.

Certainly, the most crucial difference lies in the relationship between the developing city and its society, on the one hand, and the rest of the world system, on the other. When the west experienced its enormous burst of population growth, for example, it was able to export its surplus to the New World, to Asia and even to Africa. But where is India to export hers? And when industrial production burgeoned in Europe, there were no other competitors offering the goods she produced. There were, on the contrary, large empires whose markets she controlled, whose tariffs she set, whose resources she obtained at cost, and whose cheap labor she enticed or enslaved. These were some of the advantages which the west enjoyed but which the Third World cannot expect to duplicate. In fact, many of the disabilities which the Third World has inherited were the underside of this western development.

Some of these points are recapitulated in acerbic style by Barbara Ward in the book she prepared for the United Nations Conference on Human Settlements. She reviews the intellectual and technological revolutions of the sixteenth century and beyond and shows how the creation of a world system with Europe at the core aided in the west's development. And she indicates unequivocally how many of the problems of Third World cities derive from the position they formerly occupied in that system. In her words:

> This is the background—of long subservience to an economic system designed for other nations' interests, of an infrastructure still geared to those interests, of relatively stagnant agriculture, little or no industry, and export cities dominating the urban scene—that we must bear in mind when we examine the settlements of the developing world.

We must take her advice seriously; otherwise, we shall miss important clues to understanding the subject of this book on urbanization in the Third World.

The Third World is, of course, no monolith. As we shall see in the statistical section, the subregions of Asia, Africa and Latin America are at quite different levels of urbanization and economic development, and within these vast continents there are further subregional variations which span a wide range. Some of these variations can be accounted for by the types of resources locally available; some can be attributed to differences in the density of population; and still others may be due to the policies followed by governments. But to a very significant extent, the differences derive from the past and from the role each subregion played as a peripheral satellite to the industrial core of Europe. (Important also is the current relationship of each to a widened core which now includes the so-called Second World [Soviet bloc] as well as the Asian latecomer, Japan).

Some zones (not many) were not directly colonized; even these countries, primarily Asian, did not escape the effects of colonialism for, although nominally independent, they lost control of their coastal zones and almost always of their economic autonomy (viz., the Chinese treaty ports). Most countries, however, were incorporated into the overseas empires of Spain, Portugal, Britain, Holland, France and later Germany and Italy. Clearly, differential effects of colonialism upon the cities and economies of the colonies were to be expected, depending upon who the colonizer was, when control was established and for what major purposes (settlement or extraction of resources without settlement), the extent to which prior urban traditions existed in the colony, the ratio of "natives" to settlers, and many other variables. Clearly, just as this volume cannot hope to cover fully the historical background of contemporary urbanization in the Third World, so it cannot possibly survey all the variations from region to region which are the product of that history.

Here we present only one case, that of Latin America, to demonstrate how the unique form of early colonialism there left a lasting imprint both on the spatial network of the entire urban system and on the systems of social stratification and land use within individual cities throughout that continent. The history of Latin America's colonization dates back to the sixteenth century and is therefore the longest of any Third World region. And, as Portes points out, by the end of the first century of colonization, the entire continent-wide urban network which persists to this very day was already firmly established. Not only were there to be few additions to the system of cities during the ensuing centuries but there were to be no major shifts in the basic relation of the city to its hinterland. Cities in Latin America were, from their beginning, politico-administrative centers designed for economic exploitation of the countryside. They were controlled by a land-owning elite which dominated not only the internal economy of the city and trade with Europe but also the surrounding rural areas. This elite remained insensitive to the needs of rural subordinates, whether the latter remained in the countryside or whether, as later occurred, they moved into the cities. It is striking that the themes Portes identifies as characterizing urbanization in Latin America from its earliest days are the same ones which preoccupy today's students of urbanization on that continent. One expects to find very different conditions in other parts of the Third World which experienced different historical roots.

If these are the heritages of the past, what is the situation of today and what does the future hold in store? Possibly, there will be a new reversal of positions or at least a greater balance between east and west. By the final quarter of the twentieth century, some three-quarters of the world's population were living in countries and territories still classified as "developing" (including China). Furthermore, the populations in these lesser developed countries were increasing at two or more times the rate of the developed world. Even though a sizeable proportion of Third World people are still rural, the percentage of population living in cities has been increasing dramatically. The gap between the levels of urbanization (as measured by the percentage of total population living in urban places) in the developed and developing worlds is becoming much narrower. And because the total populations are so much larger in the latter, the gap between the *number* of cities and the *number* of urbanites in both regions has already closed. Soon, before the end of the present century, most cities and most urbanites in the world will be located in the Third World.

The statistics assembled in this section, however, show that increases in the standards of living in Third World countries have in no way kept pace with the increases in the levels of urbanization. The Third World continues to be very poor and, indeed, the gap between per capita levels of income or of Gross National Product (GNP) of the wealthiest countries and the poorest ones has, contrary to ideology or wishful thinking, grown wider rather than converged in recent decades. By 1973, indeed, the developed countries (both market economies and socialist) which contained less than 30 percent of the world's population were receiving over 80 percent of the world's estimated GNP. The average North American could call upon $6,130, as contrasted with $800 in Central America, $290 in Africa, and $200 in Asia.

Given these discrepancies, it should not surprise us to find that problems in Third World cities are overwhelming in their magnitude. However, these problems must be put into the context of economic development and into the context of the world system if their true causes are to be identified and if efforts to ameliorate them are to address underlying causes rather than merely to treat their symptoms.

CHAPTER 5

Daily Life in China on the Eve of the Mongol Invasion 1250–1276

JACQUES GERNET

[I]t is possible . . . to reconstruct the town [Hangchow] as it was in 1274. The lay-out was simple: a large thoroughfare, which became the Imperial Way after the court had installed itself at Hangchow, traversed the city from north to south, terminating at the north gate of the Imperial Palace, and then, beyond the Palace, continuing southwards to the altar for the sacrifices to Heaven and Earth. This thoroughfare was crossed at right angles by others running east and west. In addition, several canals ran parallel to the Imperial Way. [Together with] the suburbs . . . it would appear that the whole built-up area must, by 1274, have covered a surface of between seven and eight square miles. . . .

The space which nature had measured out to Hangchow in such niggardly fashion was occupied in the thirteenth century by the biggest urban concentration in the world at that time. The largest cities of Europe, with a population of several tens of thousands, were nothing but petty market-towns in comparison with the "provisional capital" of China. The population of Hangchow had, by 1275, gone beyond the million mark. . . .

Hangchow itself was a city of multiple functions. It was at once the capital, since it contained the Imperial Palace and the offices of the central government, the seat of a large prefecture, and also the seat of two sub-prefectures, the offices of which were situated within the ramparts. In addition, it was an important centre of trade. . . .

"The city of Hangchow," writes one of its inhabitants, "is large, extensive and overpopulated. The houses are high and built close to each other. Their beams touch and their porches are continuous. There is not an inch of unoccupied ground anywhere." For his part, Oderic de Pordenone, who described Hangchow some years later than Marco Polo, expressed his admiration thus:

Reprinted with permission of Macmillan Publishing Co., Inc. from Jacques Gernet, H. M. Wright, trans., *Daily Life in China on the Eve of the Mongol Invasion 1250–1276* (London: George Allen & Unwin Ltd., 1962), pp. 27–55. This material has been edited for this publication.

> This city is greater than any in the world, and is quite 100 miles round; nor is there any span of ground that is not well inhabited, and often there will be a house with ten or twelve families. This city has also great suburbs containing more people than the city itself contains. It has twelve principal gates; and at each of these gates at about eight miles are cities larger than Venice or Padua might be, so that one will go about one of those suburbs for six or eight days and yet will seem to have travelled but a little way.

There was a dearth of building land, and this, together with the increase in population, necessitated the construction of dwelling-houses of several storeys.... Western travellers so frequently mention multi-storeyed buildings in Hangchow that we cannot doubt their existence, despite the lack of information in Chinese sources, which only contain vague references. It would, moreover, be difficult to understand how a town so circumscribed in area could house a population of over a million people in 1270 if the houses had mostly been single-storeyed, as in the other towns in China at that time. The multi-storeyed houses gave Hangchow a typically urban appearance, and they greatly increased the density of its population. In view of the latter factor, this mode of construction had important effects on the general style of life and on the intensity of social relations....

The hills to the south, where the Imperial Palace lay, were the residential area of the rich. The high officials lived on the hill of the Ten-thousand Pines, and the merchants who had made their fortune in the maritime trade lived on Mount Phoenix, further to the south. There, summerhouses and pavilions were scattered about amidst groves and gardens.

On the other hand, the low-lying part of the walled city, north of the palace, was overpopulated, and in the poorer districts off the Imperial Way the density was probably 324 persons per acre. It was there that the multi-storeyed houses stood, giving on to alleyways which were narrow and congested. There is a striking contrast between the spacious planning of the wide thoroughfares that cut right across the city, the monumental character of the walls and the gates, the splendour of the official edifices and of the temples, and the narrowness of the alleyways and chaotic overcrowding of the poorer districts. True, this contrast was not peculiar to Hangchow. It was common to all the towns of the empire, and it seems to reflect the political state of affairs: to be a symbol, as it were, of the juxtaposition of an omnipresent government and a vast population living in its shadow, cheek-by-jowl with those in power yet taking no part in the management of affairs....

Wherever possible, water transportation was preferred by the people of Hangchow.... It was in fact only on the Imperial Way that carts were to be seen; but then they were only light vehicles exclusively used for passengers.

"In the main street of the city," says Marco Polo, "you meet an infinite succession of these carriages passing to and fro. They are long covered vehicles, fitted with curtains and cushions, and affording room for six persons; and they are in constant request for ladies and gentlemen going on parties of pleasure. In these they drive to certain gardens, where they are entertained by the owners in pavilions erected on purpose, and there they divert themselves the livelong day, with their ladies, returning home in the evening in those same carriages."

Rich people also often went about on horseback or, if they were ladies, in chairs carried by porters. These chairs had a canopy and small folding doors. As well as the bearer-shafts, they had a third pole which rested on the shoulders of the porters.

In the streets and alleys the only method of getting goods transported was by means of porters, or, if the load was too great, donkeys and mules. . . . Porterage by men, very general and cheap because of the abundance of manpower, was done by means of a bamboo pole from which suspended cloth bundles, wicker baskets, large earthenware jars or square wooden tubs.

The Imperial Way, the city's finest thoroughfare, stretched for more than three miles, from the north gate of the Imperial Palace to the gate on the north-west ramparts. It was sixty yards wide and is thus described by Marco Polo:

> . . .the main street of which we have spoken above, which runs from one end of the city to the other, is paved like this with stones and with bricks ten paces along either side, but in the middle it is all filled with small and fine gravel, with its vaulted conduits which lead the rain waters into the canals near by, so that it always stays dry.

This was certainly a magnificent thoroughfare compared with the streets of our towns in the Middle Ages. . . .

Almost all the streets in Hangchow seem to have been surfaced with large paving-stones such as can still be seen today on the roads bordering the lake and in the environs of the town. . . .

Finally, a high standard of cleanliness was maintained in Hangchow. This was in any case essential, particularly at the hottest time of the year, in July and August, otherwise epidemics would have spread quickly and taken a heavy toll in such an overcrowded town. The streets were cleansed by the public authorities, who had the refuse removed by boat. . . .

The houses of the rich had cesspools. But the poor people who lived in the multi-storeyed houses in the poorer districts were obliged to use "horse buckets" which the scavengers came to collect each day. The nightsoil was doubtless used as manure for the gardens in the environs and for the vegetable plots in the eastern suburbs. The scavengers, commonly called "the pourers," formed a corporation. Each one had his regular customers, and would be ill-advised to take away theirs from his companions, for should he do so, notes a contemporary with amusement, the affair might grow so acrimonious that it might reach the point of being brought before the tribunal of the prefecture, and the plaintiff would not be satisfied until the case had been decided in his favour.

The establishment of the court at Hangchow seems to have been of great profit to the city; everything was done to improve its appearance and to facilitate the flow of traffic and reduce its dangers. Nevertheless, in spite of the double network of canals and streets, traffic congestion was not unknown on the main thoroughfares. It occurred chiefly at the gateways to the city, which were too narrow for the seething mass of carriages, horses, donkeys and porters, and also at the approaches to the bridges, which were often narrow and hump-backed, or, as the more poetic Chinese expression has it, "rainbow" shaped. . . .

Thanks to the river, the lake, the paved roadways leading to the city and the canals which ran through it and linked it with the big neighbouring cities, Hangchow was easily supplied with every variety of commodity. A local saying enumerating the products of daily consumption linked them with the four cardinal points: "Vegetables from the east, water from the west, wood from the south and rice from the north." And indeed firewood and timber were brought by boats which went upriver towards the interior, the vegetable gardens were numerous in the eastern suburbs where the vegetable market was held, rice came by canal from the plains which extended to the north of Hangchow and on the other side of the Yangtze, and finally, the townspeople's only drinking water was that of the lake to the west.

There were reservoirs known as the Six Wells in the northwest of the city, within the ramparts, which fed into the lake. Made fit for use in the eighth century through the cares of a governor of the city and repaired in the eleventh century, when earthenware conduits had been placed in the channels leading into the lake, they were the sole source of fresh water in the neighbourhood. Water from the river, which, because of the tides, was brackish, contaminated all the wells dug in the surroundings of Hangchow. For this reason, jealous care was lavished on maintaining the purity of the lake water. . . . It is not known how the water from the Six Wells was distributed in the town; probably pole-porters carrying buckets went about the streets and alleyways of Hangchow.

The two products consumed in largest quantity by the townspeople, and which were their basic foodstuffs, were rice and pork. "If," says a contemporary account, "one excepts the private mansions of princes, the houses of high officials, the residences and dwellings of the rich and of all salaried persons in the administration, the daily consumption of the ordinary people of the town is not lower than 70 to 140 tons of rice." Another account gives different and more precise figures derived from a source which seems reliable. "I once had the occasion," says its author, "of hearing the head of the officials employed at the prefecture say, that if one excludes the families who obtain their supplies directly from outside, there are in the city (within the ramparts) 160,000 to 170,000 people who have to buy their rice in the shops. Now, if one reckons an average of 2½ lb. daily consumption per head, it is clear that a supply of at least 210 to 280 tons of rice is required daily. And neither the suburbs to the north and to the south, nor visiting merchants, nor travellers, are included in this calculation." Probably the upper classes, who obtained their rice by direct supply, consumed less of this commodity than was the case among the lower classes, because their diet was more varied. But in all, several hundred tons of this cereal must have been brought into Hangchow every day, and the rice-barges coming from the great rice-growing regions of Chekiang and of present-day Kiangsu passed unceasingly along its canals. It was a traffic which went on uninterruptedly night and day, and it was even necessary for Hangchow to import rice from the Huai valley and, by sea, from the region of Canton, nearly a thousand miles away. The barges were unloaded at the Ricemarket Bridge and at the Black Bridge, in the northern suburbs, where their cargo was sold to the innumerable shops and restaurants of the town. . . .

But the districts where the briskest retail trade was done in Hangchow were those adjoining the Imperial Way. It was there that the luxury trade, the best shops, and most of the big taverns and fashionable tea-houses were to be found.

Let us quote Marco Polo once again. Although what he says about the location of the markets within the ramparts does not correspond with the information given by Chinese sources of around 1275, at the end of the Sung period, for the rest his account is exact:

> [There] are ten principal markets, though besides these there are a vast number of others in the different parts of the town. The former are all squares of half a mile to the side, and along their front passes the main street, which is forty paces in width, and runs straight from end to end of the city, crossing many bridges of easy and commodious approach. At every four miles of its length comes one of those great squares of two miles (as we have mentioned) in compass. So also parallel to this great street, but at the back of the market places, there runs a very large canal, on the bank of which towards the squares are built great houses

> of stone, in which the merchants from India and other foreign parts store their wares, to be handy for the markets. In each of the squares is held a market three days in the week, frequented by 40,000 or 50,000 persons, who bring thither for sale every possible necessary of life, so that there is always an ample supply of every kind of meat and game. . . .
>
> These markets make a daily display of every kind of vegetables and fruits. . . . From the Ocean Sea also come daily supplies of fish in great quantity, brought twenty-five miles up the river, and there is also a great store of fish from the lake, which is the constant resort of fishermen, who have no other business. . . .
>
> All the ten market places are encompassed by lofty houses, and below these are shops where all sorts of crafts are carried on, and all sorts of wares are on sale, including spices and jewels and pearls. Some of these shops are entirely devoted to the sale of wine made from rice and spices, which is constantly made fresh, and is sold very cheap.

. . . The intense commercial activity, the extreme density of population, and the constant influx of visitors explain why there were so many places where inhabitants and travellers alike could eat, meet and amuse themselves. The town boasted a multitude of restaurants, hotels, taverns and tea-houses, and houses where there were singing-girls. The rich met at Hangchow's celebrated tea-houses. . . .

The fashionable taverns were to be found, as were no doubt the big tea-houses also, in one-storeyed houses which did not give directly on to the street, but on to a courtyard with covered arcades. . . .

In these taverns, rice-wine of many kinds was served in little silver cups—a detail which in itself, as a contemporary remarks, shows how well-to-do the proprietors were. Various things to eat were served along with the drinks. A menu was handed to the customers, from which they could choose the dishes they wanted. . . .

The multitude and variety of commercial activities was one of the attractions of Hangchow. But not the only one. Outside the ramparts, mainly by the lakeside and in the southern suburbs, were parks and gardens to which the people had free access. On feast days these places were invaded by a holiday crowd who came to admire the rare flowers and exotic trees. Some of them, taking with them something to eat, and carrying musical instruments, made a day's excursion on the shores of the lake or on the hills surrounding it. Others hired boats for several cash coins and enjoyed the pleasure of seeing some of the most beautiful and celebrated scenery in China pass before their eyes. . . .

By 1275 the lake was over nine miles in circumference and nine feet deep. Military patrols, under the command of specially appointed officials, looked after its policing and maintenance; it was forbidden to throw any rubbish into it or to plant in it lotuses or water-chestnuts. The care bestowed over centuries on the maintenance of the lake bears witness to the extraordinary feeling for and delight in beautiful scenery shown by the Chinese of T'ang and Sung times. Its beauty spots were preserved with jealous care, and every new building had to blend in harmoniously with its surroundings. . . .

According to a description of Hangchow of 1275, there were always hundreds of boats of all shapes and sizes on the lake: small boats such as those to be seen on the canals of the city, with a large oar fixed to the stern which the boatman worked with his foot, fast boats propelled by wheel or pedals; big flat-bottomed boats 90 to 180 feet long, which could carry thirty, fifty or even 100 passengers; boats 18 to 27 feet long which could hold about twenty people. The finest workmanship had gone into the construction of these boats, and the upper parts were ornamented with fine carvings and painted in bright colours. . . .

In his description of Hangchow, Marco Polo furnishes information very close to that found in Chinese sources of the Sung period:

> On the Lake of which we have spoken there are numbers of boats and barges of all sizes for parties of pleasure. These will hold ten, fifteen, twenty or more persons, and are from fifteen to twenty paces in length, with flat bottoms and ample breadth of beam, so that they always keep their trim. Anyone who desires to go a-pleasuring with the women or with a party of his own sex, hires one of these barges, which are always to be found completely furnished with tables and chairs and all other apparatus for a feast. The roof forms a level deck, on which the crew stand, and pole the boat along whithersoever may be desired, for the Lake is not more than two paces in depth. . . . And truly a trip on this Lake is a much more charming recreation than can be enjoyed on land. For on the one side lies the city in its entire length, so that spectators in the barges, from the distance at which they stand, take in the whole prospect in its full beauty and grandeur, with its numberless palaces, temples, monasteries, and gardens, full of lofty trees, sloping to the shore. . . .

Outside the east gates the scenery was more austere: sea-going junks with square sails made of matting or of dark-coloured cloth, fishing boats and transport barges, were anchored near the banks, or sailed along the river, which is between one and two miles wide at this point. The great junks that sailed on the high seas rarely came to Hangchow because of the sandbanks which encumber the estuary of the Che river, and the boats usually seen there were of a smaller type with six to eight oars known as "wind-piercers. . . ."

This brief description of the city would not be complete if mention were omitted of the numerous entertainments which the townspeople could enjoy in the streets (jugglers, marionettes, Chinese shadow-plays, story-tellers, acrobats . . .) and in the "pleasure grounds" in which huge popular theatres were to be found where people of all conditions met and jostled together. There were daily performances there, and exhibitions of dancing, singing and instrumental music. Hangchow seems to have lived in an atmosphere of continuous feasting. The incessant activity in the streets and markets, the pleasures, the luxury, and the gaiety of the town: all this makes a vivid contrast with the poverty in the countryside and the hard, monotonous and frugal life of the peasants.

Without any question, Hangchow in the thirteenth century was, to quote again the words of Marco Polo, "the most noble city and the best that is in the world."

CHAPTER 6

The Condition of the Working-Class in England

FRIEDRICH ENGELS

Every great city has one or more slums, where the working-class is crowded together. True, poverty often dwells in hidden alleys close to the palaces of the rich; but, in general, a separate territory has been assigned to it, where, removed from the sight of the happier classes, it may struggle along as it can. These slums are pretty equally arranged in all the great towns of England, the worst houses in the worst quarters of the towns; usually one or two-storied cottages in long rows, perhaps with cellars used as dwellings, almost always irregularly built.... The streets are generally unpaved, rough, dirty, filled with vegetable and animal refuse, without sewers or gutters, but supplied with foul, stagnant pools instead .... Further, the streets serve as drying grounds in fine weather; lines are stretched across from house to house, and hung with wet clothing.

Let us investigate some of the slums in their order. London comes first, and in London the famous rookery of St. Giles.... St. Giles is in the midst of the most populous part of the town, surrounded by broad, splendid avenues.... It is a disorderly collection of tall, three or four-storied houses, with narrow, crooked, filthy streets.... A vegetable market is held in the street, baskets with vegetables and fruits, naturally all bad and hardly fit to use, obstruct the sidewalks still further, and from these, as well as from the fish-dealers' stalls, arises a horrible smell. The houses are occupied from cellar to garret, filthy within and without .... But this is nothing in comparison with the dwellings in the narrower courts and alleys between the streets, entered by covered passages between the houses, in which the filth and tottering ruin surpass all description.... Heaps of garbage and ashes lie in all directions, and the foul liquids emptied before the doors gather in stinking pools. Here live the poorest of the poor, the worst paid workers with thieves and the victims of prostitution indiscriminately huddled together, the majority Irish, or of Irish extraction....

Reprinted from Friedrich Engels, *The Condition of the Working-Class in England* (Moscow: Progress Publishers, 1973), pp. 66, 67, 71, 73, 123, 124, 125, 126. This material, first published in 1845, has been edited for this publication.

... [T]hey who have some kind of shelter are fortunate, fortunate in comparison with the utterly homeless. In London fifty thousand human beings get up every morning, not knowing where they are to lay their heads at night. The luckiest of this multitude, those who succeed in keeping a penny or two until evening, enter a lodging-house ... where they find a bed. But what a bed! These houses are filled with beds from cellar to garret, four, five, six beds in a room; as many as can be crowded in. Into every bed four, five, or six human beings are piled, as many as can be packed in, sick and well, young and old, drunk and sober, men and women, just as they come, indiscriminately.... And those who cannot pay for such a refuge? They sleep where they find a place, in passages, arcades, in corners where the police and the owners leave them undisturbed. A few individuals find their way to the refuges which are managed, here and there, by private charity, others sleep on the benches in the parks close under the windows of Queen Victoria.

Let us leave London and examine the other great cities.... Let us take Dublin first, a city the approach to which from the sea is as charming as that of London is imposing.... The city, too, possesses great attractions, and its aristocratic districts are better and more tastefully laid out than those of any other British city. By way of compensation, however, the poorer districts of Dublin are among the most hideous and repulsive to be seen in the world.... Some idea of the manner in which the poor are here crowded together may be formed from the fact that, in 1817, according to the report of the Inspector of Workhouses, 1,318 persons lived in 52 houses with 300 rooms in Barral Street, and 1,997 persons in 71 houses with 393 rooms in and near Church Street; that:

> In this and the adjoining district there exists a multitude of foul courts and alleys; many cellars receive all their light through the door, while in not a few the inhabitants sleep upon the bare floor ...; Nicholson's Court, for example, contains twenty-eight wretched little rooms with 151 human beings in the greatest want, there being but two bedsteads and two blankets to be found in the whole court.

[Engels goes on to discuss employment:]

... English manufacture must have, at all times save the brief periods of highest prosperity, an unemployed reserve army of workers, in order to be able to produce the masses of goods required by the market in the liveliest months.... This reserve army, which embraces an immense multitude during the crisis and a large number during the period which may be regarded as average ... is the "surplus population" of England, which keeps body and soul together by begging, stealing, street-sweeping, collecting manure, pushing hand-carts, driving donkeys, peddling, or performing occasional small jobs. In every great town a multitude of such people may be found. It is astonishing in what devices this "surplus population" takes refuge.... Most of the "surplus" betake themselves to huckstering. On Saturday afternoons, especially, when the whole working population is on the streets, the crowd who live from huckstering and peddling may be seen. Shoe and corset laces, braces, twine, cakes, oranges, every kind of small articles are offered by men, women, and children; and at other times also, such peddlers are always to be seen . . . with cakes and ginger beer.... Matches and such things, sealing-wax, and patent mixtures for lighting fires are further resources for such venders. Others, so-called jobbers, go about the streets seeking small jobs. Many of these succeed in getting a day's work, many are not so fortunate. . . When these people find no work and will not rebel against society, what remains for them but to beg? And surely no one can wonder at the great army of beggars, most of them able-bodied men, with whom the police carries on perpetual war. But the beggary of these men has a peculiar character. Such a man usually goes about with his family singing a pleading song in the streets or appealing, in a speech, to the benevolence of the passers-by. And it is a striking fact that these beggars are seen almost exclusively in the working-people's districts, that it is almost exclusively the gifts of the poor from which they live.... And he among the "surplus" who has courage and passion enough to resist society, to reply with declared war upon the bourgeoisie to the disguised war which the bourgeoisie wages upon him, goes forth to rob, plunder, murder, and burn!

Of this surplus population there are, according to the reports of the Poor Law commissioners, on an average, a million and a half in England and Wales [not including Scotland and Ireland where conditions are even worse].

CHAPTER 7

The Home of Man

BARBARA WARD (LADY JACKSON)

When we look at the explosion of growth and change in the twentieth century city, the first question must surely be how and why mankind ever came to reach this extraordinary upheaval in all traditional scales and types of settlement. What process of history brought him here? How did he make the quantum jump from settlements of 2,000 to the megacity of twenty million? . . .

One of the remarkable facts about the human record is the length of time during which, for all their cultural differences, their vast varieties of style and taste and worship and art, the world's great civilizations did not seem to differ very much in the actual physical arrangements and civic institutions of their settled life. Village society, which began some 15,000 years ago, with the invention of settled agriculture, could still be found in much the same physical shape in parts of twentieth-century Asia and Africa.

Cities built three thousand years before European cities in the Middle Ages had as often as not better paved streets, more elaborate sewage systems, greater convenience in bathrooms and lavatories, larger monuments, and more elaborate city walls. Mohenjo-daro, built on the Indus River around 3500 B.C. is certainly no less elaborately laid out than, say, sixteenth century Paris. The "containers" of civic life show extraordinary continuity.

It can be argued that man's activities inside these cities show a similar continuity. They all have their origins in the need to develop large-scale systems of water management in the great river valley—in Egypt, Mesopotamia, North India, China—and this more elaborate economic base brought more people and hence more strangers together in one place, varied their jobs and responsibilities, increased the exchange of goods, and demanded new institutions . . . to replace the old small-scale loyalties of the agricultural village. . . .

Reprinted from Barbara Ward, *The Home of Man* (Toronto: McClelland and Steward, 1976), pp. 11–28, by permission of W. W. Norton and Co., Inc. and the Canadian publisher, McClelland and Steward. This material has been edited for this publication.

And this process of accumulation and expansion had a further consequence. The invention of the city in the fourth millennium B.C. appears to have coincided with the invention of war. . . . We can adduce formidable material reasons for the invention of empire and the imperial city. One certainly played a part in the earlier wars—the temptation to secure manpower by capturing prisoners. . . . Another reason lay in the steady increase in international trade. . . . A vital issue in the whole unfolding of Mesopotamian trade—one which contemporary Japanese and British corporations will easily comprehend—was the shortage of raw materials. . . . It is therefore not unreasonable to guess that whether the Sumerian traders resembled the East India Company or Soviet state traders, . . . the ability to trade freely from Harappa on the Indus to Hatussas in Anatolia was a fundamental economic interest. . . . Thus, at the very start of the human city, we find an almost exact playing out of the commercial dramas of Europe's recent colonialism.

There are thus enough material explanations for the emergence of empire—the "protection of trade routes," monopoly control of resources, rounding up captives for manpower. And the reasons are quite sufficient to explain the steady recurrence of the phenomenon and its equally steady collapse. It starts up because a dominant group has these material interests. It is maintained so long as they can contain the resentment of the peoples they exploit. And, at last, they themselves, wealthy but weakened by internal discontent, are picked off by the next rising ruler with sufficient ambition and power to restart the cycle. . . .

If this is the record history presents us with—the ever-renewed round of conquest predetermined by an unshakable commitment to material ambition and greed—can it tell us anything new when our predicaments can be so easily explained in much the same terms? The city-states have become nation-states. The cities themselves have become megalopolises. One round of empire is over. . . . It is easy to conclude that the melancholy wheel is still spinning. . . .

Easy—but false: if the record is more carefully examined, the dominant themes begin to look less closed and predetermined. No one can deny the influence of material conditioning. No history of any civilization is without its cycles of war and conquest. But these are not the sole forces by which human destiny has been driven forward in the last four millennia of recorded time. What emerges from the record is much more complex, much more perplexing, infinitely more unlikely and, as a result, much more open and promising. . . .

. . . In many ways, until the nineteenth century, the physical surface of human existence does not change very much. Busy cities, artisans' quarters, market places, ships at the quayside waiting to sail for the distant Indies—they were a familiar sight in Sumeria. They were not very different in medieval Europe. The scale was increasing. We read of Venice in the fifteenth century with 50,000 workers and seamen employed in the dockyards, stores, and workshops of the arsenal. But there is nothing unfamiliar about the city's pursuits. We are still with Gordon Childe's "story of accumulating wealth, of increasing specialization of labor and of expanding trade."

Yet the shaping ideas are undergoing a steady and fateful change.... We can see the ... fateful intermingling of old and new in the concepts of wealth and property. Private property became enshrined as an absolute right in the struggles against Stuart and French absolutism. Yet its exclusion of the workers and the poor from full economic participation in society did not only contradict many medieval institutions and traditions.... It also ran counter to a deeper strain of equality and human dignity.... Concentrations of landed wealth, merchant enterprise, the growing fortunes of colonial trade overlaid but could not extinguish these uncomfortable contradictions....

The next centuries are marked above all by the steady expansion of scientific knowledge and, from the middle of the nineteenth century onward, its increasing application to man's daily work.... In intellectual terms, the intensity, the virtuosity, the incredible triumphs of experimental science in the last four hundred years are unequaled in human history.... Nothing detracts from the intellectual splendor of the discoveries nor from the rigorous and unambiguous dedication to truth entailed in scientific method. But in the end, the understanding and the power are available for any use—not necessarily "benefit," not necessarily "relief." And when we look at the human energies that in fact mobilized the whole modernizing process of science and technology and ultimately unleashed the fateful energies of the atom, we find [a] ... phrase of Bacon's perhaps more relevant. The outcome did not depend only upon the scientist's passionate search for truth. It was also driven forward by "the idols of the market, and the idols of the tribe."

During the first critical period of industrial development, the Atlantic "tribes," followed as latecomers by Japan and Russia, had as the base of their expansion a world-wide system of markets which on the one hand they fought over but, on the other, they controlled and shared. One of its earliest features was to build up supplies of raw materials in the New World—with the help of slave labor. By the nineteenth century, the goods manufactured from Europe's and North America's own resources, above all coal and iron, had knocked out much of the old luxury handicraft exports from Asia. India and China, too, became exporters of raw materials —tea, jute, cotton—Malaysia of rubber and tin, Latin America of cocoa, coffee, tin, and copper in exchange for Manchester textiles and Pittsburgh ironware. Africa was brought into the system on the same basis. Having long supplied slaves and gold, it began to be developed for new materials. Palm oil for soap became an indispensable commodity amid unspeakable filth in the new industrial order. Cocoa and coffee were added, diamonds and gold simply appropriated. In the twentieth century, the most fateful of all raw materials came into play—petroleum from Arabia and the Caribbean and the East Indies. This underlying pattern of trade—raw materials from the then colonial or near-colonial world in exchange for manufactured products from Europe and North America—provided the larger framework within which Western industrialization took place. The early centuries of trade gave it a higher base of wealth from which to start. Thereafter, since most of the mines and plantations producing the materials were based on Western investment and control, since all the value derived from processing their products was secured for Western factories, and since the final goods were sold back locally at a rate of return controlled by imperial management, the process made no small contribution to "primitive accumulation" in the West. There can be no question but that this wider framework was a prime stimulus to Western industrialization just as control of all Siberia's mineral reserves played a key role in Soviet development.

All in all, the shape of urban and industrial development in the nineteenth century was one of cities being pushed into existence by the growing requirements of thrusting economic growth. The figures bear this out. Almost invariably the percentage of the work force in industry was higher than the proportion of people living in towns of more than 20,000 inhabitants. In France in 1856, twenty-nine percent of the workers were in industry, but only 10.7 percent of the French people lived in urban areas. In 1890, Sweden had 10.8 percent of its people urbanized while twenty-two percent of the work force were in industry; for Austria the percentages were twelve percent and thirty percent. The most remarkable example is Switzerland at the same period. With cantonal decentralization and dispersed settlements, it contrived to have over forty-five percent of its work force in industry and only thirteen percent of its people in towns. . . .

It would hardly be an exaggeration to say that every historical and local circumstance that has worked to better the condition of developed settlements has had the opposite effect on the developing world. Take, first of all, the wider setting of international trade. It was precisely designed in colonial times to bring the surpluses back to the metropolitan power. Investment occurred only in those services which underpinned the export sector—routes from mines and plantations in the interior, ports growing into large coastal cities through which materials were dispatched and manufactures returned. These cities gave virtually no wider stimulus to their own hinterlands, and the lines of communication, all running to the coast, with few or no lateral links, gave an almost visual impression of what the great Dominican economist Pere Lebret called "the milch cow economy." In a very real sense, these ports—Buenos Aires, Lagos, Calcutta, Shanghai, with their modern buildings, installations, and services—were as much part of the developed world economy as today's European *bidon-villes* are a projection of the poorer countries.

It was no part of colonial policy to stimulate local manufacture. At the turn of the nineteenth century, India's textile manufacturers still could not secure protection against competition from Lancashire, and by the time the Tata family made their single-minded and determined effort to build an Indian steel industry, such vital stimulants to industrial growth as the building of the railways had already been completed—by British engineering firms and steelmakers. Nor was there much innovating activity in agriculture outside the plantation and export sector. Having introduced feudal tenure into Latin America, the Iberians maintained it. As late as 1960, some ninety percent of the agricultural land was still owned by ten percent of the people. African food production remained in the condition of communal self-sufficiency. In India, save in the large extension of irrigation, there was nothing to compare with Europe's technical revolution in agriculture. It was the great strength of Japan's independent leap into modernization in the late nineteenth century that its leaders were not tied by colonial dependence to Western controls, models, or *laissez faire.* The "Meiji" revolutionaries gave agricultural reform the first priority and built up early industrialism on small-scale enterprise. As late as the 1930s, the great majority of Japanese firms had fewer than thirty employees.

This is the background—of long subservience to an economic system designed for other nations' interests, of an infrastructure still geared to those interests, of relatively stagnant agriculture, little or no industry, and export cities dominating the urban scene—that we must bear in mind when we examine the settlements of the developing world.

CHAPTER 8

Urban Latin America: The Political Condition from Above and Below

ALEJANDRO PORTES

HISTORICAL ANTECEDENTS

The forces that have given rise to the current physical and social divisions of the Latin American city must be sought, beyond recent developments, in the early history of urban growth in the region. An impressive continuity exists between the organization and class divisions of the early colonial city and the present situation. General orientations toward use of the urban land, position and goals of the elite, role of government, and treatment accorded to the disenfranchised were crystallized early in the colonial period and, with amazing resilience, managed to persist despite the pressure of major external events in ensuing years. Imported capitalism and the institutions of Northern European bourgeois democracy never replaced the basic corporatist orientation—the framework of empire—built by Spain and Portugal in the New World cities. For this reason, the study of determinants of the current forms of urban poverty in Latin America must start with the colonial beginnings of present cities. No attempt will be made to cover in detail the history of Spanish and Portuguese urban settlements on the continent, . . . but rather we shall try to isolate the basic structural forms and cultural themes that provided the framework for present patterns of development of these cities. In the attempt to highlight such dominant trends, some violence will be done to inevitable historical exceptions. . . .

URBAN DOMINANCE

Unlike "natural" forms of urban growth, where the city emerges as a service center for an already settled hinterland, the Spanish and Portuguese cities in the New World were established in unknown and often hostile territories as centers of conquest and political control. They were

Reprinted from Alejandro Portes and John Walton, *Urban Latin America: The Political Condition from Above and Below* (Austin: University of Texas Press, 1976), pp. 7–25, by permission of the publisher and the author. This material has been edited for this publication.

founded by fiat, projected on the land as branches of an imperial strategy rather than crystallized as necessary centers for expanding commerce. . . .

A gradualistic strategy of colonization, such as the one employed by the British in North America, would have meant restricted coastal enclaves moving slowly inward after the initial territory had been settled. Casting away this model, the Spaniards moved directly inland in search of indigenous centers of political control and other points where immediate wealth could be found. In the cases of the Inca and Aztec empires, miniscule military forces subdued the leaders and capital cities to superimpose Spanish political institutions over a well-established administrative structure. The Spaniards seldom destroyed to build on their own terms but rather adapted opportunistically to what they found in order to serve their immediate economic interests. Cuzco and Tenochtitlán thus became centers of the new empire as they had been of the old, and, from their inland locations, the conquest moved toward the coast and in several directions.

This pattern of colonization, the direct opposite of the British gradualistic model, permitted Spain to conquer and control an entire continent in a few years with a very small occupying force. It was, it is true, a tenuous and fragile control at first, but it covered the entire territory and tended to solidify as the years passed. Gradually, cities abandoned the role of military enclaves for imposing European authority and became integrated as the administrative, economic, and cultural centers of vast regions. This urban-centered strategy of colonization had two immediate consequences: First, it restricted, from the start, the possible emergence of a "frontier" in the North American sense. Especially in the case of Spanish America, the major part of the territory was immediately, albeit tenuously, controlled. Subsequent colonization was directed at filling in and solidifying juridically existing holdings, seldom at creating new ones via expansion from initial coastal settlements.

Second, it established, from the start, the supremacy of city over countryside. The foundation of cities did not respond to the pressing need for urban services by established agricultural settlers or to the actions of an increasingly independent class of burghers concentrated around a marketplace. Rather it followed the strategic requirement of concentrating scarce human resources in a restricted, and therefore militarily defensible, perimeter. The city did not arise to serve, but to subdue. From it the Spaniards moved out to a hostile environment to conquer, control, and indoctrinate the surrounding populations. Conquerors lived, by and large, in the city, while conquered remained in the countryside.

This role of cities and the resulting power differential between rural and urban populations were immediately reflected in the judicial order. The *Ordenanzas de descubrimiento y población* of 1573 contained 148 articles, of which 44 were dedicated exclusively to the establishment of new cities and towns, selection of sites, designation of the form that cities should adopt, and distribution of land. . . .

Unlike that of the city in Europe and North America, jurisdiction of the Latin American city was not restricted to a specific area and did not leave the countryside in the hands of rural proprietors. Chartered Spanish American cities "owned" their hinterlands, both in the sense of economic proprietorship—since lands were granted in the king's name by city authorities—and in the sense of politico-administrative control. It was the function of cities to ensconce early settlers on the land, as strategy for consolidating imperial control and as means for fulfilling personal ambitions. The royal or viceroyal authorization granting a settlement the status of *ciudad* carried with it effective control over its hinterland, frequently with no limitation other than the extension of the jurisdiction of another city.

The urban scheme created by Spain and Portugal in the New World was remarkably comprehensive and resilient in time. Unlike the gradual pattern of urban foundations and growth in North America, the network of urban centers in Latin America emerged during the course of a single century. Between 1520 and 1580, almost all major Spanish cities on the continent were founded. The period between 1520 and 1540, following the conquest of Mexico, was especially noticeable for its feverish urbanizing activity. Cities founded during these years included Mexico City, superimposed on Tenochtitlán, and Lima, founded by Pizarro as the political center of the Peruvian viceroyalty. Also founded in this period were Puebla, Mérida, Oaxaca, Santiago de los Caballeros de Guatemala, and Quezaltenango in the Viceroyalty of New Spain. Cuzco, Quito, Trujillo, Cajamarca, Tunja, Cali, Bogotá, and Arequipa were founded in the viceroyalty of Peru. Asunción and the first foundation of Buenos Aires in what was to be the viceroyalty of the Río de la Plata also date from this period.

Between 1540 and 1580, the creation of a continent-wide urban scheme was completed with the foundations of Monterrey, Guadalajara, Caracas, Maracaibo, La Paz, Santa Cruz, Cochabamba, Potosí, and Santiago de Chile. In the Argentine, Buenos Aires was firmly established as the major harbor and Santa Fe, Córdoba, Tucumán, Corrientes, and Mendoza were founded. Early primitive harbors such as Veracruz, Piura, El Callao, Panama City, and Cartagena, developed into established city ports. Bahia was founded in the Portuguese territories, eventually becoming the first capital of colonial Brazil. Olinda-Recife on the northern coast, Saõ Paulo in the south, and Rio de Janeiro were the other major Portuguese foundations dating from this period. Among major Spanish American cities, only Montevideo and Medellín were founded after the sixteenth century. In Brazil, the foundation of Porto Alegre in 1642 completed a basic urban system modified only in recent years.

While many cities changed locations, some more than once, few casualties were registered among the early foundations. Cities struggled under adverse conditions and survived, leaving to republican Latin America a ready-made urban network. Early resilience implied, however, the rigidification of urban growth within the old colonial framework. With practically no exceptions, major cities in Latin America developed as enlarged and often distorted versions of those founded at the very start of the colonial period. . . .

ECONOMIC EXPLOITATION

While it is true that politico-administrative centers, such as Lima and Mexico City, acquired decisive importance in Spanish America, the major thrust of colonial foundations remained economic. For Crown and colonists alike, administrative centers and military enclaves were means for conquest and control of the wealth of the new countries. In the category of cities founded primarily for economic reasons, one must include mining and agricultural settlements, as well as harbors such as Callao, Cartagena, Veracruz, and La Habana. Earlier foundations searched for both fertile land and significant concentrations of docile indigenous labor. . . .

Later, discovery of mineral wealth and the concession of royalties for exploitation (*reales de minas*) became other powerful economic incentives for urban growth. These incentives were responsible for the founding of Guanajuato and Taxco in New Spain and Huancavelica and Potosí in Peru. By 1640, Potosí had 140,000 inhabitants and was by far the largest and most important city on the continent. Gold was also the cause behind the phenomenal urban growth in Southern Brazil at the beginning of the eighteenth century. Gold was discovered by *paulistas* in the territory of Minas Gerais. Villa Rica, or Ouro Preto, was the most notable of the instant cities that emerged from this discovery. Its wealth and growth were justified by the fact that between 1690 and 1770 half of the world's production of gold was extracted from Brazil. With depletion of the mines, this city and others in the region declined.

While the importance of political centers reflected the centralization of the colonial administration and while the growth of harbors was a direct consequence of the external dependence of the colonies on the Spanish and Portuguese crowns, the economic characteristics of the rest of the foundations revealed the predatory nature of the colonial enterprise. An ad hoc typology of economic motivations for conquest of new lands could be constructed that divides such efforts into two broad categories: (1) Settlers may apply their skills to organize, implement, or otherwise put to use natural resources of a new territory. In this enterprise, the main ingredients are the knowledge and organized labor of the colonists themselves. Such settlements may be termed *developmental*. (2) New territories may be occupied for the purpose of plundering, taxing, or otherwise profiting from the available resources and native populations. Colonists are less interested in developing natural resources through their direct labor than in appropriating value produced by others. This type may be labelled *exploitative.*

Spanish and Portuguese urban foundations in the New World were, by and large, of a thoroughly exploitative nature. Some were superimposed on previously existing indigenous cities; others grew with the frenzied search for mineral wealth. Cities with an agricultural hinterland differed from similar settlements in other parts of the world in that they were established prior to organized agricultural production with the explicit purpose of subduing and appropriating the labor of the surrounding populations.

The predatory nature of early urban settlements was closely linked with the interests of both the central governments and the colonists themselves. For the Spanish monarchy, the New World functioned primarily as a source of wealth, either in the form of gold and other precious metals or in the form of indirect revenues from agricultural production and commerce. While such monarchs as Charles I and Philip II made significant efforts to organize the new territories on a rational long-term basis, the main thrust from the central administration was toward maximization of immediate income from its American colonies. This effort was closely linked with the pressing need for financial resources to support the Court and army and to finance the European adventures of the kingdom.

The emphasis on short-run acquisition of wealth was reinforced by motivations of those who travelled to the New World. The ambition of early conquerors and later colonists was not to secure land they could work but rather to exploit all available opportunities to amass a rapid fortune and return to Europe. New World settlements were thus viewed, not as final destinations, but as temporary means for attainment of economic goals. This pattern corresponded to, among other factors, the prevailing system for transmission of wealth in Spain. The rule of primogeniture relegated second or later sons in noble families to a search for alternative channels through which to secure a fortune commensurate with their social rank. It became institutionalized among these families that *segundones* would leave the homeland in search of opportunities in the New World. While this source of emigration was not the only or most important one, it documents well the dominant motivations of American-bound colonists. Few in fact returned to Europe with wealth, but the original orientations persisted.

As noted by Ralph Gakenheimer [1967], Spaniards oriented themselves overwhelmingly toward the cities, regardless of whether their origins had been urban or rural. This was due in no small measure to the urban strategy of colonization of new territories, but, even in those settled areas where rural life and work were possible, few seemed disposed to move into the interior. Cities offered the only setting where an exploitative orientation could be implemented successfully. Ownership of land offered about the only effective opportunity for wealth and social position, and land was allocated by city authorities.

Founders of a new city received large tracts, with the remainder allocated among subsequent arrivals with the proper qualifications. The extensive tracts granted to early settlers were given in perpetuity and, with time, became the basis for emergence of an urban patriciate. In successful settlements, chartered land proprietors—the *vecinos*—became an elite which monopolized not only economic wealth but social prestige and access to political offices as well.

Exploitative colonization requires not only land but also subject labor. Cities also offered means for securing this resource via the royally established system of *encomiendas*. *Encomenderos* were urban settlers entrusted with a certain number of natives who, in theory, were to be protected, taught the Spanish language, and converted to the Catholic faith. In practice, the institution became an instrument for free allocation of indigenous labor. Successful colonists thus received the necessary means of production: free land and free labor with which to satisfy their economic ambitions. Most did little more than take advantage of what was available without any systematic improvement of these resources.

The strategy of urban settlements of colonization and the system of *encomiendas* thus functioned as instruments for extending the Crown's possessions in the new continent and as a means for satisfying dreams of enrichment and social position among the colonists. The result could not have been the growth of a dynamic urban bourgeoisie. Instead, a premature aristocracy, excessively preoccupied with reproducing European lifestyles in vastly different surroundings, evolved. . . .

Three aspects of the process must be stressed for their influence on later urban developments: First, cities in Spanish America were established by men whose orientations toward the settlements themselves and their hinterlands were as mere means to the achievement of economic and social goals whose final focus was in Europe. Second, this exploitative attitude toward resources was legitimized by a social order that supported it via the widespread distortion of the original purpose of *encomiendas*. Third, goals of economic wealth and social position were based on private ownership of land, an institution held sacred by colonial society. Few limitations were imposed on exploitation of land for personal enrichment since its possession was defined as a quasi-absolute right.

URBAN ELITISM

As noted above, only landowners—the *vecinos*—had access to positions of political authority. It is true that upon foundation, and for many years thereafter in the case of poor cities, the only inhabitants were the *vecinos.* In cases of important harbors, political centers, and cities in rich territories, however, the population swelled with new arrivals from Spain and natives from the interior. Spaniards of proper rank were also admitted into the landowning minority. The very size of tracts granted, however, accelerated the exhaustion of available land in the urban periphery. Without alternative channels for upward mobility, the social structure of many cities was at this point effectively closed.

Landowners remained for many years the only important city dwellers. They had the use of land and Indian labor, had the right to hold office, and bore most of the financial burdens of the municipality. In early years they were also the only source of military resources and personnel for the defense of the colonies. Thus, the first colonial censuses tended to be much more concerned with counting the number of *vecinos* than the total number of inhabitants of municipalities. . . .

The markedly elitist character of the Spanish American city was thus defined by the superior positions of the few families of the founders and early arrivals who monopolized means for acquisition of wealth, governmental positions, and prestige. Later arrivals (of high rank) formed an inferior class, termed in many cities *soldiers.* It formed a restless group in constant search of whatever opportunity became available to reach positions similar to those already occupied by the urban elite. Next came Spaniards of lower rank, foreigners, and the lower masses of mestizos and native Indians who came into cities as servants or in search of economic opportunities. To these were added, in time, imported black slaves and freed mulattoes.

Elites of rapidly growing cities governed them largely in their own interests with a measure of respect for the interests of Crown and Church, but without much concern for those of the new urban masses. Lower groups were allowed into the city, but few channels were provided for integrating them into the social structure. They had essentially the status of disenfranchised masses—permitted to fend for themselves in the city under the benign indifference of the authorities.

It is a fact seldom noted that major colonial cities quickly became acquainted with the phenomenon of internal migration. Destruction of pre-conquest patterns of agricultural production forced many natives, survivors of the *encomiendas,* to look for their livelihood in cities. Many were brought as servants; others gradually established themselves as small artisans and menial workers. These concentrated, like the Spanish population they served, in the largest cities, such as Mexico City, Lima, and Potosí, or, on a lesser scale, Cuzco and Guadalajara.

Though of much lesser magnitude than subsequent movements, early migrations confronted cities for the first time with phenomenon of impoverished groups coming in search of whatever opportunities were available.

The established city—early settlers now transformed into *señores* and political authorities—permitted the presence of disenfranchised masses and their efforts at economic survival but did not assume responsibility for their fate. Extreme cases had open to them the charity institutions of the Church, which offered the only, albeit feeble, link of aid from the powerful to the powerless.

Urban reception of the poor thus crystallized from early years into a mixture of tolerance and indifference. As cities grew, the relative initial egalitarianism among founders changed into sharp separation of the classes, dividing established groups from the masses (formed by later European arrivals), freed blacks and mulattoes, and incipient native migrants from the interior. The elitist character of urban structure and the place that rural and other migrants occupied in it emerged, therefore, quite early in the Spanish American city. It need not be emphasized how these traits were destined to serve as precedents to later urban developments.

SPATIAL PATTERNS

The ecology of the early colonial city reflected well the centralized character of its social structure. The characteristic grid pattern of streets and blocks emerged from the central plaza, which, left empty except for minor adornments, was usually a square no larger than other city blocks. The main plaza served the multiple functions of marketplace on certain days, recreational center on others, and point of military concentration in times of danger. Around the plaza were concentrated the most important buildings, such as those housing political authorities—from viceroys and *audiencias* to governors and municipalities. At the plaza was also located the cathedral or main church. . . .

The residences of the rich and powerful, the aristocracy formed by *vecinos* and political and religious authorities, were concentrated around the main plaza. This first concentric zone around the plaza contained the largest and best built houses and often had public lighting and street pavement. The convents of the most important orders were also located in this area; it was not uncommon to have small *plazoletas* facing their chapels. The next zone contained the homes of a rudimentary middle class of established artisans, government clerks, and small merchants and proprietors. Residences, which in the first zone were large one-story or even two-story buildings, became smaller and more primitive. The grid pattern persisted but lighting and pavement disappeared. The third zone, the outskirts of the city, was a mixture of the poorest residences of artisans and menial workers with the beginnings of small farms. Beyond this area only an occasional chapel or country store at a crossroads could be found. . . .

Exceptions to this model, besides earlier cities and instant mining settlements, were those situated on difficult terrain. Harbors such as Callao, Valparaíso, and Cartagena were the most common examples. In Brazil, the first important city, Bahia, followed an irregular layout of narrow, tortuous streets similar to the medieval cities dominant in Portugal. The difficult topography of Rio de Janeiro made any attempt at regular development impossible.

For the most part, however, cities in the colonial period followed the simple, regular, highly cephalic pattern delineated above. The city focused on its central plaza, and access to the centers of economic, political, religious, and recreational activity, situated around the plaza, was the most prized aspect of urban location. A rough but consistent positive correlation prevailed between the social and economic position of an individual and the physical distance of his residence from the central square: the greater the distance, the lower the social status. Elitism and centralization thus had in the colonial city their ecological counterpart. As a small aristocracy controlled access to legitimate wealth, power, and social prestige, it also came to monopolize the most desirable urban locations—those at the heart of urban life.

THE REPUBLICAN PERIOD

From the early days of independence at the beginning of the nineteenth century and until very recently, the urban scheme that Latin American countries inherited from the colony remained essentially unchanged. A *criollo* elite reared during colonial days took the place of the Spanish and Portuguese authorities but continued to emphasize urban life and to govern cities largely in its own interest. . . .

The nineteenth and early twentieth centuries brought only two changes of importance to the existing order: First, the primacy of a few cities was accentuated during this period. Lima and Mexico City, the most important colonial cities, accelerated their pace of growth, outdistancing others in the area. La Habana, as the most important remaining Spanish city, came close behind. Caracas, Bogotá, and Santiago de Chile also grew rapidly. On the Río de la Plata, Buenos Aires and, to a lesser extent, Montevideo experienced rapid growth as a consequence of massive European immigration.

The location of capitals of the new republics in the most important existing cities accentuated their predominance by superimposing central political functions on economic ones. Capitals and a few other cities benefited at the expense of others from the commercial policies followed by republican governments.

While Spanish and Portuguese colonial rules emphasized the city over the countryside, they provided existing cities with a measure of economic autonomy. Each served as a commercial center for a vast hinterland and maintained (legal or illegal) contacts with the outside. Concentration of urban development in a few cities gradually stripped the remaining ones of their economic autonomy. While they continued to grow, they came to depend more and more on directives from the primary cities. The economic and social orientation of most cities toward their own hinterland gradually changed as they became functional appendages of enterprises and authorities centered in the capital. Gigantic heads of dwarfish bodies emerged and with them the basic conditions for the regional imbalances and internal dependence characteristic of the contemporary period.

Within major cities themselves, a second change during the republican period was the gradual displacement of elite families from the center to selected portions of the urban periphery. Three related factors seem to have promoted this trend. First, by mid-nineteenth century many large cities had become relatively crowded. As density around the city center increased, wealthy families began to look for residential environments that were both more pleasant and more isolated from other social sectors. Second, better roads and means of transportation made the center of the city more accessible from distant locations than it had been. Easy access to central locations remained a crucial consideration since the political, economic, religious, and recreational life of the city continued to be concentrated in them. Third, world demand for primary products of several countries suddenly enriched many families. Coffee, sugar, or beef became the basis for revitalizing old fortunes or creating new ones. Dominant patterns of high consumption dictated that part of this new wealth should be invested in larger and more luxurious residences.

As Homer Hoyt has noted in the case of U.S. cities, upper-class displacement tended to occur consistently in a single direction. In both North and South America, the movement invariably pointed to the most desirable areas of the urban periphery in terms of geographic, climatic, and aesthetic factors. Elite displacement was preceded or followed by the extension of the highest quality of urban services, infrastructure, and means of communication available to these areas. Elite control of the mechanisms of political authority allowed it to choose its residential settings while frequently placing the burden for extending necessary urban services on the entire community.

In Lima elite displacement started relatively early. By the second half of the nineteenth century, there was a visible front of upper-class residences moving west toward the Pacific Ocean. During the first decades of this century, elegant residences reached the ocean and bordered the new Arequipa Avenue, forming the suburb of Miraflores. In Santiago de Chile the trend also started early, moving northeast along Providencia Avenue toward the Andes. The search here was for higher ground, which provided better aesthetic and, especially, climatic conditions.

In Mexico City, the movement was not visible until the late 1920s. Upper- and middle-class residences began to appear in the south, along

Insurgentes Avenue and the road of Tacubaya. The residential colonies of Lomas de Chapultepec, Mixcoac, and Tacuba, among others, date from this period. In Buenos Aires elite families moved en masse from traditional locations in the southern portion of the central city toward the north. Fear of epidemics and the generally unhealthy conditions of the old sites near the harbor lay behind this move. Wealth from beef and wheat was soon translated into magnificent new residences, which formed the nucleus of the exclusive Barrio Norte.

In Bogotá, a smaller city, elite displacement began later. During the first decades of the twentieth century, wealthy families began to move slowly toward the more desirable locations in the north of the city. The creation and growth of La Merced and La Magdalena, among other areas, date from this time. The continuous displacement of the elite toward northern locations and the growth of impoverished masses which settled in the south changed in a few years the social ecology of that city. From a simple cephalic pattern of successive concentric circles, Bogotá was transformed into a sharp physical dichotomy: the rich and powerful in the north, the poor and powerless in the south, and a feeble middle class of employees and small merchants occupying the center.

Elite displacement in these and other major cities gradually reversed the traditional correlation between residential location and socioeconomic position. Social status, previously associated with nearness to the central city, became related to distance away from it, but only in those directions chosen by elite migration.

Regardless of its location, the urban upper stratum continued to maintain control of political and economic power and to exercise it authoritatively or paternalistically, as the occasion required. In the monolithic social structure of these cities, other groups existed by virtue of, and derived their social meaning from, their relationships with the elite.

The stability of this social order was not a mere function of the elite control of wealth and military resources but was based on belief among subordinate groups in the essential legitimacy of existing arrangements. Positions of elite members were legitimized not only through their interests but also through the values shared by the rest of the population. The generalized acceptance of subordinate groups of the existing distribution of power and wealth prevented massive rebellions and permitted institutions of urban government to adopt the facile role of custodians of a consensually supported order. While serving in reality the interests of the elite, they were manifestly defined, by themselves and others, as representative of the entire city.

SUMMARY

The above outline of the early urban history of Latin America does not aim at exhaustive coverage but rather at isolating those aspects which laid the groundwork for contemporary developments. . . . [C]ontemporary urban poverty can be properly understood only against the background provided by the birth and later evolution of these cities. The argument will not be made that early historical occurrences "caused" the present situation. While direct causal links between the historical past and present can be found, the crucial point is that the colonial urban framework provided logical continuity for later structural developments.

The economic and social evolution of Latin American cities never deviated markedly from the general directions set in colonial days. It is this inertial force of early events that permitted the natural acceptance of subsequent chains of events, an almost imperceptible evolution that led, by gradual steps, to present patterns of massive poverty and structural polarization.

Present cities served as loci for the Iberian conquest, gaining from the start a dominance over the countryside, which they never lost. They were settled by men who defined the city and rural hinterland as means for social position and rapid enrichment. Their orientations promoted an exploitative culture which legitimized unregulated use of land and labor for private profit. The juridical order reflected these values by upholding the quasi-absolute character of property rights. Those who were successful within this order joined the ranks of land proprietors, which soon evolved into a self-regulating aristocracy. Their control of economic and political resources and the absence of resistance from lower groups, especially nonwhite working masses, allowed them to mold cities largely in their own image. Urban ecology reflected this elitist social order in its simple, highly focused pattern of settlement.

Independence from Spain and Portugal brought few changes except those which accentuated structural imbalances between urban and rural areas and between primary and other cities. The urban upper stratum maintained its social position but displaced itself gradually from an increasingly crowded center into more desirable locations.

Acquainted since colonial days with urban poverty and the migration of impoverished groups in search of economic opportunity, this elite permitted continuation of an inertial trend. The eventual fate of the poor, allowed to enter the city as sources of cheap labor, remained an object of total unconcern to the official and wealthy city, except perhaps as objects of Catholic charity.

It is within this general portrait of legitimatized elitism, adherence to canons of private property, growing structural imbalances, and official disregard for their consequences on popular sectors that origins of urban poverty and forms adopted by it in recent years must be understood.

CHAPTER 9

Patterns of Urbanization and Socio-Economic Development in the Third World: An Overview

RICHARD HAY, JR.

> . . . the most remarkable social phenomenon of the present century is the concentration of population in cities . . .
>
> Adna Weber, *The Growth of Cities in the Nineteenth Century,* 1899

> The fact is, therefore, that definite individuals who are productively active in a definite way enter into these definite social and political relations. Empirical observation must in each separate instance bring out empirically, and without any mystification and speculation, the connection of the social and political structure with production.
>
> Karl Marx, *The German Ideology*

The study of the levels and growth of world-wide urbanization has long reflected the differing emphases of these two statements. On one hand, an observer cannot help but be impressed by the statistics of urban growth. At the beginning of the nineteenth century, about 3 percent of the world's population lived in urban places. This figure rose to about 15 percent by 1900, presently stands at about 40 percent and is expected to reach 50 percent by the end of this century.* Corresponding figures for the North Atlantic core of western Europe and North America are much higher. This nearly-exponential increase succinctly reflects a period of

This paper incorporates excerpts from HABITAT: *United Nations Conference on Human Settlements, Global Review of Human Settlements, Item 10 of the Provisional Agenda* A/CONF.70/A/1, pp. 16–47.

*These figures are based on country-specific criteria which may vary considerably. See the U.N. extract in this article for clarification.

rapid socio-economic change which has been objectified in the built environment of today's urban societies and their constituent cities. In his pioneering study written over 75 years ago, Adna Weber noted that "the tendency towards concentration or agglomeration is all but universal in the Western world."

On the other hand, study of the magnitude of the process we call "urbanization" should not be confused with an analysis of its underlying dynamics. The urbanization process denotes a complex interplay of socio-economic, political, technological, geographical and cultural factors. While all of these factors must be taken into account, it is our view that, to varying degrees, they are all mediated through the social organization of production. Thus, variations in the level and growth of urbanization and economic development are strongly related to cross-sectional and historical variations in the socio-economic development of the world system. This is the theoretical perspective which informs much of the discussion to follow.

In this article we present an overview of the distribution of levels of urbanization and economic resources throughout the world. From these data, two predominant trends are apparent. The first is that the Third World is urbanizing at an ever-increasing rate and, although it is still populated largely by rural people, its socio-economic organization is increasingly articulated in urban systems. Current projections indicate that the "gap" in level of urbanization between the Third World and the developed world will be substantially closed in the not-too-distant future.

The second overall trend concerns the growth of socio-economic development. Put simply, urbanization in the Third World has not been accompanied by concomitant economic prosperity as it was in western nations. Quite to the contrary, it has been paralleled by increasing inequity in income and material amenities. In recent years, the developed world has become relatively "richer" while the less developed world has become relatively "poorer." Wallerstein (1974) and others account for this contradiction by noting that the Third World is presently developing within the structural constraints of a world system created and controlled by the advanced industrial nations of the west. In examining the data that follow, this systemic quality should be kept in mind.

The following report, prepared by the U.N. Secretariat as background material for the 1976 United Nations Conference on Human Settlements, discusses variations in the level and growth of urban populations on a world-wide basis. Given the limitations of space, it can only touch on the descriptive highlights of rather complex and detailed processes. Nonetheless, familiarity with this information is essential as background for later pieces in the book which elaborate on the causes and consequences of such variation.

THE URBANIZATION PROCESS: DEMOGRAPHIC ASPECTS

HABITAT: *United Nations Conference on Human Settlements, Global Review of Human Settlements, Item 10 of the Provisional Agenda* A/CONF.70/A/1, pp. 16–47.

The World Population Situation

Before surveying the demographic circumstances of urban and rural settlements, it is necessary to make a few observations about the world population situation in general.

Since about 1950 there has been an unprecedented upsurge in world population growth. Thus, between 1950 and 1975 the combined world population grew from about 2,500 million to about 4,000 million, an increase of 60 percent.

Having stated this, one is obliged at once to draw attention to the wide contrast between demographic circumstances in the more developed and the less developed regions. Birth rates in the less developed regions, on an average, are about twice as high as those in the more developed regions and most recently this gap has even shown a tendency to widen. Death rates in the less developed regions also exceed those in the more developed regions, but in this respect the gap is now much smaller and becoming narrower. Natural increase, that is the difference between the birth and death rate, is far greater in the less developed regions and has accelerated further between 1950 and 1975; it may now be near its maximum, with reductions already occurring in some countries and others likely to follow soon in others. In the more developed regions, around 1950, restoration of peace also brought the temporary "baby boom," but this has subsided since then and the moderate level of population growth is now falling off further to quite low rates. Thus it is that between 1950 and 1975 the population of the more developed regions combined grew by about one third, whereas in the less developed regions it increased about three-quarters.

The future prospects contrast in similar fashion. According to the United Nations population projections, between 1975 and the year 2000, the total population of the more developed regions may grow by about one-fifth, and that of the less developed regions again by about three-quarters despite an expected gradual slowing-down.

In less developed regions, the youthful age structure inherited from hitherto high birth rates reinforces the "demographic inertia" of rapid growth. In the more developed regions, with rising proportions of the population at advanced ages, death rates no longer decrease and may even rise somewhat. In the youthful less developed regions, meanwhile, death rates may still fall to unprecedentedly low levels.

Growth of Urban[1] and Rural Settlements in Different Parts of the World

At the beginning of the last century, the world was perhaps about 3 percent urban. By the end of the last century the concentration of population in cities was said to have been "the most remarkable social phenomenon" of that century, even though fewer than 15 percent of the world's population were then urban. The world is now almost two-fifths urban, and by the year 2000 the world may be fully one-half urban. The growth of urban and rural population between 1800 and the year 2000 is illustrated in Figure 1.

Contrasting conditions between at present more developed and less developed regions must again be borne in mind. The less developed nations, taken as a group, are not yet as highly urbanized. The more developed regions, at present, are more than two-thirds urban, and the less developed, on an average, only about one-fourth. Nevertheless, because of the enormous population of the less developed world combined, the absolute totals of urban population are now about equal between these two portions of the world. Moreover, population in the less developed regions is growing more rapidly so that, with time, they will contain an increasing majority of the world's combined urban population. It also bears emphasis that the vast majority of the world's rural population is to be found in the less developed regions. As has been calculated, by the year 2000 the less developed regions may attain an average level of urbanization equal to that of the more developed regions about the year 1930.

Regional Differences in Urban-Rural Population

In order to get beyond the extremely general categories of "developed" and "developing," it may be helpful to distinguish 24 world regions, nine of them more developed and 15 less developed. In each of the world's regions, the population of urban settlements is increasing decidedly faster than that of rural settlements, ... but it is also to be noted that increase in the combined total population is almost three times as rapid in less developed regions as compared with the more developed ones. One result of these circumstances is that even the rural population in some of the less developed regions grows with greater speed than does the urban population in some of the more developed regions.

In nearly all the developed regions rural population is declining. But in the less developed regions combined, population gains in urban areas are still exceeded by those in rural areas, the gains amounting to 143 million and 167 million, respectively. In three Asian and two African

[1] The international study of urbanization is still impeded by the fact that, in the several national statistics, "urban" settlements are diversely defined. In some of the statistics used here, "urban" places are localities with at least, say, 2,000 to 5,000 inhabitants, whereas in other instances they are distinguished by administrative criteria or by the prevalence of non-agricultural activities. This lack of a common statistical measure weakens international comparison but also reflects an actual diversity of qualitative features. ...

regions, not yet highly urbanized, the rural areas still have markedly larger population gains than do the urban areas. The 64 million gain in the rural population of Middle South Asia nearly equals the 65 million gained by the urban population of all the more developed regions combined.

As can be seen in Table 1, the more developed and less developed regions differ also in the level of urbanization attained in 1975. All the more developed regions are now at least 55 percent urbanized, and all the less developed regions less than 60 percent. In the range between 55 and 60 percent, there is a slight overlap, as the urbanization levels in Middle and Tropical South America are comparable with those in Southern and Eastern Europe. During 1950–1975, the urbanization level in the more developed regions advanced from 53.4 to 69.2 percent, that is by 15.8 points. In the less developed regions, it advanced from 15.6 to 27.3 percent, that is by 11.7 points. Yet urbanization in the less developed regions should be judged as progressing more rapidly because the initial percentage level was much lower.

Urban Concentration

Returning again to Table 1, we note that the population contained by cities with at least one million inhabitants is growing more rapidly than the combined urban population which includes numerous smaller cities and towns. This is not entirely due to population growth in the big cities themselves, since the number of million-cities also increases whenever some new city comes to surpass this size limit. For instance, the world had only 71 million-cities in 1950, as compared to 181 in 1975. During the twenty-five years, the number of million-cities rose from 48 to 91 in the more developed regions, and from 23 to 90 in the less developed regions.

The percentage of million-city population in the urban population can be regarded as a measure of urban concentration. Between 1950 and 1975, this measure rose from 30.5 to 33.5 percent in the more developed regions, and from 18.4 to 31.4 percent in the less developed. In other words, million-cities have become as important a part of the urban population in the latter regions as they are in the former. In 1975, the highest concentration of urban population in million-cities, namely 57.8 percent, is estimated for East Asia other than China or Japan. Concentrations higher than 40 percent can be noted in Northern America, Temperate South America, and Northern Europe. Among less developed regions, the concentration of the urban population is below 20 percent in seven of the less developed regions, four of them having no million-city as yet.

As a consequence of the combination of rising urbanization levels and rising concentration levels of the urban population the percentage of million-city population in the total population is rising with marked speed. Thus, in the more developed regions 15.1 percent of the total population inhabited million-cities in 1950, and 23.2 percent in 1975. Still more significantly, the percentage rose from 2.9 to 8.6 in the less developed regions. The jumps in percentage level were particularly great in Tropical South America and East Asia other than China or Japan. By 1975, roughly one third of the total population inhabited million-cities in Australia and New Zealand, Temperate South America, Northern America and Northern Europe, and more than one quarter in Japan and Other East Asia (other than Japan or China). Fewer than 10 percent of the total population lived in million-cities in the Soviet Union and in nine of the 15 less developed regions. . . .

Migration and Natural Increase

About one-half of the growth of urban population is due to natural population increase (births minus deaths) within the urban settlements themselves, and most of the other half to migratory and other transfers of population from rural to urban places. This observation applies equally to the more developed and less developed regions, each of them taken as one group, though among individual regions or countries more varied observations can be made. The impact of rural-to-urban population transfers upon the rate of change in rural population, on the other hand, can be most diverse. In a predominantly rural country, even the very fast growth of cities can draw away only a comparatively small portion of the natural increase occurring in the rural population. In a predominantly urban country, even a slow or moderate growth of cities can absorb rural migrants in excess of the rural rate of natural population increase. . . .

Perhaps the most important phenomenon . . . is that, in less developed regions, natural increase is becoming a larger component of urban growth than migration. This trend has serious implications for government policy over the next 25 years. Efforts to discourage migration, even if successful, are likely to have decreasing impact on urban growth because the already large urban population is young and growing rapidly by the force of natural increase alone.

While the relative influence of migration on urban growth seems to be decreasing with less developed countries, it appears to be increasing in the more developed regions where transfers may now be responsible for more than one-half of the urban growth. The impact of transfers on the rural population of more developed regions, however, has greatly intensified, as the rural population becomes more and more reduced to a minority. Around 1960, in the rural areas, natural increase was 14.0 per 1,000, the transfer from rural to urban areas occurred at a rate of 16.6 per 1,000 rural population and rural population decreased slightly because transfers somewhat exceeded the rural natural increase. In 1970–1975, apparently, the rural natural increase could have been 8.0 per 1,000 whereas the transfer rate, 17.7 per 1,000, may have been at least twice as high, with the consequence of a considerable rate of decrease in the rural population.

In the less developed regions, in 1960, the urban population grew at an annual rate of 45.5 per 1,000, nearly evenly divided between the rate of urban natural increase, 22.5 per 1,000, and the rate of transfers relative to the urban population, which was 23.0 per 1,000. More recently, urban population may have increased at 40.7 per 1,000 per year, of which 23.2 by natural increase and 17.5 by transfers; as compared with 1960, the urban rate of natural increase may have risen slightly, and that of transfers fallen a little so that transfers now account for somewhat less than one-half of the urban growth in the less developed regions. In the same regions, the rate of growth in rural population may have stayed nearly the same, near 17 per 1,000 at both dates. Rural natural increase may have risen slightly, but so also may the rate of transfers, relative to the rural population. In both periods, transfers from rural to urban areas offset somewhat more than one quarter of the rural natural increase. . . .

The Demographic Outlook

It remains to discuss the latest United Nations projections of urban, rural and city populations up to the year 2000. . . .

The estimated and projected growth of urban populations is illustrated in Figure 2, drawn on a logarithmic scale so that equal slopes of the graphs signify equal rates of growth. As can be seen, in Europe, Northern America and the Soviet Union urban growth, while still considerable, is now likely to slow down towards comparatively moderate rates. In East Asia, an exceptionally fast growth of urban population during the 1950s has been followed by a somewhat reduced but still very high rate which, according to the projections, may not slow down very much before the end of this century. Very high rates of urban population growth are likely to persist in South Asia, Latin America and Africa. As compared with 1975, the urban population is likely to increase, until the year 2000, by about one-third in Europe, one-half in Northern America and the Soviet Union, and two-thirds in Oceania, whereas it may double in East Asia, grow two-and-one-half times in Latin America and treble in Africa and South Asia. . . .

. . . [T]he population of large agglomerations, or cities exceeding 1 million inhabitants, will rise more steeply than the urban population as a whole, partly because of continuing growth in existing million-cities, and partly also because, with time, additional cities enter this size category. Again, it is South Asia, Latin America and Africa where foreseeable city growth will be the fastest. Between 1975 and the year 2000, million-city population is apt to increase by one half in Europe and by two thirds in Northern America, whereas it is likely to double in East Asia, to grow two-and-one-half times in Oceania and the Soviet Union, and to increase threefold in Latin America, fourfold in South Asia, and, conceivably, even sevenfold in Africa, partly because in that continent million-cities are still few but will, no doubt, become more numerous.

The growing importance of million-cities in the urban population can be inferred from these trends. Of the combined urban population, the percentage in million-cities may rise, between the years 1975 and 2000, from 16 to 27 in the Soviet Union, from 22 to 51 in Africa, from 30 to 46 in South Asia, from 31 to 34 in Europe, from 36 to 42 in East Asia, from 37 to 48 in Latin America, from 37 to 56 in Oceania and from 47 to 59 in Northern America. While more moderate in the Soviet Union and Europe, the proportion of urban population in big agglomerations is likely to approach or even exceed one-half in all the other major areas.

. . . In Europe, the Soviet Union and Northern America, rural population ceased to grow in the 1950s, is already declining and is apt to decrease with greater momentum in the future. A moderate but continuing rate of rural population growth is foreseen for Latin America and a diminishing rate also for East Asia. In South Asia and Africa rapid growth in rural population will presumably persist at least until the year 2000.

Considering the entire half century, from 1950 to 2000, one cannot help but envision that a vast transformation is in progress in the sizes and concentration of human settlements.

[Editors' note: End of U.N. document.]

URBANIZATION IN AFRICA AND LATIN AMERICA—REGIONAL VARIATION

The preceding report focused on the distribution and growth of urban settlements throughout the world, utilizing 24 regional categories. These categories, in turn, were often aggregated by geo-political criteria (e.g., North America, Soviet Union) and by level of socio-economic development (e.g., more developed, less developed). Although such aggregation is necessary when considering phenomena on a world-wide scale, it often obscures important dimensions of regional diversity.

With regard to Third World urbanization, such diversity has resulted, in large part, from the complex interaction of indigenous socio-economic structures (both urban and non-urban) with differing modes and intensities of European colonial penetration. These processes, as described in this volume by Portes, Amin, Slater and others, have clearly influenced the contemporary distribution of urban settlements. Africa, for example, is not a socio-cultural monolith and to speak of "African" urbanization subtly implies a homogeneity of social forms and historical development which just does not exist. In contrast, the pervasive influence of Spanish colonization renders the concept of "Latin American" urbanization somewhat more viable. To illustrate these points, and to provide more detailed background data about two of the areas emphasized later in the book, we present a brief discussion of the level and growth of urbanization within Africa and Latin America.

Africa

The often diverse and contradictory literature on African urbanization is usually unanimous with regard to one point: that Africa, historically one of the least urbanized areas of the world, is presently experiencing an extremely accelerated process of urbanization. Hance (1970) notes that 9 to 10 percent of the African population was living in areas of 20,000+ in 1950, and then estimates that the corresponding figure for 1967 was an astounding 15.8 percent. This marked a 76 percent increase which can be contrasted to a much lower figure for the world as a whole. Nearly every study that one encounters calls attention to this increase, though they often differ in their estimates of its absolute magnitude.

In Table 2 we present data showing the level and recent growth of African urbanization. To assure continuity with the previous discussion of world-wide urbanization levels, we have primarily presented data from publications prepared for the U.N. Conference on Human Settlements. It is often difficult to make comparisons utilizing these data, since the definition of an urban population is country-specific and may vary considerably. For example, an "urban" place in Uganda may be a trading settlement with as few as 100 people, whereas a settlement must have a population in excess of 5,000 to be classified as urban in Ghana. For this reason we have also included 1970 data collected by the Economic Commission for Africa (ECA). These figures, based on a customary analytical definition of an urban place as an agglomeration of 20,000 or more, will be utilized in the following discussion. In addition to the urbanization figures, we have provided two indices of socio-economic development—per capita GNP and the recent per capita GNP growth rate. Although we discuss these indices more fully in a later section, it is worthwhile to note that in general urban growth rates exceed economic growth rates.

Urbanization levels in Africa exhibit considerable variation, both across and within geo-political regions. Perhaps the most prominent contrast is between the relatively highly-urbanized countries of North Africa and the rest of the continent. Morocco, Tunisia, Algeria, Libya and Egypt exhibit levels ranging from 26 percent (Tunisia) to 41 percent (Egypt). Although moderate in world-wide context, their overall level (35 percent) is substantially higher than that of any other African region. These countries border the Mediterranean and have served as an easily-accessible periphery ever since Roman times. Abu-Lughod (1976) has analyzed the effect of colonization in shifting the regional urban concentration away from the interior and to the coastline.

Further southward, buffered by the expanse of the Sahara desert, West Africa was the locus of several indigenous urban systems. The walled Hausa-Fulani cities of the savannah and the dense Yoruba settlements in the rain forest still persist, but the present pattern of urban centers has been heavily influenced by the exploitation of extractive resources by the

European colonizers. Major cities grew along the coast to serve as entrepôts for the administration of colonial trade. Thus, the coastal countries such as Ghana, Ivory Coast, Senegal, Liberia and Nigeria exhibit relatively high levels (for sub-Saharan Africa) of 15 to 27 percent, while the inland countries of Upper Volta, Mali, Niger, Mauritania and others exhibit low levels of 2 to 7 percent. The sheer magnitude of Nigeria itself should be noted. Although only moderately urbanized by African standards (15 percent), Nigeria's urban population of 21 million is greater than the combined urban populations of all the other fifteen West African nations (Mabogunje, 1968). Compared to the other sub-Saharan regions, we note that West Africa is moderately urbanized (14 percent), although it possesses the largest absolute urban population.

Eastern Africa, as demarcated by the ECA, exhibits a bifurcated urbanization pattern with the mineral-producing countries of Zambia and Rhodesia being relatively moderately urbanized (20 percent) due to the presence of company towns established by the European mining companies.* The overall level of the rest of the region is, however, relatively low (6 percent). Soja (1976) notes that the eastern flank of Africa from Ethiopia to Mozambique is one of the least urbanized areas in Africa and, in fact, the world. Although some indigenous cities in this area were involved in international maritime trade, the interaction of European colonial policy with the prevailing dispersed settlement pattern did not favor the development of a large number of coastal cities. In contrast, a hierarchy of small colonial administrative centers was established under the dominance of primate colonial capitals in Kenya, Uganda and Tanzania. The growth of African communities in these urban centers was then inhibited by policies designed to control migration and guarantee the maintenance of a rural labor force (see Furedi, this volume). The contemporary pattern of a highly-concentrated urban population which is but a small proportion of the total population is a direct outgrowth of these measures.

In general, Central Africa is slightly less urbanized than coastal West Africa with the notable exception of the Peoples' Republic of the Congo (which formerly served as the capital of French Equatorial Africa) and Equatorial Guinea. Although it occupies over 30 percent of the land area of independent sub-Saharan Africa, Central Africa is sparsely populated, containing about 15 percent of the continent's population. Levels of urbanization in independent black southern Africa are very low, ranging from 2 to 5 percent and urban growth rates have not been calculated due to the small absolute numbers involved. (For greater details on African urbanization, see Gutkind, 1974.)

---

*Although the ECA includes Rhodesia, Zambia and Malawi (former members of the Central African Federation) in Eastern Africa, they are more commonly placed in Central Africa.

Latin America

The history of the colonization of Latin America is markedly different from Africa. Latin American colonization commenced at an earlier date (sixteenth century), was more protracted in duration and was typified by a predominately mercantile world economic system. The colonial histories of the two continents only begin to intersect directly toward the end of the seventeenth century when some Spanish colonists began to work their Latin American agricultural holdings with slave labor imported from West Africa. Given the differences in colonial history, it is not surprising to find notable variation in the contemporary pattern of urbanization between Latin America and Africa. In his essay in this section, Alejandro Portes discussed the historical development of urban societies in Latin America, noting that a wide-spread system of urban centers was established within the first 50 years of Spanish colonization. In the next few pages, we present a brief discussion of the level and growth of Latin American urbanization.

On the whole, Latin America is considerably more urbanized than other areas of the Third World. In 1970, approximately 38 percent of the population of Latin America (all countries south of the Rio Grande) lived in agglomerations of 20,000 or greater as contrasted with only 16 percent in South Asia and Africa (U.N., 1969). South America proper, possessing 68 percent of the population of Latin America, was 40 percent urbanized in 1970, a level comparable to that of Southern Europe (43 percent).* The three countries which comprise its southern tip, Argentina, Uruguay and Chile, are among the most urbanized societies in the world. However, these relatively high levels of Latin American urbanization are not paralleled by equally high levels of economic development. Although the level of per capita GNP ($820) is considerably higher than that of Africa ($290) and Asia ($200), Latin America remains largely underdeveloped (World Bank Atlas, 1975).

Odell (1973) and other scholars suggest that the present distribution of Latin American urban settlements was largely determined by the initial placement of the Spanish colonial cities established in the sixteenth and seventeenth centuries. Most of the major colonial cities were located in areas of relatively dense Indian settlement. Although much of the indigenous Indian population had been decimated by European-introduced diseases, the sizable number that remained constituted a source of cheap labor which was appropriated by the extractive-oriented colonial economy. Thus, the centers of the two Spanish viceroyalties, Mexico City and Lima, were located near the hearts of the former Aztec and Incan empires. Although other cities such as Panama City and Santiago de Chile were established for functional purposes (transshipment of goods and frontier

*South America proper excludes Mexico, Central America and the Caribbean.

defense, respectively), most early Spanish urban settlement remained concentrated in central Mexico and in the western and central Andean regions. In contrast to these areas, the urban structure of Brazil (colonized by Portugal) and Argentina was not well defined until later in the nineteenth century. The Spanish colonial economy was centered around the extraction of raw materials for shipment back to metropolitan Spain. The maintenance of this system of economic exchange was objectified in a pattern of urbanization which, after its early formative period, remained relatively static until the early nineteenth century.

The post-independence period (latter nineteenth century) stimulated the re-alignment of Latin American economies away from the mercantile interests of the declining metropole and toward the industrial capitalist economies of western Europe and the United States. Investment by these countries in Latin America marked the consolidation of dependency relationships which persist until this day. The effect of such investment on urbanization was to increase the marketization of local economies already centered around urban places. This, in turn, led to the accelerated concentration of resources in the the urban centers. The traditional rural subsistance economy began to disintegrate as agriculture was increasingly rationalized for export. Argentina, Chile and Uruguay even encouraged large-scale European immigration to obtain a source of tenant labor for the development of the Pampas. This immigration contributed to the high levels of urbanization exhibited by these countries today.

The preceding synopsis, however brief, is intended as a contextual background against which to view the present distribution and growth of urbanization in Latin America. The contemporary era is characterized by high rates of overall population growth and even higher rates of urban population increase. Rural to urban migration is a major component of this increase. During the period from 1940 to 1960, when overall population increase was slightly less than 3 percent per annum, the population of localities of 20,000+ increased 5 percent annually—a rate which implies a doubling within the space of 15 years. Even more remarkable is the fact that cities of 100,000+ grew at an annual rate of 11 percent during the same period (Cornelius, 1975). Although these rates appear to have declined somewhat during the last decade, they are indicative of the relatively recent urbanization of an area with a long-established urban tradition.

Although the overall level of urbanization in Latin America is considerably greater than that of Africa, both continents exhibit notable internal variation. In Table 3 we present data for Latin America similar to those presented earlier for Africa. In this table, we have substituted Kingsley Davis' estimate of the percentage of the population living in agglomerations of 100,000+ for the ECA figures which used a criteria of 20,000+. Unfortunately, this discrepancy in criteria may inhibit the direct comparison between African and Latin American countries. The Davis figures, however, seem somewhat more appropriate given the greater absolute magnitude of urbanization in Latin America. It is interesting to note that even when the larger criterion of 100,000+ is invoked, most Latin American countries still exhibit higher levels than the African countries.

As we have previously suggested, the most prominent regional contrast in Latin America is between the areas of heavy Spanish colonization (Mexico and South America) and those colonized by other powers or subject to less intense Spanish influence (Caribbean and Central America). The former countries contain the bulk of the land area and population and are more heavily urbanized. Specifically, Mexico and South America account for 87 percent of the total population of Latin America and exhibit an overall urbanization level of 36 percent. In contrast, the Caribbean and Central America contain only 13 percent of the total population, of which 19 percent is urbanized. Although we do not present data indicating the distribution of urban population within each country, all of the regions except Mexico are, in general, characterized by a high degree of concentration of urban population in one or two cities. With regard to this issue, post-revolutionary Cuba is noteworthy for adopting firm policies designed to reduce the primacy of Havana and distribute urban population among a network of settlements (Hardoy, 1970; Lynch, 1972).

In considering intra-regional variation, we note that Cuba (31 percent), Jamaica (28 percent), Costa Rica (25 percent) and Panama (25 percent) exhibit higher levels of urbanization than the rest of the countries the Caribbean/Central America which exhibit a joint level of 13 percent. In South America the highest urbanization levels are found in the southern tip of the continent: Argentina (61 percent), Chile (37 percent) and Uruguay (53 percent). The countries of Brazil, Colombia, Venezuela and Surinam are only moderately urbanized, exhibiting levels from 33 percent to 37 percent. The relation of Brazil to the rest of South America is somewhat analogous to that of Nigeria to West Africa. Although exhibiting only a moderate level of urbanization, the absolute size of Brazil's urban population is greater than the sum of the next five largest urban populations.

ECONOMIC DEVELOPMENT: LEVELS, DISTRIBUTION AND GROWTH

The growth of urban settlements in Western Europe and North America during the last few centuries was only one process out of many which converged in what is commonly called the "industrial revolution" or "modernization" of the west. The sum total of these processes is well known: western society was "transformed" along innumerable dimensions. To isolate specific "causes" and "effects" is impossible. It is specious to say that industrialization caused urbanization or vice versa. Both were manifestations of a modal change in the social organization of production—a change from "simple" or "primitive" accumulation to the expanded reproduction of industrial capitalism. (See, for example, Moore, 1960; Thompson, 1966; Marx, 1904; and Wallerstein, 1974.)

This shift was accompanied by the development of urbanized societies which accumulated heretofore unknown amounts of wealth and material comfort. These are the societies which we now refer to as "developed." But such wealth did not come easily; these societies developed only at the expense of internal (class) and external (colonial) appropriation. Marx, Engels and Harvey (1973, 1975), among others, have analyzed the relationship between urbanization and capitalist economic development in the western world. In the next few pages, we shall attempt to summarize the pattern of economic development associated with the rapid urbanization of the Third World. Our effort will be largely descriptive in nature and the reader is referred to the Portes piece in Section III of this volume for a discussion of a theoretical framework within which to interpret these data.

Before we begin, we would like to present a brief discussion on the utility of the Gross National Product in assessing Third World socio-economic development. We have relied heavily on this indicator since it is regularly reported by international agencies and is customarily used in discussions of this nature. It does suffer from several severe limitations, however. The first is that it usually fails to take into account economic activities conducted outside the national monetized accounting system. Several authors later in this volume note that such activities play an important, if not dominant, part in the lives of many people in the Third World. The retailing of goods through street trading, the construction of owner-built housing and the use of family members in economic enterprises never find their way into the national ledger. Visitors to "poor" Third World cities often expect to find moribund, stagnant poverty and are then surprised by the seeming vitality of local economic life.

Appearances not withstanding, the Third World is every bit as poor and subordinate as the GNP figures indicate. The history of capitalist development leads us to expect a tendency of increasing monetization of traditional economic systems, often in accordance with the economic interests of the metropole (or multi-national corporation). Thus, transitional local forms, although seemingly persistent from the viewpoint of one or two generations, may become quickly absorbed into the national and international monetary system should the need arise. As the proportion of productive labor entered into the national accounts continually increases in this manner, a persistently low Gross National Product may indicate increasing appropriation of that labor by external parties.

Another severe limitation of per capita GNP measures is that they do not portray inequities in the internal distribution of income within Third World countries. Although systematic evidence on this topic is difficult to obtain, we have reason to believe that wealth is more highly concentrated in a small elite in many developing countries than it is in the developed world. Such a situation is frequently associated with a highly polarized class structure of a small powerful elite at the top and large, powerless masses at the bottom. Such severe skewing in the distribution of income is not reflected, of course, in per capita measures.

With these caveats in mind, we now turn to a brief discussion of the data presented in Tables 4, 5 and 6. Once again, familiarity with the overall patterns evident in these data will aid the reader in following subsequent arguments in this book.

In considering what has been called the "development of underdevelopment" in the Third World, the economist Rosenstein-Rodan (1972) denotes three periods in recent world economic history. The first period, 1800–1940, was characterized by a century of relative political and economic stability from the Congress of Vienna in 1815 until the outbreak of World War I in 1914. This century saw the emergence and expansion of corporate industrial capitalism—first in Western Europe and North America, and then throughout the world (see Lenin, this volume). World income began to grow at a rate of 2.5 percent to 3.0 percent per annum as contrasted with its previous rate of 1 percent per annum. However, the income began to be more and more concentrated in those core countries where capitalism originated. At the beginning of the nineteenth century, the difference in per capita income between the rich and the poor countries was approximately in a ratio of 2 to 1; today it is around 40 to 1 in nominal terms and 20 to 1 in real terms.

Growth rates in the developed and developing worlds approached a parity of around 4 percent per annum during the second period (1950–1962) but, of course, by this point the absolute differences in income were quite large. The gap in per capita income continued to grow due, in part, to a slower population growth rate (1.5 percent per annum) in the developed world and a higher population growth rate (2.5 percent per annum) in the developing world. During the third period (1963–1968), economic growth rates again began to diverge. Rates in the developed countries rose to 4.8 percent while those of the underdeveloped countries remained stationary at 4 to 4.5 percent.

Taken as a whole, the post-war period of 1952 to 1972 saw a relative widening of the gap between developed and underdeveloped countries both in aggregate and per capita terms. During these twenty years the per capita income of the developed world rose by $2,000 while the corresponding increase in the developing world was only $125 (United Nations Council on Trade and Development, 1976a). Thus, the increment in per capita income growth in the developed world was 16 times that of the developing world. By 1973, the developed market economies had a per capita GNP of $4,200—nearly 14 times that of the developing countries' per capita GNP of $300. (See Table 4.)

The effects of these unequal growth rates are cogently stated in a report prepared by the Secretariat of the U.N. Conference on Trade and Development:

> Taking the post-war period as a whole, the increments in *per capita* real income have proved to be merely marginal in a considerable number of developing countries, increments which have been totally inadequate to make any significant impact on the economic and social problems which they confront. Moreover, more often than not, the benefits of growth, such as they were, did not percolate to the mass of the population, but resulted rather in a widening of existing social disparities and a heightening of internal political tensions. These are, in some respects, matters for which the developing countries themselves are answerable, they cannot be entirely divorced from the manifold ways in which the domestic systems in these countries are linked to the international economic order. The widening of the economic gap between developed and developing countries over the past 20 years or more, as indicated by the real income figures, can also be traced more concretely, in terms of the consumption of basic foods, clothing, housing and essential services, such as health care and educational provision. In many of these elements making up the individual's standard of living, it would seem that only marginal progress has been achieved in a considerable number of developing countries. Moreover, the underlying problems of hunger and malnutrition, even famine, of unemployment and underemployment, of rural poverty and urban degradation, are even more pressing today than they were a quarter of a century ago at the beginning of the phase of rapid economic expansion in the developed countries.
>
> United Nations, Council on Trade and Development, 1976a:5

It is often difficult to grasp the profound effect that changes of only a few percentage points in growth rates might have. Time serves as a powerful amplifier and the continued maintenance of unequal growth rates over a long duration results in greatly attenuated differences. Tables 4, 5 and 6 present data indicating the current manifestation of the historical processes of development that we have been considering. Portrayed in absolute terms, the income inequality between the "haves" and the "have-nots" is indeed disquieting.

The data reveal that the developed market economy countries

(primarily western and southern Europe, North America and Japan) account for only 20 percent of the world's population and yet accumulate 68 percent of the world's total Gross National Product. In contrast, the 47 percent of the world's population which lives in developing market economy countries (the non-socialist Third World) receives only 12 percent of the total GNP. Ignoring the type of economic system for the moment, we find that the 70 percent of the world's population who live in developing countries receive only 16 percent of the total GNP whereas the 30 percent who inhabit the developed countries receive 84 percent of the total GNP. This overall distortion in the distribution of the world income is echoed in Table 6. Seventy-five percent of the world's people live at levels well below the U.S. poverty line and receive only 21 percent of the world total GNP.

Narrowing our attention to the Third World itself, we find a certain amount of variation occurring within the confines imposed by its subordinate economic position. In examining these data (Table 4), we should keep in mind the considerable spreads in urbanization within Africa and Latin America. The sources and extent of such diversity are often obscured when large aggregates of continental dimensions are compared. Nonetheless, the developing world can be divided roughly into two groups on the basis of per capita GNP. Latin America, Oceania and the oil-producing countries of the Middle East range from $700 to $950 while the rest of the developing world (Asia and Africa) exhibits levels from around $160 (market economy Asia) to $250 (Africa). The fact that the vast majority of the Third World lives in this latter region accounts for its overall per capita GNP of approximately $300.

It is noteworthy that the per capita GNP of socialist Asia ($260) is somewhat higher than that of market economy Asia ($160). Although socialist nations must function externally within a capitalist world system, they seem better able to mobilize internal resources for development purposes.

In closing, we would like to observe that the existence of such gross inequalities in the distribution of world income is a widely-acknowledged fact. The development of theory to account for these differences is, however, marked with strife and disagreement (see Portes, Section III). Though it is beyond the scope of this article to elaborate on the origins and development of such inequality, we would like to note one recent piece of research which considers these questions.

It is often contended that the inequitable distribution of socio-economic resources is due to the failure of developing countries to adopt the "modern" institutions and mechanisms which produced prosperity in the west. Specifically addressing this issue, Meyer, Boli-Bennet and Chase-Dunn (1975) conclude that the last two decades have seen a trend toward increasing equalization in the worldwide distribution of such indices as education, electricity consumption, urbanization and size of the non-agricultural labor force. Notably, the trend toward diffusion of these mechanisms *is not* associated with an equalizing trend in the distribution of income. To the contrary, the developing countries now find themselves with the costly rudiments of a western infrastructure and a lower proportionate share of the Gross World Product.

FIGURE 1. The Growth of the World's Urban and Rural Population, 1800-2000 (In Millions)*

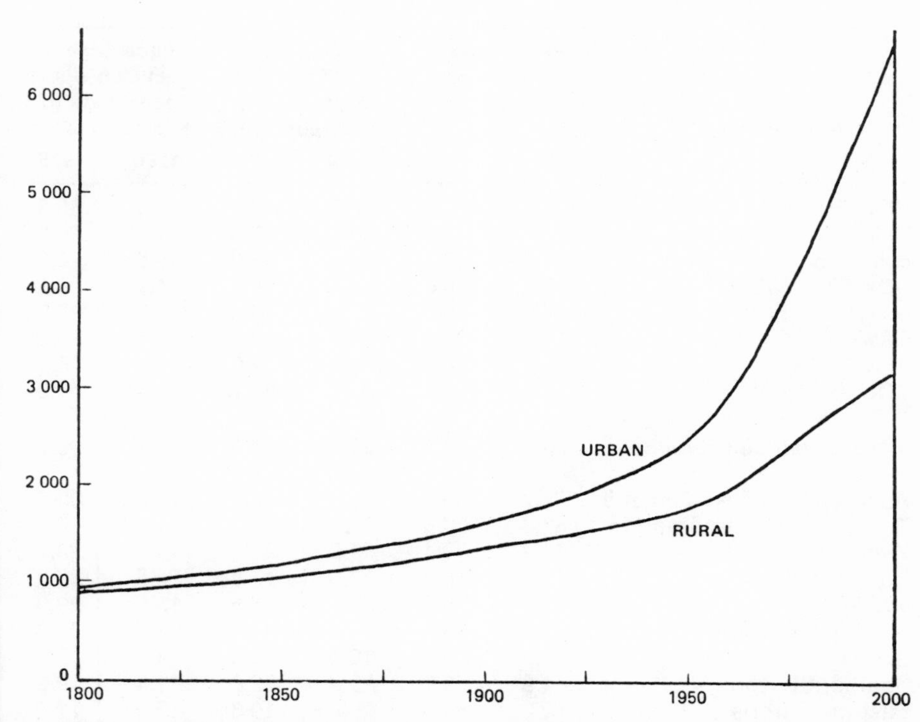

SOURCES: Data for 1800, 1950, 1900 adapted from estimates made by Kingsley Davis and Hilda Hertz as published in P. M. Hauser, ed., **Urbanization in Asia and the Far East** (Calcutta: UNESCO, 1957), p. 56. Data for 1950 from United Nations Population Division.
*Figure 1 reprinted from HABITAT: **United Nations Conference on Human Settlements, Global Review of Human Settlements, Item 10 of the Provisional Agenda** A/CONF.70/A/1, p. 18.

TABLE 1. Percentages of Urban in Total Population, Million-city Population in Urban Population, and Million-city Population in Total Population, 1950 and 1975, in Eight Major Areas and 24 regions of the World*

Area or region	Percentage of urban in total population		Percentage of million-city population in urban population		Percentage of million-city population in total population	
	1950	1975	1950	1975	1950	1975
NORTHERN AMERICA	63.6	76.5	37.9	47.5	24.5	36.3
OCEANIA	64.5	71.6	36.6	37.0	23.6	26.5
EUROPE	54.8	67.2	28.4	31.2	15.5	20.9
SOVIET UNION	39.4	60.5	10.5	16.4	4.1	9.9
LATIN AMERICA	40.9	60.4	22.5	36.9	9.2	22.3
EAST ASIA	16.6	30.7	27.6	36.0	4.6	11.0
AFRICA	13.2	24.4	8.2	22.0	1.1	5.4
SOUTH ASIA	15.5	23.0	15.3	29.7	2.4	6.8
More developed regions	**53.4**	**69.2**	**30.5**	**33.5**	**15.1**	**23.2**
Australia and New Zealand	78.7	85.5	37.4	39.2	29.4	33.6
Temperate South America	62.8	80.8	34.9	44.6	21.9	36.0
Western Europe	63.2	77.1	23.2	30.7	14.7	23.7
Northern America	63.6	76.5	37.9	47.5	24.5	36.3
Japan	50.3	75.2	27.5	38.6	13.8	29.0
Northern Europe	70.8	75.1	44.8	41.3	31.7	31.0
Soviet Union	39.4	60.5	10.5	16.4	4.1	9.9
Southern Europe	44.9	59.2	25.3	32.6	11.4	19.3
Eastern Europe	42.2	56.6	20.4	19.8	8.6	11.2
Less developed regions	**15.6**	**27.3**	**18.4**	**31.4**	**2.9**	**8.6**
Tropical South America	36.5	59.3	17.0	38.9	6.2	23.1
Middle America	39.5	57.1	20.3	32.2	8.0	18.4
Other East Asia[a]	23.2	50.0	33.7	57.8	7.8	28.9
Caribbean	33.0	48.2	20.1	17.4	6.6	8.4
Southern Africa	36.5	46.2	0.0	31.2	0.0	14.4
Western South Asia	23.3	43.7	0.0	30.7	0.0	13.4
Northern Africa	23.2	39.5	19.8	32.0	4.6	12.6
Micronesia and Polynesia	20.6	32.1	0.0	0.0	0.0	0.0
Middle Africa	8.1	24.6	0.0	18.4	0.0	4.5
China	11.1	23.5	26.9	31.8	3.0	7.5
Eastern South Asia	13.4	22.1	13.3	34.2	1.8	7.6
Middle South Asia	15.6	21.1	18.0	27.6	2.8	5.8
Western Africa	9.6	18.5	0.0	9.7	0.0	1.8
Melanesia	2.0	13.7	0.0	0.0	0.0	0.0
Eastern Africa	5.3	12.3	0.0	7.7	0.0	0.9

[a]East Asia other than China or Japan.
SOURCE: United Nations Population Division.
*Table 1 reprinted from HABITAT: **United Nations Conference on Human Settlements, Global Review of Human Settlements, Item 10 of the Provisional Agenda** A/CONF.70/A/1, pp. 21–22.

FIGURE 2. Urban Population, 1950-2000, in Seven Major Areas*

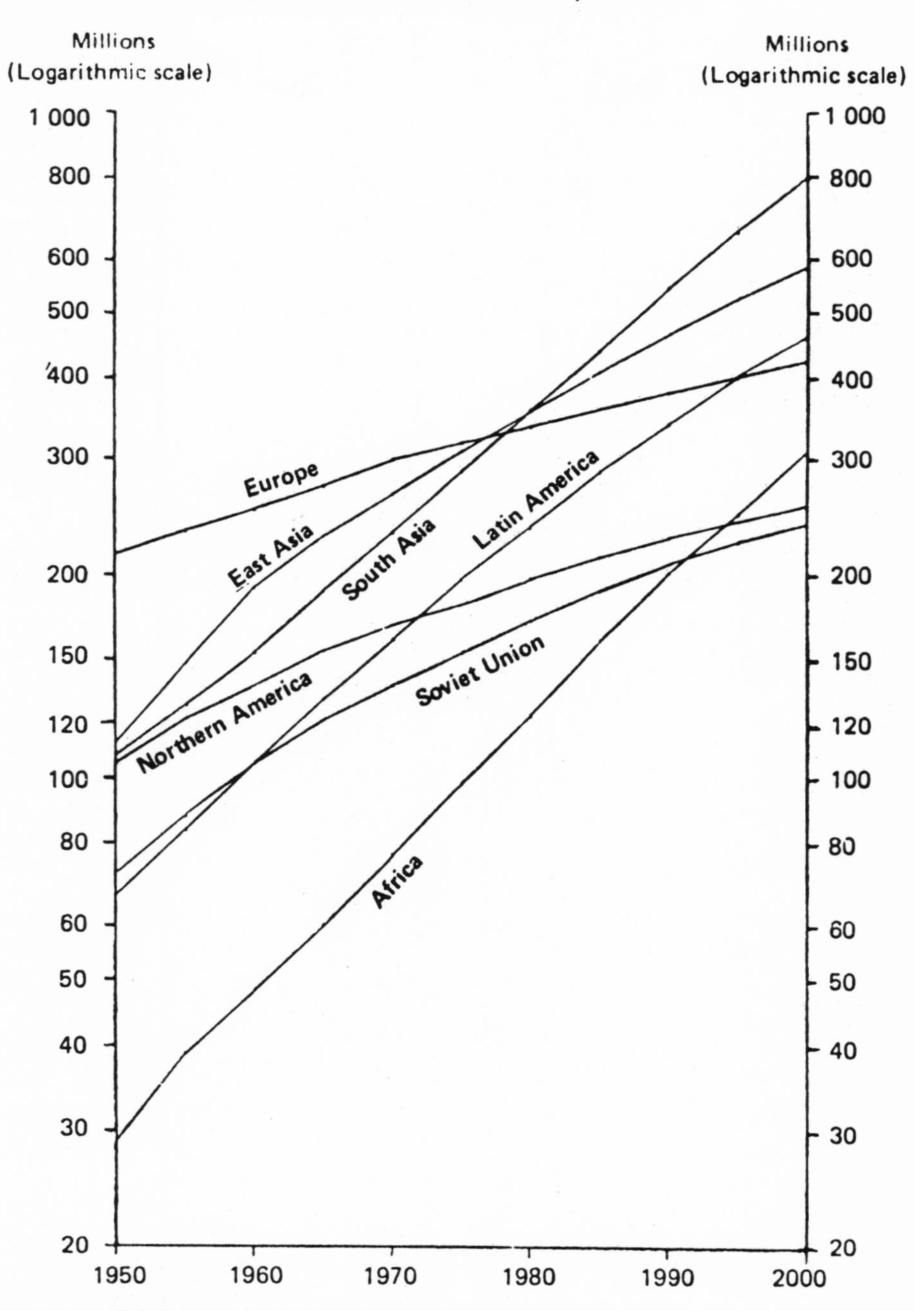

SOURCE: Statistical annex (A/CONF.70/A/1 Annex, table 1.
*Figure 2 reprinted from HABITAT: **United Nations Conference on Human Settlements, Global Review Review of Human Settlements, Item 10 of the Provisional Agenda** A/CONF.70/A/1, p. 45.

TABLE 2. Africa: Level and Growth of Urbanization and Socio-Economic Development.

COUNTRY	--1950 POP-- (THOUSANDS) TOTAL	URBAN	--1975 POP-- (THOUSANDS) TOTAL	URBAN	AVG. ANNUAL URBAN GROWTH RATE 50-75	ECA PCT. URBAN 1970	GNP PER CAPITA 1973	GNP/CAP GROWTH RATE 60-73
** AFRICA **	204672	23360 (11%)	373439	84562 (23%)	5.5%	16%	240	2.1%
--NORTH AFRICA--	42726	11355 (27%)	79841	36327 (45%)	4.9%	35%	440	3.7%
EGYPT	20461	6527 (32%)	37543	17906 (48%)	4.0%	41%	250	1.5%
MOROCCO	8953	1846 (21%)	17504	6655 (38%)	5.1%	30%	320	1.6%
ALGERIA	8753	1877 (21%)	16792	8385 (50%)	5.9%	30%	570	1.7%
TUNISIA	3530	913 (26%)	5747	2694 (47%)	4.3%	26%	460	3.4%
LIBYAN ARAB REP.	1029	192 (19%)	2255	687 (30%)	5.1%	24%	3530	10.5%
--WEST AFRICA--	64382	6181 (10%)	115168	21310 (19%)	5.3%	14%	200	1.6%
NIGERIA	34331	3556 (10%)	62925	11418 (18%)	4.7%	15%	210	3.6%
GHANA	5024	727 (14%)	9873	3199 (32%)	5.8%	17%	300	0.0%
UPPER VOLTA	3769	149 (4%)	6032	500 (8%)	4.9%	4%	70	-.4%
MALI	3426	276 (8%)	5697	766 (13%)	4.2%	7%	70	1.0%
IVORY COAST	2822	178 (6%)	4885	994 (20%)	6.8%	19%	380	3.1%
NIGER	2291	109 (5%)	4592	430 (9%)	5.4%	4%	100	-1.9%
SENEGAL	2600	492 (19%)	4418	1252 (28%)	3.8%	27%	280	-1.8%
GUINEA	2687	149 (6%)	4416	859 (19%)	6.9%	11%	110	.1%
BENIN	1732	102 (6%)	3074	553 (18%)	6.7%	12%	110	1.0%

TABLE 2 continued

COUNTRY	--1950 POP-- (THOUSANDS) TOTAL	URBAN	--1975 POP-- (THOUSANDS) TOTAL	URBAN	AVG. ANNUAL URBAN GROWTH RATE 50-75	ECA PCT. URBAN 1970	GNP PER CAPITA 1973	GNP/CAP GROWTH RATE 60-73
SIERRA LEONE	1779	175 (10%)	2983	448 (15%)	3.9%	13%	160	1.6%
TOGO	1201	95 (8%)	2248	304 (14%)	5.1%	11%	180	4.4%
LIBERIA	1066	61 (6%)	1708	259 (15%)	5.8%	18%	310	2.2%
MAURITANIA	796	39 (5%)	1283	142 (11%)	4.9%	2%	200	4.1%
GUINEA-BISSAU	511	47 (9%)	525	121 (23%)	--	5%	330	3.6%
GAMBIA	347	26 (7%)	509	65 (13%)	--	10%	130	3.7%
--CENTRAL AFRICA--	26198	2113 (8%)	45230	11112 (25%)	5.7%	13%	220	2.1%
ZAIRE	13055	1027 (8%)	24485	6402 (26%)	7.1%	15%	140	2.6%
UNITED REP. CAMEROON	4092	368 (9%)	6398	1525 (24%)	5.6%	7%	250	3.2%
ANGOLA	3969	230 (6%)	6353	1160 (18%)	6.4%	10%	490	3.8%
CHAD	2461	98 (4%)	4023	558 (14%)	6.7%	8%	80	-2.1%
CENTRAL AFRICAN REP.	1145	121 (11%)	1790	643 (36%)	6.5%	16%	160	.4%
PEOPLE'S REP. CONGO	815	185 (23%)	1345	535 (40%)	4.3%	31%	340	1.7%
GABON	434	49 (11%)	526	149 (28%)	4.0%	11%	1310	4.4%
EQUATORIAL GUINEA	227	35 (15%)	310	140 (45%)	4.6%	33%	260	.4%
--EASTERN AFRICA--	69926	3696 (5%)	130893	15636 (12%)	5.9%	7%	170	1.6%
ETHIOPIA	16675	801 (5%)	27975	3137 (11%)	5.4%	5%	90	2.4%

TABLE 2 continued

COUNTRY	--1950 POP-- (THOUSANDS) TOTAL	URBAN	--1975 POP-- (THOUSANDS) TOTAL	URBAN	AVG. ANNUAL URBAN GROWTH RATE 50-75	ECA PCT. URBAN 1970	GNP PER CAPITA 1973	GNP/CAP GROWTH RATE 60-73
SUDAN	9067	673 (7%)	18268	2411 (13%)	5.1%	8%	130	-.9%
UNITED REP. TANZANIA	7892	286 (4%)	15438	1043 (7%)	5.2%	6%	130	2.8%
KENYA	6018	335 (6%)	13251	1499 (11%)	6.0%	8%	170	3.1%
UGANDA	5969	175 (3%)	11353	952 (8%)	6.7%	5%	150	2.1%
MOZAMBIQUE	5743	130 (2%)	9239	584 (6%)	6.0%	6%	380	3.3%
MALAGASY REP.	4330	339 (8%)	8020	1426 (18%)	5.8%	11%	150	.3%
RHODESIA	2276	233 (10%)	6276	1236 (20%)	6.5%	17%	430	1.7%
ZAMBIA	2473	273 (11%)	5022	1838 (37%)	7.5%	24%	430	1.7%
MALAWI	3033	96 (3%)	4916	314 (6%)	4.9%	4%	110	3.5%
RWANDA	2189	20 (1%)	4200	158 (4%)	7.4%	--	70	.3%
BURUNDI	2435	34 (1%)	3765	141 (4%)	5.9%	2%	80	1.3%
SOMALIA	1826	301 (16%)	3170	897 (28%)	4.4%	12%	80	-.2%
--SOUTHERN AFRICA--	1440	15 (1%)	2307	177 (8%)	--	3%	190	5.1%
LESOTHO	766	6 (1%)	1148	36 (3%)	--	2%	100	3.8%
BOTSWANA	421	5 (1%)	691	74 (11%)	--	--	230	4.7%
SWAZILAND	253	4 (2%)	468	67 (14%)	--	5%	330	6.9%

NOTES:

1. Columns 1-4, Population figures, 1975—
SOURCE: **Global Review of Human Settlements: Statistical Annex,** United Nations, Department of Economic and Social Affairs.
COMMENT: As noted in the text, the criterion for an urban population is country-specific and may vary considerably. Percent urban was calculated by the author.
2. Column 2, Average Annual Urban Population Growth Rate, 1950-1975—
SOURCE: Same as cited in Note 1.
COMMENT: This figure was calculated by taking the mean of the 1950-1960, 1960-1970 and 1970-1975 growth rates as reported in Note 1. If the growth rate was not reported for a given interval(s), then the mean of the reported rate(s) was used.
3. Column 6, Urban Population (criterion=20,000+) as a Percent of Total Population, 1970—
SOURCE: **Statistical Yearbook, 1970,** United Nations, Economic Commission for Africa.
COMMENT: These figures are based on standard criteria (urban=agglomeration of 20,000+) and are presented to facilitate comparison between specific countries.
4. Columns 7 and 8, GNP per Capita and GNP per Capita Growth Rate (%) 1960-1973—
SOURCE: **World Bank Atlas, 1975,** International Bank for Reconstruction and Development.
COMMENT: The data are calculated on the basis of 1973 market prices. In certain instances, the interval employed in the calculation of growth rates is different from that shown.
5. The regional groupings of North, West, Eastern, Central and Southern Africa were adopted (with minor revisions) from the Statistical Yearbook of the Economic Commission for Africa. (See section of text on African urbanization.) Regional summary measures were calculated by the author from the information in the table. Growth rates, especially, may be somewhat inaccurate.
6. We have limited our discussion to independent African countries. Thus, we have not included minority-ruled South Africa and the territories which it presently controls. In light of recent promising political developments, we have included Rhodesia.
7. The ECA figure for Botswana was omitted since it appeared to be inaccurate.

TABLE 3. Latin America: Level and Growth of Urbanization and Socio-Economic Development

COUNTRY	--1950 POP-- (THOUSANDS) TOTAL	URBAN	--1975 POP-- (THOUSANDS) TOTAL	URBAN	AVG. ANNUAL URBAN GROWTH RATE 50-75	DAVIS PCT. URBAN 1970	GNP PER CAPITA 1973	GNP/CAP GROWTH RATE 60-73
** LATIN AMERICA **	160495	65796 (41%)	319226	192960 (60%)	3.7%	34%	820	2.5%
-- MEXICO --	26606	11207 (42%)	59204	37400 (63%)	4.8%	34%	890	3.3%
-- CARIBBEAN --	13408	4321 (32%)	22434	10316 (46%)	3.4%	21%	900	2.7%
CUBA	5752	2866 (50%)	9481	5839 (62%)	3.2%	31%	540	-1.0%
DOMINICAN REPUBLIC	2313	539 (23%)	5118	2253 (44%)	5.7%	18%	520	2.7%
HAITI	3097	376 (12%)	4552	945 (21%)	3.6%	8%	130	-.3%
JAMAICA	1403	311 (22%)	2029	915 (45%)	4.2%	28%	990	3.6%
TRINIDAD + TOBAGO	632	145 (23%)	1009	253 (25%)	2.2%	--	1310	2.1%
BARBADOS	211	84 (40%)	245	111 (45%)	1.3%	--	1000	5.5%
-- CENTRAL AMERICA -	9119	2898 (32%)	19264	7420 (39%)	3.9%	17%	560	2.8%
GUATEMALA	3023	893 (30%)	6129	2154 (35%)	3.6%	15%	500	3.3%
EL SALVADOR	1931	701 (36%)	4108	1647 (40%)	3.0%	14%	350	1.9%
HONDURAS	1390	340 (24%)	3037	851 (28%)	3.8%	10%	320	1.3%
NICARAGUA	1109	396 (36%)	2318	1111 (48%)	4.2%	18%	540	3.3%
COSTA RICA	866	284 (33%)	1994	795 (40%)	4.1%	25%	710	2.7%
PANAMA	800	284 (35%)	1678	862 (51%)	4.4%	30%	920	4.4%

TABLE 3 continued

COUNTRY	--1950 POP-- (THOUSANDS) TOTAL	URBAN	--1975 POP-- (THOUSANDS) TOTAL	URBAN	AVG. ANNUAL URBAN GROWTH RATE 50-75	DAVIS PCT. URBAN 1970	GNP PER CAPITA 1973	GNP/CAP GROWTH RATE 60-73
-- SOUTH AMERICA --	111362	47370 (43%)	218324	137824 (63%)	3.8%	37%	800	2.3%
BRAZIL	52901	18699 (35%)	109730	65253 (59%)	4.9%	34%	760	3.6%
COLOMBIA	11689	4202 (36%)	25890	15989 (62%)	5.2%	39%	440	2.4%
ARGENTINA	17150	11011 (64%)	25384	20300 (80%)	2.4%	61%	1640	2.7%
PERU	7915	3268 (41%)	15326	8742 (57%)	4.0%	23%	620	2.1%
VENEZUELA	5145	2838 (55%)	12213	10061 (82%)	4.9%	37%	1630	2.0%
CHILE	6091	3541 (58%)	10253	8507 (83%)	3.7%	37%	720	1.7%
ECUADOR	3224	915 (28%)	7090	2952 (42%)	4.7%	21%	380	1.9%
BOLIVIA	3019	778 (26%)	5410	2013 (37%)	3.8%	15%	230	2.5%
URUGUAY	2194	1434 (65%)	3108	2505 (81%)	2.1%	53%	950	-.2%
PARAGUAY	1371	455 (33%)	2647	991 (37%)	3.2%	19%	410	1.9%
GUYANA	423	116 (27%)	791	265 (34%)	3.3%	27%	410	1.5%
SURINAM	215	100 (47%)	422	212 (50%)	3.1%	33%	870	4.2%
FRENCH GUIANA	25	13 (52%)	60	34 (57%)	--	--	1360	3.8%

NOTES:

1. Columns 1-4, Population figures, 1975—
SOURCE: **Global Review of Human Settlements: Statistical Annex,** United Nations, Department of Economic and Social Affairs.
COMMENT: As noted in the text, the criterion for an urban population is country-specific and may vary considerably. Percent urban for 1950 and 1975 was calculated by the author.
2. Column 5, Average Annual Urban Population Growth Rate, 1950-1975—
SOURCE: Same as cited in Note 1.
COMMENT: This figure was calculated by taking the mean of the 1950-1960, 1960-1970 and 1970-1975 growth rates as reported in Note 1. If the growth rate was not reported for a given interval(s), then the mean of reported rate(s) was used.
3. Column 6, Urban Population (criterion=100,000+) as a Percent of Total Population, 1970—
SOURCE: **World Urbanization 1950-1970, Volume I: Basic Data for Cities, Countries, and Regions,** Kingsley Davis.
4. Columns 7 and 8, GNP per Capita and GNP per Capita Growth Rate (5) 1960-1973—
SOURCE: **World Bank Atlas, 1975,** International Bank for Reconstruction and Development.
COMMENT: The data are calculated on the basis of 1973 market prices. In certain instances, the interval employed in the calculation of growth rates is different from that shown.
5. The regional groupings of Mexico, Caribbean, Central and South America were adopted (with minor revisions) from Ruddle and Barrows (1974), **The Statistical Abstract of Latin America.**
6. Regonal summary measures were calculated by the author from the information in the table. Growth rates, especially, may be somewhat inaccurate.

TABLE 4. Distribution of Total Gross National Product by Region, 1973

	Population (Thousands)	% Total World Pop.	Annual Pop. Growth Rate(%) 70-73	Total GNP (Millions)	% World GNP	GNP per capita
MARKET ECONOMIES	2 585 254	67.7	1.8	3 814 645	79.9%	1480
I. Developing Countries	1 802 301	47.2	2.5	543 400	11.4%	300
Asia	1 097 037	28.7	2.1	170 980	3.6%	160
Oil-Producing Middle East	52 280	1.4	3.2	49 470	1.0%	950
Africa	355 400	9.3	2.6	87 500	1.8%	250
Oceania	4 200	0.1	1.6	3 000	0.1%	700
Latin America	293 384	7.7	2.8	232 450	4.9%	790
II. Developed Countries	782 953	20.5	1.0	3 271 245	68.5%	4200
Western + Southern Europe	399 100	10.4	0.9	1 345 400	28.2%	3370
Japan, Israel, South Africa	135 253	3.5	2.4	448 145	9.4%	3300
Australia, New Zealand	16 100	0.4	1.6	63 900	1.3%	3970
North America	232 500	6.1	0.9	1 413 800	29.6%	6080
SOCIALIST ECONOMIES	1 219 016	31.9	1.3	951 750	19.9%	780
I. Developing Countries	862.116	22.6	1.7	230 150	4.8%	270
Asia	853 200	22.3	1.7	225 300	4.7%	260
Cuba	8 916	0.2	1.7	4 850	0.1%	540
II. Developed Countries	356 900	9.3	0.9	721 600	15.1%	2020
Eastern Europe	107 153	2.8	0.9	215 110	4.5%	2000
USSR	249 747	6.5	1.0	506 490	10.6%	2030
WORLD *	3 820 100	---	1.9	4 776 900	---	1250
I. Developing Countries	2 664 417	69.7	2.1	773 550	16.2%	290
II. Developed Countries	1 139 853	29.8	1.0	3 992 845	83.6%	3500

SOURCE: **Handbook of International Trade and Development Statistics, 1976,** United Nations Council on Trade and Development UNTD/STAT.6.
Percentages and sub-totals calculated by author.
Summary growth rates may be inaccurate.
*World total for all items includes data not allocated by economic class or region.

TABLE 5. Gross National Product, by Major Regions, (1973) (Ranked by GNP per Capita)

Region or country	GNP per capita (US $)	GNP (US$000 millions)	% of total GNP	Pop (millions)	% of total Pop
North America	6,130	1,425	30	233	6
Japan	3,630	393	8	108	3
Oceania	3,400	71	1	21	1
Europe (excluding USSR)	2,990	1,516	32	507	13
USSR	2,030	506	11	250	7
Middle East	1,080	82	2	76	2
South America	840	170	4	203	5
Central America	800	79	2	99	3
Africa	290	114	2	392	10
Asia (excluding Japan)	200	399	8	1,947	51

SOURCE: **World Bank Atlas, 1975.**
Percentages computed by author.

TABLE 6. Gross National Product, by Income Group (1973)

Income Group	No. of Countries	Pop (millions)	% of total Pop	Cum. %	GNP (US$000 millions)	% of total GNP	Cum. %	Average GNP per capita (US $)
Less than $200	43	1,151	30	30	136	3	3	120
$200 to $499	52	1,184	31	61	332	7	10	280
$500 to $1,999	53	531	14	75	530	11	21	1,000
$2000 to $4,999	28	654	17	92	1,871	39	60	2,860
$5000 and over	12	316	8	100	1,886	40	100	5,970

SOURCE: **World Bank Atlas, 1975.**
Percentages computed by author.

SECTION III

NEW THEORIES ON THIRD WORLD DEVELOPMENT

INTRODUCTION

In the preceding section we noted that economic development and urbanization, which were presumed to be linked together in the process of industrialization and modernization, have split apart in the Third World. Rapid urbanization there is; economic development, on the other hand, lags behind and indeed involutes. How are we to account for this anomaly? As exceptions to the theory? As temporary deviations or lags in a process which eventually will correct itself? As products of human failure and inadequate planning which can be corrected with greater will and knowledge? Or have our theories erred? Are the regularities observed in so many parts of the Third World, referred to by one of the authors (Arrighi) in a later chapter as "growth *without* development," the logical consequences of a process which we have failed to understand? What alternate theories exist which might help to explain the increasing gap in development for so large a portion of the Third World? And if these alternate theories apply, how shall we have to change our view of cities and urbanization in the light of the newer theories?

We pose these questions in Section III, but can do no more than present a brief overview of the theories and a single sample analysis based upon them. It must be recognized that theories are not "true" in the way "facts" or physical events appear to be. Theories must be judged in terms of their usefulness in guiding research and in terms of how well they seem to account for observed regularities. We contend that the older theories leave us unprepared for the situation observed in country after country in the Third World and leave us with too many anomalies, too many deviations and failures. Perhaps it is time to change our theory.

Some twenty years ago, most academicians working on questions of economic development and urbanization would have subscribed to the following (admittedly oversimplified and travestied) description of the process whereby Third World countries were to close the gap between themselves and the developed west:

1. Such countries were called backward or underdeveloped because they were chiefly agricultural and rural. Development would occur with industrialization. Such industrialization would involve (require) a massive movement of the labor force out of farming and primary industries into the processing and manufacturing industries. Since industries were associated with urban centers, this would mean a physical redistribution of population from rural to urban areas. Therefore, industrialization and urbanization would move together, each supporting and facilitating the other.

2. Industrial development requires high capital investment and the shift from animated to nonanimated sources of energy. Both of these shifts permit higher productivity which is directly translatable into higher incomes for workers. Capital investment is made possible through savings (deferred consumption) which are invested in factories, machines and infrastructure (roads, rails, other communication facilities, etc.). There are major economies of scale to be achieved through centralization. This reinforces the tendency for development and urbanization to vary together.

3. Industrial development requires a population with higher skills, literacy, and

inventiveness than does agricultural production. Therefore, one of the major requirements for economic growth is universal education and a system of values supportive of industrial growth: namely, the qualities necessary for a disciplined labor force, such as punctuality, hard work, capacity to respond to monetary incentives by working longer and harder, stability, etc.; and the characteristics necessary for an entrepreneurial class, rational, sober and puritanical yet adventurous enough to take risks. Since these are characteristics associated with "urbanism as a way of life," in contrast to the "folk culture" of illiterate, traditional agricultural communities, it was assumed that not only would urbanization accompany industrialization but it would indeed assist in bringing about the latter.

4. Just as the causes of underdevelopment lie within the underdeveloped country itself, so the major solutions to the problems lie with internal economic, social and cultural transformations. The role of the developed countries, to be undertaken chiefly for altruistic motives or as "enlightened self-interest," is to extend assistance, in the form of technical expertise, in the extension of credit and loans to help capital accumulation, and in the training of local entrepreneurs, technicians and industrial workers.

5. Implicit in all of this is the idea that social and economic change takes place by a process of diffusion from the core to the periphery. Innovations and advances are generated in the developed world (the Metropole) and diffuse to the lesser developed countries; these innovations are introduced into the cities of the underdeveloped countries and from there diffuse, in turn, to the countryside. In this model, the cities of the less developed countries play an important creative role in integrating the country and in stimulating its parallel transformation.

This scenario of development—predicted by the theories prevalent for the past few decades—has not occurred. Industrialization has not accompanied urbanization and many now doubt whether the wide-scale substitution of industry for agriculture is wise or even possible. Population growth, as well as the subordination of agriculture to the monoculture demands of the world market which often require *less* labor, has led to a massive migration from rural areas, without a commensurate expansion of employment opportunities in the cities. High capital investment, whether achieved through state spending or imported via multinational firms, has led to enormous indebtedness, the draining of the surplus by the repatriation of profits or the interest payments on the debt, and has failed to create sufficient jobs because of technological innovations which minimize the need for labor. Western-trained technicians and industrialists (in or out of local governments) define development as imitation of western solutions and seek a consumption style both in public and private life which favors luxury for the few rather than basic necessities for the many. Scarce international exchange credits are squandered on their importation. Education is diffused, but the educated cannot be used by the economy. Monetary incentives are created, indeed forced, but the opportunities to achieve them are not there. And in all of this, the role of the city as a generator of change and as a fertile source of ideas and innovations exported to the countryside has failed to materialize. Indeed, the most frequently observed scenario is an almost complete reversal of this role; the major cities serve chiefly to drain the resources of the countryside, just as the Metropole economy drains the resources of the underdeveloped country in a chain of exploitation which increases discrepancies rather than bridging and equalizing them.

These are some of the anomalous findings which are causing social scientists to reexamine their theories of development and their evaluations of the role of cities in that process. While new theories about urbanization per se have not been refined, there is obviously a need to rethink the entire question, particularly if one accepts some of the premises of dependency theory.

This section begins with an article by Alejandro Portes which takes up the question of sociological theories of modernization and economic-cultural development. He identifies two theories which for some time have tended to dominate sociological thinking about development. These he calls the theory of "development as social differentiation" and the theory of "development as the enactment of values." After defining each he proceeds to criticize them and to show how they have internal flaws in reasoning and lack external congruence to the facts of Third World development. He then introduces the theory of dependency which, while it can be criticized in its application as often too rhetorical and mechanical, at least does appear to explain many observed conditions in the Third World. He suggests that it offers a new set of leads for future research, not only into development per se but into the role of cities in that process.

Dependency theory, however, is far from a simple and agreed upon set of interrelated propositions, as is evident from the article written by Ronald Chilcote. The latter reviews the history of the development of this theory and the variety of "schools" of thought within it. As can be seen, the theory was largely generated by scholars working in Latin America where the conditions deviate in very significant ways from other parts of the Third World. While Latin America was initially colonized, political independence was achieved quite early. Despite this political independence, however, autonomy and development failed to materialize. The theory of internal colonialism and of economic dependency was an attempt to explain the persistence of subordination to the goals and interests of Britain and the United States despite nominal independence.

It is clear that this theory is not easily transferable in its original form to other parts of the Third World. Within Africa, certainly, more raw and brute exercises of power are obvious and it has hitherto been unnecessary to search for subtler forms of domination since more direct ones could be found. Variations over time and space, however, can be found in the history of Africa, as Samir Amin's analysis indicates, and these variations have left different regions of post-independent Africa with somewhat different types of underdevelopment. If, indeed, concrete historical cases are the ultimate test of the value of a theory, such a theory should be able to explain the changing forms of dependency and should be able to predict the current shift in Africa from the earlier control under colonialism to the newer forms of dependency which begin to approach the Latin American type; Amin's application demonstrates such a test.

Asia presents another variation but one which has not yet been incorporated into the new theory, although McGee's two books (1967, 1971a) come close. Further study and analysis of this complex case are needed if we are to discover the validity of dependency theory for understanding developments in the Far East.

CHAPTER 10

On the Sociology of National Development: Theories and Issues

ALEJANDRO PORTES

. . . The study of national development occupies a paradoxical position within sociology. From classical times to the present, it has had a central place in the minds of theorists concerned with the transition toward more advanced social stages. At the same time, familiarity with the concrete historical experiences of countries in the "underdeveloped" world has remained a tangential preoccupation. This is especially true with regard to the actual dilemmas faced by nations attempting to break away from their past and move toward different models of the future.

A major gap appears to exist between theoretical perspectives chosen by modern sociology and recurrent dilemmas and concrete restrictions faced by the nonindustrialized world. In part, the paradox which makes of "development" both a central and an esoteric concern within sociology stems from the confluence of two different major themes. One is the century-and-a-half-old recapitulation of major processes of change which occurred in Europe beginning in the 16th century. The other is the more recent comparison between countries that are "developed"—wealthy, industrialized, technologically advanced, militarily powerful, politically stable, etc.—and those that are at different stages of "underdevelopment."

The first theme has generated a vast literature around the question, What were the forces which impelled Europe to evolve such drastically new social forms out of a rather conventional feudal order? The second has been concerned with the issue, What forces, present in some societies, have been absent in others and hence prevented their rapid advance toward industrialism? . . .

The goal of this paper is to further explore relationships between the experiences of underdeveloped countries, as they attempt to bridge the gap separating them from advanced nations, and contributions made by modern sociology to an understanding of these phenomena.

Reprinted from the *American Journal of Sociology* 82, no. 1 (July, 1976): 55–85, by permission of The University of Chicago Press and the author. This paper has been edited for this publication.

DEVELOPMENT: PRELIMINARY CONSIDERATIONS

When scholars or leaders of countries in Africa, Asia, or Latin America refer to "development," they consistently sound one or more of the following major themes, which, taken together, form a working definition of the concept:

1. Economic transformation, in the direction of sustained and rapid increases in the national product and the conquest of "decision centers" in manufacturing, which give the country a measure of autonomy for guiding its future growth (Furtado, 1964).
2. Social transformation, in the direction of a more egalitarian distribution of income and widespread access of the population to "social goods" such as education, health services, adequate housing, recreational facilities, and participation in political decision making (Weiner, 1966).
3. Cultural transformation, in the direction of reaffirmation of national identity and traditions. Emergence, in elites and masses alike, of a new self-image which dispels feelings of second-rate nationality and external subordination (Lagos-Matus, 1963).

There are three important aspects of the relationships between wealthy and impoverished nations on which there seems to be growing consensus. They are introduced here because they serve as appropriate background against which to discuss contributions of sociology to the problem of development.

Growing Inequality

Little more than a decade ago the line between economically advanced and underdeveloped countries was set by the U.N. at U.S. $500–$600 of GNP per capita. Table 1 presents some recently compiled data on GNP per capita and annual growth rates for the decade from 1963 to 1973. According to the compiler (U.S. Department of State, 1975), the demarcation line now stands at roughly $1,500 (1972 dollars). Apart from the effects of inflation, the tripling of the figure stems from the growing dynamism of economic, especially industrial, production all over the world.

Such growth has not been distributed equally, however, nor has it contributed to narrowing the gap between economically advanced and underdeveloped societies. Figures from the State Department can hardly be suspected of exaggerating that gap. Yet those appearing in Table 1 indicate that the average GNP per capita of developed countries increased from 12.88 times that of underdeveloped nations to 14.13 times that amount in 10 years. Average annual growth rate in developed countries during this period stood at 3.9 percent as compared with 3.2 percent in underdeveloped ones. The latter figure is inflated by inclusion of the oil-producing countries, which experienced an extraordinary growth during the decade. The OPEC countries as a whole were estimated to grow

TABLE 1. Gross National Product per Capita (In 1972 Dollars), 1963-1973

Year	Developed Nations*	Underdeveloped Nations*	World
1963	2,381	185	821
1964	2,509	194	857
1965	2,619	202	887
1966	2,742	207	917
1967	2,832	208	934
1968	2,967	212	967
1969	3,074	223	996
1970	3,164	236	1,020
1971	3,256	244	1,043
1972	3,378	251	1,071
1973	3,561	252	1,112
Average annual growth	3.9	3.2	2.9

SOURCE: Adapted from U.S. Department of State (1975).
*"Developed nations" include: United States; Canada; all European NATO countries except Greece and Turkey; all Warsaw pact countries except Bulgaria; and Austria, Finland, Ireland, Sweden, Switzerland, Australia, New Zealand, Japan, and South Africa. Twenty-eight countries were classified as developed. "Underdeveloped nations" include: all those in Latin America; in the Near East including Egypt; in East Asia except Japan; in South Asia; in Africa except South Africa; and also Albania, Bulgaria, Greece, Malta, Spain, Turkey, and Yugoslavia. One hundred eight countries were classified as underdeveloped.

at an average rate of 4.7 percent per year (U.S. Department of State, 1975).

The situation may be further clarified by considering individual countries. They are cross classified in Table 2 by 1963 per capita GNP and by average annual growth rates from 1963 to 1973. There is obviously a great deal of dispersion in these data. Nevertheless, dominant trends exist which may be summarized as follows: Among countries considered "developed," none failed to grow at annual rates of at least 2.00 percent. Among the least developed countries (those with per capita GNP of less than $300 in 1963), more than half failed to reach the 2.00 percent annual growth figure, and only 10 percent exceeded growth rates of 4.00 percent. In the intermediate categories (per capita GNPs between $300 and $999 in 1963), slightly more than a third of the countries had growth rates of 4.00 percent or higher; a third, however, fell below the minimal criterion of 2.00 percent per year. . . .

State Intervention

A second, related point of consensus is the need for decisive state intervention in initiating and sustaining a process of development. This applies to economic growth as well as to social redistribution and cultural transformation. Examples of "late development"—countries which crossed the threshold after the dichotomy development-underdevelopment became part of international reality—furnish perhaps the most useful point of reference for analysis of the process. Japan and the Soviet Union are

TABLE 2. Gross National Product per Capita and Growth Rates, 1963-1973

Average Annual Growth Rate 1963–73 (%)	GNP per Capita, 1963 (in 1972 Dollar Equivalents)							
	Less than 100	100–199	200–299	300–499	500–999	1,000–1,999	2,000–2,999	3,000 or More
Less than 1.00 ...	Afghanistan, Burma, Chad, India, Nepal, Somalia, Upper Volta	Cambodia, Haiti, Niger, Sudan, North Vietnam, South Vietnam, Yemen	Egypt, Ghana, Jordan, Senegal	Zambia	Cuba, Lebanon, Mongolia, Uruguay	...	...	Kuwait
1.00–1.99	Burundi, Dahomey, Ethiopia, Mali, Rwanda	Central African Republic, Guinea, Malagasy Republic, Sierra Leone	El Salvador, Honduras, Morocco	Albania, Congo, Nicaragua, Peru	Chile	Venezuela	...	...
2.00–2.99	Ceylon	Bolivia, Uganda	Colombia, Iraq, Paraguay, Philippines, Syria		South Africa	...	Iceland, Luxembourg, United Kingdom, New Zealand	Sweden, Switzerland, United States
3.00–3.99	Indonesia, Laos, Tanzania	Cameroon, China (People's Republic), Kenya, Liberia, Nigeria, Togo, Zaire	Ecuador, Ivory Coast, North Korea, Tunisia	Algeria, Costa Rica, Dominican Republic, Guatemala, Malaysia	Argentina, Jamaica, Mexico, Trinidad-Tobago	German Democratic Republic, Ireland	Australia, Denmark, Norway	Canada, German Federal Republic
4.00–4.99	Mauritania, Pakistan	Thailand	...	Brazil, Turkey	Panama	Austria, Czechoslovakia, Finland, Italy, Hungary, Poland, Soviet Union	Belgium, France, Netherlands	...
5.00–5.99	...	...	...	...	Spain, Yugoslavia	Israel	...	...
6.00 or more	...	South Korea	China (Taiwan), Iran	Gabon, Saudi Arabia	Bulgaria, Cyprus, Greece, Rumania, Lybia, Portugal	Japan	...	...

SOURCE: Adapted from U.S. Department of State (1975).

generally regarded as the most remarkable examples of late development. Countries like Argentina, Greece, Portugal, Spain, Yugoslavia, and Poland experienced, at some point or other, rapid processes of social or economic change and, hence, partially attained developmental goals (O'Donnell, 1972; Linz, 1973). Finally, nations ranging from Cuba to Brazil and from China to Peru are currently engaged, under differing political regimes, in sustained efforts at economic and/or social transformation. . . . .

The main lesson emerging from these experiences is that development does not just happen. Instances of accelerated economic growth or social transformation among modern nations have generally been preceded by deliberate policies initiated and sustained by national governments. This process, in turn, has been dependent on arrival to positions of power of new ruling groups. . . .

Early industrial development in northern Europe and the United States differed from late or contemporary development, for the former lacked the definite teleological component of the latter. Ever since the international division between technologically advanced powerful nations and technologically backward poor ones became fact, rapid development has generally been the result of a consciously guided process. The presence of development-oriented ruling groups, their effective control of the state, and their deliberate attempts to transform the social structure have been necessary, albeit not sufficient, conditions for national development (Apter, 1971). Governments under the control of such groups have aimed at reversing those "natural" processes of the international economy which, left to themselves, would increase disparities between their countries and developed ones.

Emergent Realities

Despite the goal of bridging the gap and the analogy of movement from a backward to a modern situation, there is agreement that contemporary development cannot be a mere repetition of earlier models. The international context in which national transformations occur has been altered drastically in the course of a century. International trading alliances—the growth of common markets and producer organizations—and especially the rise of a "transnational" corporate economy in recent years have produced an immensely more complex situation (Sunkel, 1974).

In addition, advanced countries do not simply furnish a static end point for the process of development but are themselves rapidly evolving. Both technologically advanced and backward nations are integral parts of a changing "world-system." Within it, societal development does not occur as a simple linear advancement toward a static goal but rather as a dialectic exchange between national and transnational actors (Wallerstein, 1974). . . .

The preliminary remarks above are intended as introduction to the general topic of this paper and as generally accepted points of reference for guiding the ensuing discussion.... The purpose is to assess the standing and limitations of current sociologies of development. The strategy consists of presenting the major perspectives on the problem in rough chronological order. A brief outline of each is followed by comments. I believe that the three viewpoints to be examined exhaust major ones advanced in sociology. Thus, discussion of them should clarify current limitations of the field and, I hope, point in alternative and more promising directions.

DEVELOPMENT AS SOCIAL DIFFERENTIATION

Theory

Nineteenth-century scholarly efforts to apprehend processes leading to industrialism culminated in the various theories of social evolution. Borrowing from the biological sciences, the concept of "development" was applied to the continuous transformations experienced by human societies. The latter were envisioned as growing organisms which passed through a series of ordered and inevitable stages. Such stages culminated in the highest levels of societal complexity, represented by advanced European nations....

The status of evolutionism as the dominant perspective in the social sciences did not suffer because of its Europe-centered character. Instead, its popularity floundered on two formal shortcomings: first, the assumption that development of human societies was unilinear and its major "stages" were universal, and second, the failure to specify fully the causal mechanisms and processes of change which led to the transition from one "stage" to another (Eisenstadt, 1964).

The demise of evolutionary theory was not total, for it left behind a series of notions and hypotheses influential to our days. Three such sets of ideas can be mentioned: first, a wide assortment of bipolar theories which compress the former evolutionary stages into ideal-typical extremes; second, the notion of "social differentiation" as a continuous and largely irreversible process; and third, the more recent listing of "evolutionary universals." Each idea eventually acquired a double role: as part of general social theory, each continued the tradition of classic European thought, preoccupied with the historical transformation of the West. Each was also made to apply, eventually, to comparisons between contemporary "developed" and "underdeveloped" societies and the processes of change occurring within the latter.

Bipolar theories of society trace their origins to classic sources of modern social thought. Theorists in this tradition were less concerned with encompassing the entire history of mankind than with apprehending that moment of European transition from a feudal-agricultural to a capitalist-industrial order. Their theories necessarily adopted a "before-and-after" model, in which beginning and end stages were described in different terms and with different value tonalities (Bendix, 1967). . . .

Ferdinand Toennies's "community" and "society" and Emile Durkheim's "mechanic" and "organic" solidarities are perhaps the best-known classic examples. Toennies (1957) came to deplore the absorption of the spontaneity and intimacy of "community" into the impersonal automatism of "society." Making systematic use of the notions of social differentiation and integration, Durkheim (1933) viewed the process more optimistically as a transformation toward higher social forms. Charles H. Cooley (1962) opposed "primary" to "secondary" social attachments in a manner similar to Toennies's. Henry Maine (1907) distinguished between societies based on status and those based on contract. Robert Redfield (1965) came to oppose "folk" to "urban" cultures on the basis of his anthropological research. Howard Becker theorized on differences between "sacred" and "secular" social orders. In this context, the contemporary distinction between "tradition" and "modernity" (see Lerner, 1965; Levy, 1966) can be easily conceptualized as a latter-day counterpart of a theme present since the beginning of sociological thought.

One of Talcott Parsons's best-known contributions is the attempt to break down this time-honored polarity into a series of more specific and exhaustive "value orientations." The resulting pattern variables (Parsons, 1964a) were originally intended as a contribution to general social theory. It did not take long, however, for writers of the functionalist school to transform the scheme into a tool for the comparative diagnoses of societies.

For Hoselitz (1960), the pattern variables constitute a paradigm of the sociological aspects of economic growth: underdeveloped (traditional) societies are those in which roles are ascribed, functionally diffuse, and oriented toward narrow particularistic goals. Developed (modern) societies are characterized, on the other hand, by clearly delineated specific roles, which are acquired through achievement criteria and oriented toward universal norms (Chodak, 1973). The transformation of classic bipolar theories into the pattern variables scheme and of the latter, in turn, into a theory of societal development is perhaps the clearest instance of the application of notions emerging from European history to the contemporary situation of underdeveloped societies.

Among residues left by evolutionism, none receives a greater degree of acceptance at present than the concept of "social differentiation." The theory of society based on this notion is one in which pressures faced at a given point are eliminated through increasing specialization and differentiation, which, in turn, give rise to problems of integration, which are solved through emerging networks of interdependence. The whole process results in an ever-growing societal "complexity" or "systemness."

Social differentiation is another concept emerging from general social theory and made to apply to contemporary comparisons of development and underdevelopment. For Smelser (1968, p. 138), "Development proceeds as a contrapuntal interplay between differentiation (which is divisive of established society) and integration (which unites differentiated structures on a new basis)." . . . Finally, the notion of "evolutionary universal" is Parsons's major late contribution to social theory. The concept is defined as "any organizational development sufficiently important to further evolution that, rather than emerging only once, it is likely to be 'hit upon' by various systems operating under different conditions (Parsons, 1964b, p. 339). . . .

In sum, the sociology of development which emerged as a by-product of classic theory is built on polarities provided by ideal-typical forms of social organization. Development is conceptualized as gradual, qualitative passage from less to more differentiated social forms. This occurs through processes of ever more complex specialization and functional interdependence. Through them, social roles are transformed to approach modern standards of universalism, specificity, and achievement. As societal development proceeds, certain "adaptive features" are incorporated which increase the capacity of the system to survive in its environment. A money economy, formal rationality in the administration of justice, and, finally, the democratic association are among such structural features.

Discussion

Critiques of the definition of development as evolutionary transformation have been made before (Frank, 1969b; Bendix, 1967). They have not dealt with attempts at operationalizing the concept of differentiation (through measurements of division of labor, etc.) but rather with the theoretical statements reviewed above. The discussion below aims less at raising new points than at systematizing those already made. In this regard, four major aspects must be considered:

Abstractness Notions derived from general evolutionary theory may prove useful as a comprehensive tool for the analysis of human societies. They could also be accepted as a broad perspective on social history. Their utility, however, breaks down when they are applied to the concrete "middle-range" problem of national development.

Students of developing societies have had difficulty in applying concepts stemming from the general evolutionary perspective to the analysis of specific historical instances. . . .

To make the diagnosis, for example, that social transformations occur in the passage from underdevelopment (backwardness, undifferentiation, ruralism, etc.) toward modernization (industrialism, social complexity, urbanization, etc.) is to beg the entire question. Such descriptions (assuming for the moment their accuracy) are not the end point of scientific inquiry into the problem but their beginning. The question is why such transformations occur in some societies and not others, why they take place at different rates and in different forms, under which conditions such processes are paralyzed or even reversed, and under which they successfully overcome structural obstacles. . . .

Similarly, the statement that development proceeds through the "counterpoint of differentiation and integration" or that it constitutes a process of increasing "social systemness" can be readily accepted without advancing in any way our understanding of the phenomenon. . . . Inquiries, for example, into the composition of elites oriented toward development, their struggles for control of the state, and the dilemmas they face in allocation of scarce national resources do not flow naturally from this paradigm nor can they be readily integrated into it (Kerr, et al. 1960; Furtado, 1973b).

Abstractness, in sum, means that the effort to encompass processes of change in all societies has rendered the evolution-differentiation perspective virtually useless for close-range analysis of individual ones. . . .

Gradualism The imagery conveyed by the concepts of differentiation and integration is one of a relatively smooth, gradualistic, and irreversible societal process. Problems to be solved by greater functional specialization are part of the natural internal evolution of the social system in its march toward greater complexity. Such a view may again be applicable in the abstract to all societies in "evolutionary time." . . . Yet it can hardly accommodate the many contingencies of contemporary social change in the underdeveloped world.

These contingencies involve the composition and orientations of different political and economic elites, the situation and distribution of rural and urban masses, the uncertain outcomes of struggles for control of the state, the sudden thrusts, convulsions, and reversals suffered by the social fabric during processes of national transformation. Descriptions of development as automatic, growing societal systemness offer few conceptual tools with which to understand the structural constraints and failures that form an intrinsic part of any developmental effort (Malloy, 1971; Seers, 1969). . . .

Above all, the gradualistic imagery of development as social differentiation provides no insight into the role of social actors in social change. In underdeveloped countries, that role involves the commitment of strategic groups—elites and masses—to overcoming past structural barriers (Horowitz, 1966). Developmental efforts are aimed precisely at reversing, not encouraging, the "natural" evolution of the preceding social structure (see Scott, 1973). As seen above, natural processes of differentiation in the international economy have tended to increase inequalities between wealthy and impoverished nations. Development, as experienced in the latter, is not a gradual, automatic, and impersonal process but a consciously guided, difficult, and often drastic alteration of existing social trends.

External factors A criticism leveled more frequently at this approach is its neglect of interrelationships among nations and the impact of such factors in the internal structures of each. Societies are conceived as autonomous units which change according to internal forces. Room is made for some intersocietal exchanges and consequent processes of "diffusion." However, the introjection of the economy and polity of some societies into the internal structure of others and the supranational dynamism of an organized world-economy are not comprehended.

It is perhaps for the above reason that this perspective on development is able to conceptualize the phenomenon as movement from an initial (backward) state to a final (modern) one. Within the analogy, autonomous societies occupy more or less advanced positions, with the "winners" coming to resemble the United States and the industrialized western European nations. As seen above, this imagery neglects the fact that the internal configuration of each competing unit is profoundly affected by its interactions with the others, especially the more powerful ones.

Though abstract and partial, the evolutionary viewpoint provides at least a statement on the internal structure and change of societies. It fails completely, however, to provide a framework for understanding the insertion of individual countries in an evolving international system. Distinctions between core and peripheral economic regions are foreign to the theory. Nor does it grasp the possibility that "autonomous competitors" in the developmental race may be integral parts of transnational units in which weaker states are kept in place by a context of overwhelming political and economic forces (Sunkel, 1972). Contemporary development is not a matter of autonomous change but one composed largely of exchange and confrontation in an integrated world-system, in order to alter the position of individual nations (Wallerstein, 1974).

Conceptual misplacement Limitations of the passage from a Europe-centered theory of change into a theory of contemporary development are reflected in identification of the "wrong" dimensions as indicative of or strategic for development. Theoretical writings may thus contain extensive discussions of aspects which have not been central to historical experiences of change in the Third World. Deflection of attention from important problems occurs because the theory is based, not on extensive knowledge of underdeveloped societies, but on reflection on the European past.

Thus, for example, development may be conceptualized as an attempt to attain missing "evolutionary universals." Minimal familiarity with Third World nations would indicate that, by these standards, most of them are already "developed." Such features as money and markets, extensive bureaucratic regulation, and formal legal systems have been long known and present in underdeveloped countries. It is perhaps for this reason that when the abstract discussion of "universals" reaches for concrete examples, it selects primitive tribal societies as poles of contrast to modern Western nations. . . . Third World nations, regardless of stereotypes dear to theorists, are not in the tribal stage.

Here again, the problem lies in conceptualizing underdevelopment as a feature intrinsic to autonomous societies rather than as part of an international system. Structural characteristics like currency and markets, formal legal regulation, and bureaucracies exist in both developed and underdeveloped nations, for both kinds are part of an integral world-order. The functions of these traits vary, however, between the two types of societies because their "adaptive" potential is employed in accordance with the particular insertion of the country into the international economy.

Since colonial times, extensive legal and bureaucratic regulation of dependent territories has been employed by metropolitan centers as a means of insuring their hegemony. The historical dialectics by which "modern" structural features serve to perpetuate weak and stagnant societies are not understood by proponents of the evolutionary perspective. . . .

DEVELOPMENT AS ENACTMENT OF VALUES

Theory

Myron Weiner (1966) notes that, for many scholars, the starting point of any definition of development is not the character of the society but the character of individuals. The same author observes that "although there are differences among social scientists as to how values and attitudes can be changed, it is possible to speak of one school of thought that believes that attitudinal and value changes are prerequisites to creating a modern society, economy, and political system" (Weiner, 1966, p. 9).

For Chodak (1973, p. 11), writers of this school do not ask, What is development? or What happens in its course? but rather why it happened and what specifically caused the breakthrough from traditional into modern societies. Where such factors are present, development happens; where they are absent, stagnation prevails. The distinctive factor is then sought in the sphere of value orientation.

Unlike the first perspective, whose emphasis was on describing evolutionary trends, this one is essentially explanatory. The search in this case is for those "mental viruses" (McClelland, 1967) changing the "spirit" (Inkeles, 1969) of men so that they come to adopt and promote a modern society. This perspective derives its impetus from the general emphasis in United States sociology on value-normative complexes (Parsons, 1964a) as opposed to the structure of material interests in society. . . . More specifically, the value approach to the problem of development lays claim to, and often labels itself as a direct continuation of, the thesis Weber developed in *The Protestant Ethic and the Spirit of Capitalism.* . . .

Weber's argument was, however, securely embedded in a body of research which clearly brought forth the importance of structural forms and the politico-economic interests of groups and classes. Emergence of an urban burgher class out of the feudal "oikos" and the relative vulnerability of feudalism, as opposed for example to the "prebendary" system of China, are subjects examined at length in his work (Weber, 1951). The combined effects of the political assault by the central state and the economic assault by rising classes on the weakened feudal order meant an increasingly "open" structure for capitalist expansion (Weber, 1958). . . . Only because of the growing predicament of a lordly class incapable of defending its position by enforcing old prerogatives could the Protestant "spirit" of capitalism, or any other spirit for that matter, transform the economic order to its own advantage (Wallerstein, 1974).

Psychological theories of development, such as those proposed by McClelland (1967) and Hagen (1962), have chosen to ignore the Weberian treatment of historicostructural issues and concentrate on the primacy of ideas in society: "This is just one more piece of evidence to support the growing conviction among social scientists that it is values, motives, or psychological forces that determine ultimately the rate of economic and social development. . . . *The Achieving Society* suggests that ideas are in fact more important in shaping history than purely materialistic arrangements" (McClelland, 1963, p. 17).

Since ideas inhere in individuals, these theories result in an "additive" image of societal development in which the larger the number of people "infected" by the strategic psychological ingredient, the greater the economic growth of the country. This arithmetic approach is concerned neither with differences in positions in the stratification system nor with existing arrangements of economic and political power. Theorists of this persuasion subscribe to the proverb, Where there is a will, there is a way; their voluntarism is, in turn, predicated on the creation of sufficiently high levels of motivation. . . .

More recent, and perhaps more accepted among sociologists, is the theory of "modernity" as a psychosocial complex of values. "Modern man" is characterized, internally, by a certain mental flexibility in dealing with new situations and, externally, by similarity to the value orientations dominant in industrial Western societies (see Lerner, 1965). The spirit of modernity is regarded by these writers both as a precondition for societal modernization and as a major consequence of it: "Indeed, in the end, the ideal of development requires the transformation of the nature of man—a transformation that is both a *means* to the end of yet greater growth and at the same time one of the great *ends* itself of the development process" (Inkeles, 1966, p. 138).

Inkeles identifies nine major attitudes and values distinguishing modern man: (1) readiness for new experience and openness to innovation, (2) disposition to form and hold opinions, (3) democratic orientation, (4) planning habits, (5) belief in human and personal efficacy, (6) belief that the world is calculable, (7) stress on personal and human dignity, (8) faith in science and technology, and (9) belief in distributive justice.

This list is not complemented by a similar description of "traditional man." The latter tends to be defined by default: whatever is not properly modern must be traditional. It is difficult to understand, however, why "traditional man" does not stress "personal and human dignity" or believe in distributive justice. . . .

In addition to the dimensions quoted above, the following are frequently encountered in characterizations of modern man: (1) participation: motivation and ability to take part in organizations and electoral processes; (2) ambition: high mobility aspirations for self and children and willingness to take risks; (3) secularism: limited religious attachments and low receptivity to religious and ideological appeal; (4) information: frequent contact with news media and knowledge of national and international affairs; (5) consumption orientation: desire to own new goods and technologically advanced recreation and labor saving appliances; 6) urban preference: desire to move to or remain in urban areas; (7) geographic mobility: experience of moving and/or willingness to move from original residence in search of better opportunities (Lerner, 1965; Kahl, 1968; Schnaiberg, 1970; Horowitz, 1970; Portes, 1974).

Discussion

Some critical observations have already been made in the course of outlining this second perspective on development. The value-enactment viewpoint has probably been the most controversial of all three perspectives. Critics have noted, for example, its neglect of international economic and political linkages. As Frank (1969b, p. 47) states: "Those engaging in [the value-enactment] mode of analysis resolutely avoid the study of the international structure of development and underdevelopment of which the domestic structure of underdevelopment is only a part."

Discussion of this second sociology of development must consider, however, several important aspects. A systematic presentation of these will cover three major points:

Structural constraints No matter how compelling the image of highly motivated entrepreneurs racing to break the barriers of stagnation, the fact remains that individual action is highly conditioned by external social arrangements. Despite the frequent application of "tribal" imagery to underdeveloped nations, the reality of such societies is not one of an open frontier awaiting conquest by an entrepreneurial elite. Indeed, a complex structure of economic and political interests penetrates every aspect of them. To think that more modernity, achievement motivation, or status withdrawal will automatically transform these structures is, at best, naive. Regardless of what psychologists may think, societies are not the simple "additive" sum of individual members.

An active set of individuals, motivated by whatever psychological mechanism one may wish to posit, must still cope with existing economic and political arrangements. . . .

Individual motivations for achievement can be absorbed, fulfilled, and utilized without changing a basic situation of economic subordination and social maldistribution. The issue is not how much individual motivation there is, or what its sources are, but rather to what goals it is directed. The fundamental individualism apparent in theories of achievement and entrepreneurial motivation may be either irrelevant or inimical to struggles for national transformation. Elites committed to the task of development are not formed by "moderns" but by "modernizers"—individuals committed to achievement of collective economic and social change (Kerr et al., 1960).

Consumption-oriented values A second, related aspect has to do with some of the values defined as "modern." That most modern of traits—"empathy"—is usually described as ability to comprehend and place oneself symbolically in the midst of urban-industrial life (see Lerner, 1965). This, in turn, is directly linked with a "demonstration effect" which raises demands for consumption beyond what a poor country can realistically afford. . . .

The market is certainly broadened, often for the benefit of multinational enterprises, strains are placed on the country's capacity to import, and new "possibilities" are taught which often bear no relationship to local conditions. Such modern values—premature wants, imported needs and tastes, excessive consumption—are not the values of development. . . .

Historical fiction Theories of modernity share with those of evolutionary differentiation the belief that development proceeds from an early traditional stage toward a terminal "advanced" one. It is proper at this time to complete analysis of the character of this analogy.

As seen above, tradition is described in terms which are only logical counterparts to those embodied in modernity. There is no existing nation in the Third World which can be labeled "traditional" in this sense. The fictional character of the initial stage of the process is due to the fact that it is not based on observation of actual societies but on reflection on the features of the "terminal" stage. . . .

DEVELOPMENT AS LIBERATION FROM DEPENDENCY

Theory

The third and newest perspective on the sociology of development is one not original to the United States or western Europe but to countries of the

Third World. In a sense, the theory of "dependency" is the counterpart of earlier theories of imperialism (Lenin, 1939; Hobson, 1965), seen from the standpoint of the subordinate nations.

This perspective views development and underdevelopment not as two different stages in the history of mankind but as integral parts of the same "world-economy." The forms that social poverty and economic stagnation take in the Third World at present are largely creations of the process of capitalist world expansion. It is argued that rapid industrial growth in the West could not have occurred without the conditioning of a "periphery" from which an economic surplus is extracted and necessary raw materials secured (see Wallerstein, 1974).

Thus underdevelopment is not a "backward" state prior to capitalism but a variant of the latter and a necessary consequence of its evolution. . . .

Earlier writings, such as those of Baran (1957) and the initiators of the U.N. Economic Commission for Latin America (Prebisch, 1950), addressed a situation in which international forces kept peripheral economies stagnant and restricted their role to that of producers of raw materials. . . . Import-substitution industrialization was expected to lead the way toward greater economic autonomy, less reliance on raw export commodities, progressive weakening of the power of the landowning aristocracy, and the diffusion of economic benefits to the impoverished masses (O'Brien, 1975).

Writings in English, such as those of Gunder Frank (1967), have created the erroneous impression that "dependency" analyses continue to be concerned with the same quasi-colonial situation of economic stagnation and foreign control of export enclaves. On the contrary, contemporary dependency studies address a situation in which domestic industrialization has occurred along with increasing economic denationalization; in which sustained economic growth has been accompanied by rising social inequalities; and in which rapid urbanization and the spread of literacy have converged with the ever more evident marginalization of the masses (Sunkel and Paz, 1970).

The complex dialectics of economic successes that lead to social failures and of strategies for national autonomy which produce external subordination lie at the core of the situation analyzed by proponents of this third perspective. Efforts to untangle the dialectics have centered on the structure of foreign-controlled multinational corporations and their role in underdeveloped societies.

The presence of multinational enterprises is credited with the bulk of the dismal outcomes of earlier import-substitution industrialization. In a sense, multinational corporations stood earlier goals of national autonomy on their heads by employing protectionist barriers as instruments for the more effective control of domestic markets and internal economic activity. . . .

[D]ependency analyses generally identify two major consequences of the activities of multinational enterprises within import-substitution industrialization: externally, from international centers, these enterprises have maintained and expanded the economic dependence of underdeveloped nations; internally, they have led to emergence of new privileged groups and the acceleration of social inequality.

External subordination comprises, in turn, two aspects. First, use of complex imported technology by local corporate affiliates and the inputs they require ties the dependent economy ever more closely to that of the centers. Second, the costs of capital intensive technology, added to profit remittances abroad, intercompany transfers, and so forth, amount to a constant drain of scarce foreign exchange resources. . . .

The internal impact of multinational corporations has been described as a virtual restructuring of the domestic social order. At one extreme, superior resources of foreign subsidiaries create "instant" obsolescence and progressive elimination of preexisting national enterprises. Those which survive must accept a satellized role, and the small number of domestic enterprises created to satisfy input demands of the multinationals are entirely dependent on the interests and policies of the latter. Thus, a progressively increasing portion of elite economic activity comes to revolve around the interests and priorities of foreign subsidiaries, in turn obedient to their metropolitan offices.

At the other extreme, a capital intensive technology not only is costly to import but also fails to absorb labor surpluses and, in some areas, increases them. While the working sector associated with the multinationals may evolve into a veritable labor aristocracy, the masses remain in a subsistence situation where unemployment, scarcity of bare essentials, and lack of access to expanding social and economic benefits are the norm. Industrialization under the aegis of foreign-owned corporations promotes a structure of increasing inequality where those working in the "modern" sector and able to purchase its products are a self-contained minority, progressively divorced from the plight of the majority (Furtado, 1973b). The basic cleavage in these new quasi-industrial societies does not occur between traditional peasant and landowning classes or even between domestic bourgeoisie and urban proletariat but between groups linked with the foreign-owned sector (workers and employers alike) and the rest of the population (Sunkel, 1972).

A situation of contemporary dependency is thus quite different from one of foreign colonialism. For underdeveloped societies, dependency is a condition not only external but also internal. Foreign domination is not imposed by an "army of occupation," as naive versions have asserted; instead, it is initially implemented by a willing local bourgeoisie (Quijano, 1967). As Torres Rivas (1974, p. 195) has asserted concerning Central America: "[Such] dependent power is not only that which confronts and submits itself to imperialist power, but the behavior of dominant classes which, even prior to that confrontation, are already submitted to it. This is the consequence of anticipatory socialization, product of the very weakness of these groups."

At later stages, multinational enterprises are able to evolve their "own" bourgeoisie and proletariat, both allied in the defense of a privileged situation against the majority. This remolding of social structure serves well the short-term interests of the companies by creating artificially high demand in a docile market. It runs counter, however, to the interests of the nation as a whole. In this context, development consists logically of liberation from external control and from the internal structure of inequality which it promotes.

Discussion

Rhetoric The theory of dependency does not represent a system of logically interrelated propositions. Rather, it refers to a historical "vision" (Sunkel and Paz, 1970), a sociohistorical "model" (Roth, 1975) meant to apply to a plurality of situations. Such theoretical strategy helps define the scope of a field of study, sets priorities for empirical investigations, and provides an abstract framework under which more concrete hypotheses can be articulated. Like all historical models, however, it can also be applied so mechanistically as to suggest that there is nothing new under the sun.

In the hands of some writers, dependency has become a *deus ex machina* explanation for everything that is wrong with Third World societies. . . .

The rhetoric of dependency characterizes primarily the earliest and crudest versions. Their excesses are largely responsible for the two criticisms leveled most frequently against the perspective as a whole: First, some writings have equated national development with liberation from foreign domination (see Frank, 1967). . . .

Second, denunciations of imperialism and dependency as global phenomena leave no room for analysis of national variants. Countries like Canada are profoundly "dependent," in the sense of penetration of their economies by foreign-owned subsidiaries, and yet exhibit a much higher per capita income, a better distribution of wealth, and more efficient health and educational services than most "Third World" countries. . . . While the theoretical framework may encompass comparisons across a wide variety of national situations, the fervor of earlier writers managed to limit them to those documenting only a particular viewpoint. . . .

Internal dynamics A focus on the international economy in the study of underdevelopment has certainly been the major contribution of dependency studies. At the same time, it has led to neglect of the analysis of internal dynamics. . . .

The limitation—which is shared by all types of dependency theory—

lies in a deductive approach to internal conditions, making them conform to what is logically expected on the basis of external forces. The data base for dependency studies is usually limited to aggregate economic figures on production and distribution, exports and imports, capital flows, inflation, and the like. From these, inferences are made concerning the impact of the situation on different classes and sectors and about their subsequent behavior.

The empirical weakness of dependency studies, as pointed out frequently by critics, is based precisely on the absence of autonomous data for each country to validate these deductive inferences. Without them, assertions derived from an abstract framework are always suspect. . . .

A final point is that a view of social structure centered on the international economy fails to take account of autonomous internal forces and their potential role in processes of change. All historical evidence points to the existence of certain "degrees of freedom" for national governments and their ability to carry out, under certain circumstances, fairly drastic policies of internal and external transformation. While analyses of dependency often end by calling for precisely this kind of action, the forces which could promote it and the dilemmas and options that it would confront are not explicitly integrated into the theory. . . .

CONCLUSION

. . . One thing seems certain—the almost complete exhaustion of earlier modernization theorizing as a source of insights into the situation of underdeveloped nations or of significant questions for research. Newer perspectives—elaborated at closer range to the realities of underdevelopment —represent a potentially more useful guide for future investigation.

Still, much remains to be done before the sociological approach to the problem can acquire a clearer sense of priorities. The passage from earlier evolutionary theories to modernization theories and thence to dependency studies has mostly clarified what a sociology of development should not be concerned with. It should avoid, for example, a social philosophical perspective in which processes of change are described in such abstract terms as to be of little use for the analysis of concrete situations. . . .

What then are likely trends to be followed by this field in the future? A first and predictable concern will be the continuing analysis of external linkages and their projection into the social structures of underdeveloped societies. Much of the dependency literature has been written from the perspective of political economy rather than sociology proper. While analyses of external and internal economic transactions are likely to continue as a central focus, there is an increasingly diversified concern with the internal manifestations of dependency.

Such studies will include ones of trends in spatial distribution of the population, urban growth and rural-urban migration, the changing ecology of cities, partition of the labor force into an "internationalized" sector and its "domestic" counterparts, the growth of public sector enterprises and their relationships with multinational firms, emergence of a "national bourgeoisie" of entrepreneurs and civil servants and their efforts to cope with or resist external penetration, the insertion of foreign-based corporate bureaucracies into local power structures, the extent to which cultural themes and patterns of consumption are affected by diffusion from international centers, and so forth. . . .

A second likely trend stems from limitations of past dependency studies, noted above. It will have to do with conditions and constraints affecting national development efforts. An abundant literature on many of these topics already exists, but it has not been successfully integrated into a coherent framework. Analysis of past and current efforts and strategies of development in countries of Latin America, Africa, and Asia furnished the major historical source for these studies.

Topics will include the origins and socialization of political elites; the uses of ideology for mass legitimation; the demographic, educational, economic, and cultural constraints which different development strategies face; the interplay of coercion and charisma in achieving acceptance of consumption sacrifices; the limitations of democratic, totalitarian, and authoritarian political systems at different stages of development; allocative decisions in "balanced" and "unbalanced" strategies of economic growth; choices between "pure" economic development and "social capital" investment; identification of those sectors tapped as sources of support for development and of those "made to pay" its costs; consequences of "growth poles" versus decentralized investment decisions; processes of exchange among the state, transnational finance, and national entrepreneurial groups; the role of organized labor as source of redistributive demands or instrument for increasing productivity; and so forth.

These topics form a partial and crudely outlined agenda of matters which development scholars are likely to undertake in the future. Many have precedents in the past, but the marked shift in direction represents a significant departure from earlier experience. Students of Third World societies tended in the past to enter this field in a somewhat casual manner. Underdeveloped societies appeared to be appropriate settings in which to test theories evolved in more advanced contexts or on which to try explanatory frameworks based on different historical experiences. At other points, they have appeared to be ideal recipients of charitable intentions, and the sociology of development has been limited to analysis of the welfare aspects of the process.

Colonies first and weak nations second had to accept "tribal" comparisons and stereotypes in lieu of analysis of their own situation. Times have changed, but fulfillment of the current promise offered by this field is contingent on abandonment by sociologists of the naive notions and extrapolations of the past. For the sociology of development to contribute in proportion to current expectations, it must revert to those who, in one way or another, have acquired close familiarity with the internal problems of "peripheral" societies and the external constraints under which they must survive.

CHAPTER 11

Dependency: A Critical Synthesis of the Literature

RONALD H. CHILCOTE

Over the past decade a new perspective of development and underdevelopment has emerged. Labelled by its advocates as dependency theory, this perspective focuses on the problem of foreign penetration in the political economies of Latin America. Generally, this theory explains underdevelopment throughout Latin America as a consequence of outside economic and political influence. More specifically, the economy of certain nations is believed to be conditioned by the relationship to another economy which is dominant and capable of expanding and developing. Thus the interdependence of such economies assumes contrasting forms of dominance and dependence so that dependent nations might develop as a reflection of the expansion of dominant nations or underdevelop as a consequence of their subjective relationship. This explanation approximates the definition of dependency offered by Dos Santos (1968:6) who states:

> By dependence we mean a situation in which the economy of certain countries is conditioned by the development and expansion of another economy to which the former is subjected. The relation of inter-dependence between two or more economies, and between these and world trade, assumes the form of dependence when some countries (the dominant ones) can expand and can be self-sustaining, while other countries (the dependent ones) can do this only as a reflection of that expansion, which can have either a positive or a negative effect on their immediate development.

Reprinted from *Latin American Perspectives* I (1974): 4–29, by permission of the publisher and the author. This paper has been edited for this publication. The bibliography which has been entirely deleted for this publication appears in the original, above-cited source.

In more specific terms, Osvaldo Sunkel . . . elaborates on this interpretation [in *Foreign Affairs,* April 1972: p. 519]:

> . . . foreign factors are seen not as external but as intrinsic to the system, with manifold and sometimes hidden or subtle political, financial, economic, technical and cultural effects inside the underdeveloped country. . . . Thus, the concept of "dependencia" links the postwar evolution of capitalism internationally to the discriminatory nature of the local process of development, as we know it. Access to the means and benefits of development is selective; rather than spreading them, the process tends to ensure a self-reinforcing accumulation of privilege for special groups as well as the continued existence of a marginal class.

These definitions are central to an understanding of dependency, for like other theory in an infant stage, dependency theory has spawned a plethora of interpretations and applications and has been adopted by ideologues on all sides of the political spectrum. However, initial comprehension of the theory should revolve around the relationships of nations, one to the other in terms of dominance versus dependence.

Dependency theory quickly caught the attention of Latin Americans and more recently it has come under wide acceptance by others, both in Europe and the United States. There have been criticisms of the theory, from left and right circles. Perhaps more important, a number of thrusts have emerged in the everexpanding literature that has ensued. . . .

THEORETICAL DIRECTIONS

In all the writings of dependency only two offer a synthesis of the many directions and positions. Fernando Henrique Cardoso (1973c) provides the most recent of these. He finds the foundation for the concept in the writings of Lenin and Trotsky, then attempts to relate these classical formulations to the literature of the past decade. He notes three tendencies in the recent literature. One concentrates on analysis which critiques the obstacles to national development, a good example being the publications of the Instituto Superior de Estudos Brasileiros (ISEB), established in the middle fifties to study and introduce to Brazilian society a new conception of nationalism and development. Helio Jaguaribe (1970), a founder of ISEB, has carried this tradition forward in his view that Latin America faces three alternatives: dependency, autonomy or revolution. Dependency will be overcome, he argues, through autonomous national development and non-revolutionary change. This view of development falls into what . . . I called the diffusion model; it has been criticized by Frank (1967), Theotônio Dos Santos (1970), and Cardoso (1965). A second tendency incorporates analysis on international capitalism in its monopolistic phase. The thrust of this tendency springs from Marx and Lenin, especially the latter. Refinements and elaboration of the early ideas were offered by Paul Baran and Paul Sweezy (Baran, 1957; Baran and Sweezy, 1966); and Harry Magdoff (1969) effectively ties theory to fact in contemporary world affairs. A raging debate from differing perspectives is found in recent issues of *Socialist Revolution;* particularly noteworthy are the views of Robert Fitch who is critical of Baran and Sweezy's corporate model. The third tendency identified by Cardoso attempts to describe "a historical structural process of dependency in terms of class relations, tying the economy and international politics to corresponding local factors which in turn generate internal contradictions and political struggle. . . ." Cardoso's own contributions fall into this tendency.

Claire Savit Bacha (1971) has contributed the other synthesis of the literature in a study that embraces theory and relates to the Brazilian experience. She examines five conceptions of dependency as elaborated in the writings of Vasconi, Lenin, Frank, Dos Santos, and Cardoso. Let us briefly examine each conceptualization.

[Savit] Bacha describes the effort by Tomás Vasconi . . . as oriented to a "systematization of the concept of dependency." Vasconi's conceptualization of dependency relates to distinctions between underdevelopment and development, on the one hand, and between the center and the periphery, on the other. Dependency, he argues, permits one to see the center and the periphery as parts of a capitalist structure, this structure being a system of relations of international interdependence. Accordingly, a central economy expands as it reaches the peripheral economy, incorporating it within that system. Vasconi then suggests a number of propositions. First, dependent nations may or may not develop, but the process of development can lead to a rupture in the ties of dependence. Second, dependency is determined historically. Third, dependency includes all internal and external forces that historically affect a nation, forming its structures in relation to its historical and international position. Vasconi elaborates. During the development of capitalism, he says, dependent nations are isolated from the center. These peripheral nations remain dependent until they break their dependent relations, but in either case they may or may not experience development. Changes in dependent relationships, however, are tied to historical forces.

Another major conceptualization of dependency is closely tied to imperialism. Both concepts deal with relations between the center and the periphery and both explain underdevelopment. Drawing upon Hobson and others, Lenin refined the concept of imperialism as the consequence of capitalism itself. Monopoly capital, he argued, needed to export its surplus of capital, to search for new external markets, and to expand profit-making opportunities. Lenin identified two types of nations: imperialist and dominated nations and he referred in his work to the concept of dependency:

> Since we are speaking of colonial policy in the epoch of capitalist imperialism, it must be observed that finance capital and its foreign policy, which is the struggle of the great powers for the economic and political division of the world, give rise to a number of *transitional* forms of state dependence. Not only are there two main groups of countries, those owning colonies, and the colonies themselves, but also the diverse forms of dependent countries which, politically, are formally independent, but in fact, are enmeshed in the net of financial and diplomatic dependency. . . .
>
> (Lenin, 1967, Vol. 1:742–743)

Thus, dependentistas can turn to Lenin for the theoretical underpinnings of their argument. Lenin makes clear the external imposition that imperialist nations force upon many nations, and by also focusing on dependency, he is able to combine internal with external forces in interpreting the national reality of a dependent nation.

Economist André Gunder Frank offers a third conceptualization of dependency. In his early work (1967), Frank affirms that "it is capitalism, both world and national, which produced underdevelopment in the past and which still generates underdevelopment in the present." His analysis centers on the metropolis-satellite structure of the capitalist system as he traces throughout the history of certain countries the development of underdevelopment. He identifies the internal contradictions of capitalism as "the expropriation of economic surplus from the many and its appropriation by the few, the polarization of the capitalist system into metropolitan center and peripheral satellites, and the continuity of the fundamental structure of the capitalist system throughout the history of its expansion and transformation. . . ." His central thesis focuses on these contradictions; capitalism, he argues, has "generated underdevelopment in the peripheral satellites whose economic surplus was expropriated, while generating economic development in the metropolitan centers which appropriate that surplus." With this thesis, Frank suggests a series of hypotheses which contend with some literature which explains backwardness through a dualist model of society and advocates change through a progressive national bourgeoisie. His critique of these ideas set in motion new thinking and provoked a multitude of criticisms. . . . Further, Frank concentrates attention on exploitation, thereby turning attention to the internal consequences of nations caught up in dependence.

A fourth conceptualization offers a further refinement. Known as the "New Dependency" and elaborated by Brazilian sociologist Theotônio Dos Santos (1968), this conceptualization differs from colonial dependency, based on trade export, and from financial-industrial dependency, characterized by the domination of big capital in the hegemonic centers at the end of the nineteenth century. The new dependency is a recent phenomenon, based on multinational corporations which after the Second World War invested in industries geared to the internal market of underdeveloped countries. Dos Santos characterizes it as a "technological-industrial dependence."

Finally, [Savit] Bacha turns to the early work of Cardoso and Faletto (1969) on dependency. They stress internal structure. For example, classes or groups are analyzed in relation to the structure of outside domination. Dependency, therefore, is viewed not only as an external variable but within "a system of relations among different social classes in an environment characteristic of dependent nations." Like Frank and Dos Santos, they trace dependency through history. They also criticize the economic emphasis of these writers, and attempt to elaborate on theory by suggesting that politics and the internal forces are more decisive than economics and external forces in determining forms of dependency. Their own approach embraces these four levels of analysis: internal and external, political and economic. . . .

My own synthesis of the dependency theory builds upon the earlier efforts of Cardoso and Bacha. I shall recast the categories to reflect the evolution of the literature from the early conceptualization until the present time. Rather than focus on particular writers, I concern myself with two thrusts and six formulations which seem to stand out in the literature. The thrusts revolve around distinctions between the diffusionist and dependency models mentioned earlier while the formulations relate to one or the other model. These formulations are not necessarily mutually exclusive. Indeed they overlap, but they are representative of particular theoretical directions. I now turn to a discussion of each in an effort to identify major theoretical works, attempts to implement the theory, and critical assessment.

THE DIFFUSION MODEL AND THE ECLA AND INTERNAL COLONY FORMULATIONS

The diffusion model embraces a number of fundamental premises. Progress comes about through the spread of modernism to backward areas. Inescapably these areas evolve from a traditional toward a modern state as technology and capital are introduced. Underdevelopment is a condition which all nations have experienced at one time. Some nations have managed to develop, while others have not. In some underdeveloped nations, modern cities have arisen through contact with the developed world, while the countryside maintains a system of unproductive agriculture of large feudal estates.

These premises lead to two controversial propositions. One is that developing nations are structured into dual societies, one advanced and modern and the other backward and feudal. The other proposition suggests that in the advanced society there will emerge a new bourgeoisie, commercial and industrial in character. This bourgeoisie may become progressive and a supporter of national interests as capitalist development diffuses itself into rural areas and as economic and political policies restrict the domination and penetration of foreign interests. Both propositions are embraced, at least partially, by two formulations which sometimes are linked to the foundations of dependency theory. These formulations were proposed, on the one hand, by the United Nations' Economic Commission for Latin America (ECLA) and, on the other by advocates of internal colonialism.

The ECLA school of thought evolved after the Second World War. It was nationalist and sometimes anti-imperialist but non-Marxist in orientation. Its analysis sprang from Latin American economists grouped around the Argentine, Raúl Prebisch. Their philosophy was shaped by beliefs and principles set forth in a manifesto on development (United Nations, ECLA, 1950). The history of the ECLA movement dates from its manifesto and breaks into three phases: from 1950 to 1953 when its ideology was formed, elaborated, and tested; from 1953 until 1958 when intensive studies were made of individual Latin American countries with the objective of proposing plans for their future development; and since 1958 when attention shifted to the study and promotion of regional integration through formation of a common market. The ECLA thesis divides the world into an industrial center and a primary producing periphery, both of which should benefit from the maximizing of production, income, and consumption. However, unrestrained competition tends to result in appropriation to the center of most of the increment in world income. In short, the thesis correctly links Latin American underdevelopment to the international economic system, and thus affirms an underlying assumption of dependency theory. But analysis also is limited. For example, the thesis neglects an adequate examination of the conscious policies and specific needs of the nations of the center; it mistakenly attributes Latin American backwardness to traditional or feudal oligarchies; it inappropriately assumes that development would be promoted by a progressive, nationalist bourgeoisie, an assumption thus far negated by historical experience; and its stress on import substitution as a solution to consumptive dependence on the outside world has resulted in even greater dependence on the international system and in economic stagnation.

Theories of internal colonialism relate to dependency. The early work of the Mexican sociologist, Pablo González Casanova (1970), proposed a framework for analysis of internal colonialism. With the elimination of traditional forms of colonialism, characterized by foreign domination over nations, he suggests that the same conditions of the past colonialism may be found internally: "With the disappearance of the direct domination of foreigners over natives, the notion of domination and exploitation of natives by natives emerges." He describes the forms of internal colonialism, focusing on monopoly and dependence (the metropolis dominates the isolated communities, creating deformation of the native economy and decapitalization); relations of production and social control (exploitation plunders lands and discriminates everywhere); and culture and living standards (subsistence economies accentuate poverty, backward techniques, low productivity, lack of services, and traditionalism). . . .

THE DEPENDENCY MODEL: FOUR FORMULATIONS

The dependency model distinguishes underdeveloped Latin America from pre-capitalist Europe. It does not view underdevelopment as an original condition, but instead assumes that nations may once have been undeveloped but never underdeveloped and that the contemporary underdevelopment of many parts of Latin America was created by the same process of capitalism that brought development to the industrialized nations. Latin America is underdeveloped because it has supported the development of Western Europe and the United States. When the center of the expanding world economic system needed raw materials, it was supplied by Latin America. This relationship has not basically changed, even though the United States has replaced Great Britain as the metropolis which dominates over the area, resulting in a strengthening of dependency through foreign corporate and governmental penetration of banking, manufacturing, retailing, communications, advertising, and education. Within each country the pattern of metropolis-periphery relations is replicated as the economic surplus of the countryside drains into urban areas.

These premises lead dependentistas to a number of propositions. First, they argue that while feudalistic conditions and relationships exist, the backwardness of the countryside cannot be explained by the image of a dual society. Rural areas are poor not because of feudalism but because they have been responsive to urban and international market influences. The consequence has been the enrichment of the cities and the dominant nations. Second, dependentistas assert that the capitalist link between the city and the countryside is characterized by commerce between landowners and merchants who form an agro-commercial bourgeoisie which is subject to the market forces of a national and international capitalist economy. Empirical evidence verifies that agriculture, financial, and industrial interests are often found in the same economic groups, the same firms, and even in the same families. Thus, the capital of archaic latifundia may be invested by their owners in lucrative enterprise in the cities; or the grand families of the city, associated with foreign capital, may also be the owners of the backward latifundias. Thus, the landowning aristocracy and the urban commercial bourgeoisie often align with the manufacturing bourgeoisie. Third, dependentistas believe that dominant class interests are dependent on world imperialism for the manufacture of some goods, for foreign currency, and for foreign capital. Even if a segment of this class manifests nationalist xenophobia or resentment against imperialism, it has no other choice than to accept its condition as a dependent bourgeoisie. (The clearest case for these propositions is in Quijano's analysis of contemporary Peru—see 1971). Let us now examine four formulations of dependency theory which relate to the above propositions. These formulations relate to directions in the literature which we might label (1) the development of underdevelopment; (2) the new dependency; (3) dependency and development; and (4) dependency and imperialism.

The Development of Underdevelopment

The bulk of writing during the past decade has focused on the development of underdevelopment. The thesis was most explicitly set forth in the early writing of André Gunder Frank (1966 and fully elaborated in 1967). He emphasized commercial monopoly rather than feudalism and precapitalist forms as the economic means whereby national and regional metropolises exploit and appropriate the economic satellites. Thus, capitalism on a world scale produces a developing metropolis and an underdeveloped periphery. This same process can also be found within nations between a domestic metropolis (a capital city, for example) and the surrounding satellite cities and regions.

Frank's theoretical perspective has been neatly summed up and critiqued by Ernesto Laclau (1971). The summary includes the following theses: First, development does not occur through a succession of stages, and today's developed countries were never underdeveloped, although they were once undeveloped. Second, underdevelopment is part of the historical product of relations between the underdeveloped satellites and the present developed metropolises. Third, the dualist interpretation must be rejected because capitalism has effectively and completely penetrated the undeveloped world. Fourth, metropolitan-satellite relations are found within countries as well as in the imperialist world order. Fifth, Frank hypothesizes that development of satellites is limited by their dependent status; satellites experience their greatest growth only when their links to the metropolis are weakened, say during depression or world war; the most underdeveloped regions are those which were closely linked to the metropolis; originally latifundia were capitalist enterprises responsive to the growing demand in the national and international market. These ideas emerged in his earlier works cited above and were refined in a series of essays, reprinted as an anthology (Frank, 1969b)....

The New Dependency

Much of the thrust of dependency theory emanates from the notion of the new dependency. Types of dependence are identifiable through periods of history, according to Dos Santos (1970:232). Colonial dependency characterized the relations between Europeans and the colonies whereby a monopoly of trade complemented a monopoly of land, mines, and manpower in the colonized countries. Financial-industrial dependency consolidated itself at the end of the nineteenth century with, on the one hand, domination of capital in hegemonic centers and, on the other, investment of capital in the peripheral colonies for raw materials and agricultural products which in turn would be consumed by the centers. A new dependency based on investments by multinational corporations emerged after the Second World War. Dos Santos labels this a technological-industrial dependency. An elaboration of theory on the new dependency is found in several of his writings (especially Dos Santos, 1968 ...). The thrust of his argument is directed against prevailing bourgeois assumptions about development in Latin America ...; and it attempts to relate traditional notions of imperialism to the internal situation of the Latin American countries. Let us explore this latter concern.

The new dependency places limits on the development of Latin American economics. Industrial development is dependent on exports which generate foreign currency to buy imported capital goods. Exports in turn are usually tied to traditional sectors of the economy which are controlled by oligarchies. Often the oligarchies are tied to foreign capital; and they remit their high profits abroad. Thus, it is not surprising that foreign capital controls the marketing of exported products, even though dependent countries have attempted to impose policies of exchange restrictions and taxes on foreign exports and have leaned toward the nationalization of production. Industrial development then is conditioned by fluctuations in the balance of payments which in dependent countries often leads to deficits caused by trade relations in a highly monopolized international market, the repatriation of foreign profits, and the need to rely on foreign capital and aid.

These conditions and relations of the new dependency have been related to the colonial heritage of Latin America by Stanley and Barbara Stein who affirm that "in backward, underdeveloped, or dependent areas of the globe, the heritage of the past has shaped and is shaping current widespread poverty" (1970:189)....

Dependency and Development

The notion that capitalist development takes place within dependent situations has evolved primarily in the writings of Fernando Henrique Cardoso. Let us trace his line of argument (1972). Cardoso begins with the assumption that modern capitalism and imperialism differ from Lenin's earlier conceptions. Capital accumulation, for example, is more the consequence of corporate rather than financial control. Investment by multinational corporations in Latin America is moving away from raw materials and agriculture to industry. More often than not these corporations comprise "local and state capital, private national capital, and monopoly international investment (but in the last analysis under foreign control)" (Cardoso, 1973b:11). Thus monopoly capitalism and development are not contradictory terms; and dependent capitalist development has become a new form of monopolistic expansion in the Third World. This development is oriented to a restricted, limited, and upper class-oriented type of market and society. At the same time, the amount of net foreign capital in dependent economies is decreasing. New foreign capital is not needed in some areas where there are local savings and reinvestment of profits in local markets; further, dependent economies during times of monopolistic imperialistic expansion are exporting capital to the dominant economies.

This analysis leads Cardoso to a critique of other dependentistas. First, analysis "based on the naive assumption that imperialism unifies the interests and reactions of dominated nations is a clear oversimplification of what is really occurring" (Cardoso, 1972:94). Second, the notion of development of underdevelopment and the assumption of a lack of dynamism in dependent economies because of imperialism are misleading (Cardoso, 1972:94). On the one hand, new trends in international capitalism have resulted in increased interdependence in production activities at the international level and in a modification in the patterns of dependence that limit developmental policy in the peripheral countries of the international capitalist system (Cardoso, 1973a:146). On the other hand, international capitalism has gained disproportional influence in industry. Whether or not industrial firms are owned by foreigners or nationals, in either case "they are linked to market investment, and decision-making structures located outside the dependent country" (Cardoso, 1973a:146). . . .

Dependency and Imperialism

As mentioned above, Lenin related imperialism to dependency. A recent synthesis elaborates on this relationship. According to Benjamin Cohen (1973:15), imperialism refers to "any relationship of effective domination or control, political or economic, direct or indirect, of one nation over another. . . ." This relationship involves dominance and dependence among nations which are large and small, rich and poor. Three principal forms of imperialism are evident through history. First, during the sixteenth and seventeenth centuries European mercantilism characterized the "old imperialism." Second, the European empire building of 1870 and thereafter represented a shift from informal to formal mechanisms of control and influence in the colonies during a period known as the "new imperialism" (see Fieldhouse, 1961, for a useful review of ideas on the literature on the old and new imperialisms). Third, the breakup of empires was accomplished by analysis of neo-colonialism and what today might be called "modern imperialism." Analytically, the theory of modern imperialism moves in two directions. One emphasizes the view from the metropolis and argues that imperialism is necessary for the advancement of capitalist economies. The other stresses the view from the periphery and focuses on the detrimental consequences of capitalist trade and investment in the poorer economies of the world.

Both these theoretical directions incorporate analysis of dependency, although the view from the periphery has provoked a variety of perspectives, some being non-Marxist and others Marxist. Some of these perspectives have attempted to relate imperialism to dependency, while others have refuted dependency altogether in favor of an interpretation based solely on imperialism. . . .

Among the many recent efforts to relate dependency and imperialism to the Latin American experience are writings by Hinkelammert (1970), Quijano (1972), and essays in the reader by Rhodes (1970). While Hinkelammert examines classical imperialism and underdevelopment, Quijano attempts to relate imperialist domination to the class struggle. Esteban (1961) analyzes imperialism and capitalist development within the context of dependency in his case study of Argentina. Bodenheimer (1970) distinguishes between non-Marxist and Marxist interpretations of imperialism in moving toward her position that a Marxist theory of imperialism can complement a theory of dependency. . . .

CASE STUDIES AND ATTEMPTS TO APPLY DEPENDENCY THEORY

There has been little effort empirically to verify the assumptions of dependency theory. Tyler and Wogart (1973) undertake a modest test along lines of an international comparison. They inquire into Sunkel's "general notion of the relationship between increasing international integration of less-developed countries and their national disintegration" and conclude that "there is not sufficient evidence to reject the dependency hypothesis" (p. 42). While rejecting "the normatively charged theorizing which credits *dependencia* with causing virtually all of Latin America's ills" Schmitter (1972:99), feels that "the basis for a more subtle, differentiated, and empirically testable theory have been laid" (p. 100) and in an exploratory manner he has attempted to measure external dependence and its impact on political outcomes (Schmitter, 1971).

Case studies supported with new information and descriptive analysis are more prevalent, however. In this writer's opinion, Quijano's (1971) class analysis of Peruvian society is the most sophisticated, yet not definitive, effort to link class structure and actions to the outside world. . . .

The relevance of dependency theory is apparent in studies of other areas outside Latin America. Samir Amin (1972) writes about underdevelopment and dependency in Africa. He studies "how the dialectic reveals itself between the major colonial policies and the structures inherited from the past" (1972:505). Completed forms of dependence, he argues, appeared when Africa was made the periphery of the world capitalist system in an imperialist stage. William Minter . . . examines the impact of dependency in Angola. Certainly the most exhaustive account is provided by Walter Rodney (1972) who offers conceptualizations of development and underdevelopment, then relates them to his thesis that European colonialism led to the underdevelopment of Africa. Finally, Esseks (1971) relates economic dependence to political development in Africa while Barbara Stallings (1972) ties Africa to Latin America, basing her analysis essentially on the contributions of Latin American writers. These writers and their ideas provoked Samir Amin (1974) to critique the theory of underdevelopment. He notes that economic growth is an uneven process, analyzes the inequality of international specialization, and assesses the consequences of the international flow of capital for the center and the periphery. The concern with underdevelopment also extends to Asia, Gunnar Myrdal's work serving as an example. Rudebeck . . . critiques that work on the grounds that it provides a framework for analysis of internal aspects relating to underdevelopment but that it ignores external aspects, especially as related to dependency and international capitalism. . . .

CHAPTER 12

Underdevelopment and Dependence in Black Africa—Origins and Contemporary Forms

SAMIR AMIN

Contemporary Black Africa can be divided into wide regions which are clearly different from one another.... The unity of Black Africa is, nonetheless, not without foundations.... [T]he striking similarities of social organisation make a living unity of Black Africa. This physical reality, extensive and rich, did not wait for colonial conquest to borrow from, or give of itself to, the other wide regions of the Old World—the Mediterranean in particular, but also Europe and Asia. The image of an ancient, isolated and introverted Africa no longer belongs to this age: isolation—naturally associated with a so-called "primitive" character—only corresponded to an ideological necessity born out of colonial racism. But these exchanges did not break the unity of Africa; on the contrary, they helped to assert and enrich the African personality. The colonial conquest of almost the whole of this continent strengthened this feeling of unity in Black Africa. Seen from London, Paris, or Lisbon, Black Africa appeared to European observers as a homogenous entity, just as the North Americans regard Latin America as a continent which extends south of the Rio Grande.

Looked at from the opposite point of view, that is to say from inside, Black Africa, like Latin America, evidently appears as extremely variegated. It is true that the present states are the result of an artificial carve-up, but almost nowhere does this constitute the sole or even the essential basis of their diversity....

Between these two extremes—African unity and micro-regional variety—the continent can be divided into a few wide macro-regions. I propose to identify three, and shall discuss the basis for such a distinction.

Traditional West Africa (Ghana, Nigeria, Sierra Leone, Gambia, Liberia, Guinea-Bissau, Togo, former French West Africa), Cameroun, Chad, and the Sudan together constitute a first macro-region, which I wish to

Reprinted from the *Journal of Modern African Studies* 10, no. 4 (1972): 503–24, by permission of Cambridge University Press. This paper has been edited for this publication.

describe as *Africa of the colonial trade economy.* I shall give a precise definition to this term, which, unfortunately, is too often treated lightly. This integrated whole is clearly divisible into three sub-regions; (i) the coastal zone, which is easily accessible from the outside world, and which constitutes the "rich" area; (ii) the hinterland, which mostly serves as a pool of labour for the coast, and as a market for the industries which are being established there; and (iii) the Sudan, whose particular characteristics will be examined later.

The traditional Congo River basin (Congo-Kinshasa, Congo-Brazzaville, Gabon, and the Central African Republic) form a second macro-region, which I wish to define as *Africa of the concession-owning companies.* Here also it is necessary to explain how, over and above the difference in the policies and practices of the French and Belgian Governments, genuine similarities in the mode of colonial exploitation characterise the whole of the region, and this justifies its demarcation.

The eastern and southern parts of the continent (Kenya, Uganda, Tanzania, Rwanda, Burundi, Malawi, Angola, Mozambique, Zimbabwe, Botswana, Lesotho, Swaziland, and South Africa) constitute the third macro-region, which I wish to call *Africa of the labour reserves.* Here also, apart from the varied nature of each country, the region was developed on the basis of the policy of colonial imperialism, according to the principle of "enclosure acts" which were applied to entire peoples.

Ethiopia, Somalia, Madagascar, Réunion, and Mauritius, like the Cape Verde islands on the opposite side of the continent, do not form part of these three macro-regions, although here and there are to be found some aspects of each. However, they also display features of other systems which have played an important part in their actual development: the slavery-mercantilist system of the Cape Verde islands, Réunion, and Mauritius; and the "pseudo-feudal" system of Ethiopia and Madagascar. Obviously questions of frontiers between the regions remain: Katanga, for example, belonged to the area of the labour reserves, and Eritrea to that of the colonial trade.

TOWARDS A DEFINITION OF PERIODS IN AFRICAN HISTORY

My proposed distinction is deliberately based on the effects of the *last* period in the history of Africa: that of colonisation. It will be necessary to study how the dialectic reveals itself between the major colonial policies and the structures inherited from the past. To do so, we have to go back in time, and to distinguish four separate periods.

The pre-mercantilist period stretches from the earliest days until the seventeenth century. In the course of this long history, relations were forged between Black Africa and the rest of the Old World, particularly from both ends of the Sahara, between the savannah countries (from Dakar to the Red Sea) and the Mediterranean. Social formations emerged which cannot be understood if they are not placed, here as elsewhere, within the context of all the multitude of other social systems and their relationships with one another. During that period, Africa, by and large, does not appear as inferior, or weaker than the rest of the Old World. The unequal development within Africa was not any worse than that north of the Sahara, on both sides of the Mediterranean.

The mercantilist period stretches from the seventeenth century to 1800. It was characterised by the slave trade, and the first retrograde steps date back to this time. It was not only the coastal zone which was affected by this trade: there was a decline in productive forces throughout the continent. There were two distinct slave-trading areas: the Atlantic trade (by far the most harmful, due to the great numbers involved), which spread from the coast to the whole of the continent, from Saint-Louis in Senegal to Quelimane in Mozambique; and the Oriental trade operating from Egypt, the Red Sea, and Zanzibar, towards the Sudan and East Africa. This second type of mercantilist activity was carried beyond 1800, because the industrial revolution which shook the foundations of society in Europe and North America did not reach the Turkish-Arab part of the world.

The next period lasted from 1800 to 1880–90, and was characterised by attempts—at least in certain regions within the influence of Atlantic mercantilism—to establish a new form of dependence with that part of the world where capitalism was firmly entrenched by industrialisation. These attempts, however, had very limited backing. . . . The area of influence of Oriental mercantilism was not affected.

The fourth period, that of colonisation, completed the work of the previous period in Western Africa, took over from Oriental mercantilism in Eastern Africa, and developed with tenfold vigour the present forms of dependence of the continent according to the models of the three macro-regions mentioned above. The present throws light on the past. The completed forms of dependence—which only appeared when Africa was actually made the periphery of the world capitalist system in its imperialist stage, and was developed as such—enable us to understand, by comparison, the meaning of previous systems of social relations, and the way in which African social formations were linked with those of other regions of the Old World with which they had contact.

I. THE PRE-MERCANTILIST PERIOD: UP TO THE SEVENTEENTH CENTURY

During this time, Black Africa was not on the whole more backward than the rest of the world. The continent was characterised by complex social formations, sometimes accompanied by the development of the state, and forged between Black Africa and the rest of the Old World, particularly from both ends of the Sahara, between the savannah countries (from Dakar community. . . .

This is why the trans-Sahara trade was so significant. It enabled the whole of the Old World—Mediterranean, Arab, and European—to be supplied with gold from the main source of production in Upper Senegal and Ashanti, until the discovery of America. The importance of this flow can hardly be adequately stressed. For the societies of tropical Africa, this trade became the basis of their organisation. The mining of gold under the orders of the king provided the ruling classes of the countries concerned with the means to obtain across the Sahara, on the one hand, rare luxury goods (clothes, drugs, perfumes, dates, and salt), and on the other, and in particular, the opportunity to establish and strengthen their social and political power by the acquisition of horses, copper, iron bars, and weapons. This trade thus encouraged social differentiation, and the creation of states and empires, just as it promoted the improvement of instruments, and the adaptation of techniques and products to suit local climatic conditions. In return, Africa supplied mainly gold, a few other rare products, notably gum and ivory, and some slaves [Bovill, 1958]. . . .

For centuries the Mediterranean societies and those of tropical Africa were united by a bond, for better or for worse. The vicissitudes of one area had quick repercussions on the other, just as wealth and glory reached them all simultaneously. Thus the gradual shifting of routes from west to east found a parallel shift in the civilisation and power of the nations both in North Africa and in the West African savannah lands—reflected, for example, in the successive might of the ancient Empires of Ghana and Mali, the Hausa cities, Bornu, Kanem, and Darfur. This also explains why there was a crisis in Africa when the centre of the newly born European mercantile capitalism moved from the Mediterranean towards the Atlantic. This shift, studied by Fernand Braudel [1973] with his usual talent and care for detail, heralded the decline, in the sixteenth century, of the Italian towns which, since the thirteenth century, had opened the way for a decisive evolution in the future history of mankind. Similarly we can say that this change was to cause the downfall of both the Arab world and the Sudan-Sahel regions of Black Africa. Soon afterwards the presence of Western Europe along the coasts of Africa was to become a reality. This shift of the centre of gravity of trade in Africa, from the savannah hinterland to the coast, was a direct consequence of the change of commercial emphasis in Europe from the Mediterranean to the Atlantic. But the new trade between Europe and Africa was not to play the same role as that of the preceding period; henceforth it was to take place under mercantile capitalism.

2. THE MERCANTILIST PERIOD: THE SEVENTEENTH AND EIGHTEENTH CENTURIES

As I have pointed out elsewhere, the mercantilist period saw the emergence of two poles of the capitalist mode of production: (i) the creation of a proletariat resulting from the decline of feudal relationships, and (ii) the accumulation of wealth in the form of money [Amin, 1974: ch. 2, sect. 3]. During the industrial revolution the two became united; money wealth turned into capital, and the capitalist mode of production reached its completed stage. During this long period of incubation covering three centuries the *American* periphery of the Western European mercantile centre played a decisive role in the accumulation of money wealth by the Western European bourgeoisie. Black Africa played a no less important role as *the periphery of the periphery.* Reduced to the function of supplying slave labour for the plantations of America, Africa lost its autonomy. It began to be shaped according to foreign requirements, those of mercantilism.

Let us finally recall that the plantations of America did not constitute autonomous societies, in spite of their slave-based form of organisation. As I have argued previously, this mode of production was here an element of a non-slave-based society, i.e., it was not the dominant feature of that society. The latter was mercantilist, and the dominant characteristic of the plantation economy was the trade monopoly which, under its control and for its benefit, sold the products of these plantations on the European market, thus quickening the disintegration of feudal relations. The peripheral American society was thus an element in the world structure whose centre of gravity was in Western Europe. . . .

The mercantilist slave trade had similar devastating effects on all the regions of Africa where it took place. Along the coast, from Saint-Louis to Quelimane, it affected almost the whole of the continent, except the northeastern area of the Sudan, Ethiopia, Somalia, and East Africa. . . . There were wars and anarchy almost everywhere on the continent, and the flight of peoples towards regions of shelter which were difficult to reach and also very often poor—such as those of the paleo-negritic peoples in the over-populated mountains of West Africa. It all ended with an alarming decrease in the population. The processes of integration were stopped, as well as the construction of large communities, begun in the pre-mercantilist period. Instead there was an incredible fragmentation, isolation, and entanglement of peoples, and this, as we know, is the root cause of one of the most serious handicaps of contemporary Africa. . . .

3. INTEGRATION INTO THE FULL CAPITALIST SYSTEM: THE NINETEENTH CENTURY

The slave trade disappeared with the end of mercantilism, that is to say essentially with the advent of the industrial revolution. Capitalism at the centre then took on its complete form; the function of mercantilism—the primitive accumulation of wealth—lost its importance, and the centre of gravity shifted from the merchant sector to the new industry. The old periphery of the plantation of America, and its African periphery of the slave trade, had now to give way to a new periphery whose function was to provide *products* which would tend to reduce the value of both constant and variable capital used at the centre: raw materials and agricultural produce. The advantageous terms under which these products were supplied to the centre are revealed by the theory of unequal exchange. . . .

4. INTEGRATION INTO THE FULL CAPITALIST SYSTEM: COLONISATION

The partitioning of the continent which was completed by the end of the nineteenth century multiplied the means available to the colonialists to attain capital at the centre. We must remember that their target was the same everywhere: to obtain cheap exports. But to achieve this, capital at the centre—which had now reached the monopoly stage—could organise production on the spot, and there exploit both the cheap labour and the natural resources, by wasting or stealing them, i.e., by paying a price which did not enable alternative activities to replace them when they were exhausted.* Moreover, through direct domination and brutal political coercion, incidental expenses could be limited by maintaining the local social classes as "conveyor belts." Hence the late development in Africa of the peripheral model of industrialisation by import substitution. It was not until independence that the local élites who took over from the colonial administration constituted the first element of a domestic market for "luxury goods" [see Amin, 1971]. . . . Hence also the markedly bureaucratic nature of the "privileged classes."

However, although the target was the same everywhere, different variants of the system of colonial exploitation were developed. These did not depend, or only slightly, on the nationality of the coloniser. The contrast between French direct and British indirect rule, so frequent in the literature, is not very noticeable in Africa. It is true that a few differences are attributable to the nationality of the masters. British capital, being richer and more developed, and having additionally acquired the "best pieces" of land, carried out an earlier and more thorough development than French capital [Amin, 1967]. Belgium, which had been forced to come to terms with the Great Powers, and had to accept the competition of foreign goods in the Congo, did not have the direct colonial monopolies which France used and abused to her advantage. Portugal similarly agreed to share her colonies with major Anglo-American capital.

*This problem of the looting of natural resources is beginning to be studied with the present-day awareness of "environmental problems," although the term is ambiguous.

In the region which I have called "Africa of the labour reserves," capital at the centre needed to have a large proletariat immediately available. This was because there was great mineral wealth to be exploited (gold and diamonds in South Africa, and copper in Northern Rhodesia), and an untypical settler agriculture in the tropical Africa of Southern Rhodesia, Kenya, and German Tanganyika. In order to obtain this proletariat quickly, the colonisers dispossessed the African rural communities—sometimes by violence—and drove them deliberately back into small, poor regions, with no means of modernising and intensifying their farming. They forced the "traditional" societies to be the supplier of temporary or permanent migrants on a vast scale, thus providing a cheap proletariat for the European mines and farms, and later for the manufacturing industries of South Africa, Rhodesia, and Kenya. [See, *inter alia,* Arrighi, 1967.]

Henceforth we can no longer speak of a traditional society in this part of the continent, since the labour reserves had the function of supplying a migrant proletariat, a function which had nothing to do with "tradition." The African social systems of this region, distorted and impoverished, lost even the semblance of autonomy: the unhappy Africa of *apartheid* and the Bantustans was born, and was to supply the greatest return to central capital. . . .

Until recently there was no known large-scale mineral wealth in West Africa likely to attract foreign capital, nor was there any settler colonisation. On the other hand, the slave trade was very active on this coast, and caused the development of complex social structures which I have analysed above. The colonial powers were thus able to shape a system which made possible the large-scale production of tropical agricultural products for export under the terms necessary to interest central capital in them, i.e., provided that the returns to local labour were so small that these products cost less than any possible substitutes produced in the centre itself.

The net result of these procedures, and the structures to which they gave rise, constituted what I have called "Africa of the colonial trade economy". . . . These processes were, as always, as much political as economic, and included the following: (i) the organisation of a dominant trade monopoly, that of the colonial import-export houses, and the pyramidal shape of the trade network they dominated in which the Lebanese occupied the intermediate zones while the former African traders were crushed and had to occupy subordinate positions; (ii) the taxation of peasants in money which forced them to produce what the monopolists offered to buy; (iii) political support to the social strata and classes which were allowed to appropriate *de facto* some of the tribal lands, and to organise internal migrations from regions which were deliberately left in their poverty so as to be used as labour reserves in the plantation zones; (iv) political alliance with social groups which, in the theocratic framework of the Muslim brotherhoods, were interested in commercialising the tribute they levied on the peasants; and last but not least, (v) when the foregoing procedures proved ineffective, recourse pure and simple to administrative coercion: forced labour.

Under these circumstances, the traditional society was distorted to the point of being unrecognisable; it lost its autonomy, and its main function was to produce for the world market under conditions which, because they impoverished it, deprived the members of any prospects of radical modernisation. This "traditional" society was not, therefore, in transition to "modernity"; as a dependent society it was complete, peripheral, and hence at a dead end. It consequently retained certain "traditional" appearances which constituted its only means of survival. The Africa of the colonial trade economy included all the subordination, domination relationships between this pseudo-traditional society, integrated into the world system, and the central capitalist economy which shaped and dominated it. Unfortunately the phrase "colonial-type trade" has been used so frequently that its meaning has been reduced to a mere description: the exchange of agricultural products against imported manufactured goods. Yet the concept is much richer: it describes analytically the exchange of agricultural commodities provided by a peripheral society shaped in this way, against the products of a central capitalist industry, imported or produced on the spot by European enterprises.

The results of this colonial-type trade have varied according to different regions of this part of Africa. To give honour where honour is due, it was British capital which initiated a perfectly consistent formulation of aims and procedures. At the beginning of colonisation, when Lever Brothers asked the Governor of the Gold Coast to grant concessions which would enable them to develop modern plantations, he refused because "it was unnecessary." It would be enough, the Governor explained, to help the traditional chiefs to appropriate the best lands so that these export products could be obtained without extra investment costs. Lever then approached the Belgians and obtained concessions in the Congo. . . .

I have analysed elsewhere [1974] the conditions for the success of this colonial-type trade, but these may be summarised as follows: (i) an "optimum" degree of hierarchy in a "traditional" society, which is exactly the case in those zones formed by the slave trade; (ii) an "optimum" population density in the rural areas of 10–30 inhabitants per square kilometre; (iii) the possibility of starting the process of proletarianisation by calling upon immigrants foreign to the ethnic communities of the plantation zone; (iv) the choice of "rich" crops, providing a sufficient surplus per hectare and per worker, at the very first stage of their development; and (v) the support of the political authority, making available to the privileged minority such resources—political and economic, especially agricultural credit—as would make possible the appropriation and development of the plantations.

The complete model of this colonial-type trade was achieved in the Gold Coast and German Togoland by the end of the nineteenth century, and was reproduced much later in French West and Equatorial Africa. This lateness reflected that of French capitalism, and was attributable to the attempts at quasi-settler colonisation even under unfavourable conditions—for example, French planters in the Ivory Coast and in Equatorial Africa—and the corresponding maintenance of forced labour until the modern period, after World War II.

This colonial economy took two main forms. Dominant in the Gulf of Guinea, where conditions enabled this kind of trade to develop, was the *kulak* class of indigenous planters of rural origin, who employed paid labour, and secured virtually exclusive appropriation of the land. On the other hand, in the savannah zone from Senegal through Northern Nigeria to the Sudan, the Muslim brotherhoods permitted another type of colonial trade: the production and export of groundnuts and cotton in vast areas subject to a theocratic power—that of the Mourid brotherhoods of Senegal, the Emirates of Nigeria, and the Ansar and Ashiqqa in the Sudan. They kept the form of a tribute-paying social system, but this was integrated into the international system, because the surplus appropriated in the form of tribute levied on the village communities was itself marketed. It was the Egyptian colonisation in the Sudan which created the most advanced conditions for the development of this type of organisation, which in that country tended towards a pure and simple latifundia system of large estates. The British merely gathered the fruits of this evolution. The new latifundia owners accepted the colonial administration after 1898, and grew cotton for the benefit of British industry. Powerful modern techniques were made available to them, notably large-scale irrigation in the Gezira.

There was a "second transformation of Islam" in West Africa, after the colonial conquest opened the way to the same kind of evolution, although less definite and slower. We have already seen that Islam in this region underwent a first transformation: from being the religion of a minority caste of merchants in the pre-mercantilist period, integrated into an animist society (hence similar to Judaism in Europe), it became the ideology of popular resistance to the slave trade in the mercantilist period. This second transformation made Islam—"restored" by the aristocracy and the colonial authorities—the guiding ideology of peasant leaders for the organisation of the export production which the colonisers desired. The Mourid phenomenon of Senegal is probably the most striking example of this second transformation. The fact that the founders of the brotherhood, and some short-sighted colonial administrators, felt hostile to each other for some time, does not matter. Ultimately the brotherhood proved to be the most important vector for the expansion of the groundnut economy, and for the submission of the peasants to the goal of this economy: to produce a large amount, and to accept very low and stagnating wages despite progress in productivity.

To organise this colonial-type trade it was necessary to destroy the pre-colonial pattern, and to reorganise the flows in the direction required by the externally orientated nature of the economy. For there had been, before, regional complementarities with a broad, natural forest-savannah base, strengthened by the history of the relations between the West African societies. The domestic trade between herdsmen and crop farmers, and in kola and salt, as well as the outflow of exports and the dissemination of imports, constituted a dense and integrated network, dominated by African traders. The colonial trading houses had to gain control of these flows and to direct them all towards the coast; that was why the colonial system destroyed African domestic trade and then reduced African traders—when they were not eliminated—to the role of subordinate primary collectors. The destruction of the trade of Samory, like that of the people of mixed blood in Saint-Louis, Gorée, and Freetown, like that of the Hausa and Ashanti of Salaga, and of the Ibo of the Niger delta, bear witness to this other crippling socio-economic effect of *l'économie de traite.*

Thus the colonial trade necessarily gave rise to a polarisation of dependent peripheral development at the regional level. The necessary corollary of the "wealth" of the coast was the impoverishment of the hinterland. Predisposed by geography and history to a continental development, organised around the major inland river arteries (thus providing for transport, irrigation, electric power, and so on), Africa was condemned to be only "developed" narrowly along the coast. The exclusive allocation of resources to the latter zone, a planned policy of colonial trade, accentuated this regional imbalance. The mass emigration from the hinterland to the coast forms part of the logic of the system: it made cheap labour available to capital where capital required this, and only "the ideology of universal harmony" can see in these migrations anything other than their impoverishment of the departure zones.* The culmination of the colonial trade system was a balkanisation, in which the "recipient" micro-regions had no "interest" in "sharing" the crumbs of the colonial cake with their labour reserves.

Thus the bounties of the colonial trade were highly relative. However, it was impossible to implement this system in Central Africa, the third

*Elliot J. Berg, "The Economics of the Migrant Labor System," [in Kuper, 1965] . . . reflects better than anyone else this non-scientific ideology. The conventional assumption is that migrations "redistribute" one factor of production (labour) which originally was unequally distributed. If that were so, migrations would tend to equalise the rates of growth of the economies of the various regions, but we can see that they are everywhere accompanied by a growing disparity between rates of growth: the acceleration of growth *per capita* in the immigration zones, and its reduction in the emigration zones.

micro-region of the continent. Here, ecological conditions had to some extent protected the peoples who took refuge from the ravages of the slave trade in zones unlikely to be penetrated from the coast. The low density of population, and the lack of sufficient African hierarchies made the colonial-trade model non-viable. Discouraged, the colonial authorities gave the country to any adventurers who would agree to try "to get something out of it" without resources—since adventure does not attract capital. The misdeeds of the concessionary companies have been duly denounced: between 1890 and 1930 they ravaged French Equatorial Africa with no result except a trivial profit. As for the Congo, it will be remembered that Lever Brothers were welcomed by the Belgians, after the firm's unsuccessful attempt to establish itself in the Gold Coast. But it was only after World War I, when the solution was adopted of having industrial plantations established directly by the major capitalists, that a small-scale colonial-type trade infiltrated as an extension of the plantation zones belonging to foreign capital. As for French Equatorial Africa, this area had to wait until the 1950s before seeing the first symptoms. Thus the negative impact of this period, still omnipresent, justifies the name which I have given to the region—"Africa of the concessionary companies."

In all three cases, then, the colonial system organised the African societies so that they produced exports—on the best possible terms, from the point of view of the mother country—which only provided a very low and stagnating return to local labour. This goal having been achieved, we must conclude that there are no traditional societies in modern Africa, only dependent peripheral societies. . . .

SECTION IV

PROBLEM MANIFESTATIONS IN URBANIZATION

REGIONAL IMBALANCES

INTRODUCTION

Urbanization is, by definition, a differentiation and specialization in the utilization of *space,* whereby certain locations receive a disproportionate share of residents and the facilities by which they live. In this sense, urbanization *is* inequality. But inequality in itself is not imbalance. What is implicit in the idea of "imbalance" as it is used by scholars to denote a pathological situation in urbanization is the concept that certain levels of inequality are normal and necessary and result in mutual benefit to the parties (as in a healthy symbiosis between a trading center and its agricultural hinterland) whereas other levels are excessive, resulting in an unequal exchange between the parties and an "unfair" exploitation of one by the other. From this description it should be clear that scholars are making all sorts of value judgments when they talk about regional imbalances and the issue is hardly one of "scientific" character.

Nevertheless, common sense does support the notion that there is some range of optimal spatial distribution which lies somewhere between a concentration so great that it becomes totally unmanageable and unliveable and a dispersal so scattered that the units cannot be integrated through communication into a coherent national order. It is this nebulous level which is often the hidden norm of "balanced" geographic pattern against which specific urban systems are judged "imbalanced."

In the literature about Third World cities one often comes across statements that imply (or state outright) that the major cities are too large and growing too fast in relation to the rest of society or in relation to the smaller cities and towns which exist within the same country. When the former comparison is made, the country is said to be "overurbanized"; when the latter comparison is made, the country is said to be suffering from primacy or macrocephaly. In both instances, a judgment is being made that the growth is abnormal. Occasionally, this abnormality is defined as being deviant from some mathematical relationship (the rank-size rule) which appears to fit the urban hierarchies in a number of large developed countries. There is, as we have suggested above, however, no way to prove this scientifically and in the abstract.

Nevertheless, it is possible in individual cases to evaluate the consequences of particular patterns of spatial concentration and inequality, to search for the underlying causes—both natural and politico-historical—which have given rise to these patterns, and to explore possible future changes in the system of cities which might occur with economic development, with or without spatial planning. The readings included in this section address themselves to just these issues and therefore move substantially beyond the more traditional approaches to the question of regional imbalance which have so often merely been hidden value judgments. Cities *are* spatial nodes in a geographically-based network as well as social, economic and political systems. The spatial aspects, however, intersect with the social, economic and political and cannot, therefore, be understood as "natural" events of the environment; they are planned and created, either centrally or not, either consciously or not.

The first article in this section, by Soja and Tobin, describes how geographers are now beginning to look at urban systems in the developing world. While on occasion the authors seem to imply that the deviation-amplifying system (i.e., the tendency of the areas which benefit from early concentration to continue to grow the fastest during early stages of development) is a natural or automatic process, this would be a misreading of their position. They merely suggest that, in the absence of policies to the contrary, there are compelling reasons why concentration should result.

This initial tendency, however, is often fostered by elites who stand, of course, to gain from a heightened inequality which concentrates resources in the city where they live and which are, therefore, available to the class to which they belong. That is why it is often difficult to separate "natural" processes from those which are designed. In Slater's history of the development of the urban system in Tanzania under colonialism, it is clear how the urban system was transformed, by what mechanisms, and for whose benefit. The regional system which resulted in that country and which has been inherited by Tanzania after independence is quite typical for many other colonized countries, whether in Africa or elsewhere. The focus of the railroads and other parts of the transportation system from resource extraction points in the interior toward the colonial-enclave port city designed to transship them, the lack of transversal lines linking parts of the interior with one another, the heavy concentration of the accoutrements of "modernity" (factories, schools, government offices, etc.) in a few major cities linked as much to Europe as to their hinterlands—all these are recurring spatial imbalances in many Third World countries, even when they have not been colonized directly. Nor does the pattern simply vanish with independence, as can be seen from Bryan Roberts' discussion of post-revolution Peru, where the central government continues to exert its control over portions of the hinterlands. Roberts argues that not only is this not possible, for it implies greater control than the central government can marshall, but it is counter-productive since there is considerable vitality and talent in the smaller cities and towns which could be utilized more constructively if the political-economic system were more decentralized.

The issues raised in this section will reappear in Section V when we investigate policies being followed by various countries in their attempt to reach balanced regional development. We end here simply on the note that while science cannot resolve the question of what balanced regional development is, politics certainly attempts to. In the last analysis, an urban system is "unbalanced" when the distribution of income and goods that results from the spatial inequality is judged unsatisfactory to those who are able to express their will in the political system; the question of balanced spatial distribution is therefore intimately connected to the question of distributive justice as this is defined by a society.

CHAPTER 13

The Geography of Modernization: Paths, Patterns, and Processes of Spatial Change in Developing Countries

EDWARD W. SOJA
RICHARD J. TOBIN

INTRODUCTION

Over the past two decades, the academic discipline of geography has experienced a conceptual and methodological transformation which rivals that of any other social science. From an essentially descriptive discipline content primarily with the collection and categorization of data about countries and regions, modern geography has become more quantitative and theoretical in orientation and technically more dependent upon mathematical and statistical modes of analysis. For those who find comfort in historical precedent, it is possible to trace the roots of this transformation deeper into the history of geographic thought. But it has only been since 1950 that the isolated strands of change have become interwoven and sufficiently consolidated to redirect the context and objectives of geographical research.

If one central theme can be said to have emerged from the revolutionary changes of the last twenty years, it is *the spatial organization of human society*. Much of contemporary geography rests upon the assumption that there exists an inherent geographic order in human society, a spatial "anatomy" of human behavior and societal organization which has regular and discoverable characteristics. The bulk of modern geographical research has been associated with the search for order and regularity in spatial systems—in the distribution and size of settlements, the patterns of industrial location and agricultural land use, the growth and form of cities, the geometry of spatial patterns, and the structure of communications and human interaction in space. . . .

Reprinted from G. D. Brewer and Ronald Brunner, eds., *Political Development and Change: A Policy Approach* (New York: Macmillan, 1975), pp. 197–208, by permission of the publisher and the authors. This material has been edited for this publication.

SOME DYNAMICS OF SPATIAL DEVELOPMENT

Spatial development involves a set of processes working within a system of regions to transform structurally the spatial organization of society. This transformation can be viewed as a consequence of innovation diffusion in which the values, attitudes, and material attributes of "modernity" are spatially disseminated through a population; differentially adopted; and eventually incorporated within a territorially defined social system as a primary basis for social, economic, political, and cultural organization.

Modernity, as it is used here, is a relative term referring to a cluster of associated innovations which, at a given point in time, represents a predominant paradigm for societal organization. Although strands of continuous change can be identified in the evolution of human society—progressive increases in the "complexity" of human relations and organization, expansion in societal scale, increases in societal differentiation—modernization is perhaps most accurately viewed as a discontinuous, but cumulative, process keyed to the dynamics of human invention.

Every historical period is thus characterized by its own prevailing form of modernity, linked to what the economist Simon Kuznets [1966] has called "epochal innovations." These epochal innovations are capable of transforming social values and generating new world views to express altered norms. Examples in early human history would include the still poorly known cultural inventions which promoted sharing in human groups (e.g., the social control of sexual relations, the extension of kinship); the massive ecological revolution associated with the domestication of plants and animals; and the political and economic changes linked to the growth of cities and the emergence of the centralized territorial state.

The major epochal innovation of the contemporary period, according to Kuznets and many others, is the extended application of science to problems of economic production. More broadly viewed, this involves a variegated cluster of convergent social, political, technological, and economic innovations which have made the production and consumption of energy more efficient and produced social systems more highly adapted to continuous innovative change than any others in human history. *Twentieth-century modernization,* therefore, represents the conjoining of contemporary innovation diffusion with a series of structural transformations—the former associated with the application of modern science, technology, and organizational theory; the latter characterized by the emergence of attitudes, values, and behavior which support sustained or continuous change and innovation.

Like all societal processes, modernization has a "geography," a salient spatial component. Put most simply, spatial development refers to the *geography of modernization,* the evolutionary changes in spatial organization, behavior, and perception arising from the impress and diffusion of modernity. Moreover, it is assumed that whatever recognizable order or regularity may exist in the clustered processes of societal modernization will be associated with changing patterns and processes of human spatial organization. This view and the tasks inherent in an attempt to construct a spatial theory of development are succinctly expressed by John Friedmann [1972].

> Society is spatially organized in the sense that human activities and social interactions are *space-forming* as well as *space-contingent.* It follows that as a society undergoes development, its spatial structure will be transformed, but the development process will also be influenced by the existing patterns of spatial relations and the dynamic tensions that will result from them. . . . In order to state a spatial theory of development, therefore, it is necessary to establish a linkage between the separate but correlative theories of social change and spatial organization. [Italics added.]

Most existing theories of regional or spatial development have been derived either from development economics or from related structural-functionalist concepts in the other social sciences. As a result, spatial systems, like economic or social systems, are usually viewed as being in a state of dynamic equilibrium. Although never fully integrated, the systems are presumed to contain built-in mechanisms which tend to assure a long-run adaptability to externally generated change and to reduce the impact of dysfunctional stress and internal deviations from equilibrium conditions. A fully integrated equilibrium state is never reached, but the developing spatial system is viewed as moving toward such a goal in a process of gradual and self-adjusting change.

In the discussion which follows, a significantly modified perspective on the development process is introduced, one which emphasizes not the stability and structural maintenance of evolving spatial systems, but the growth and change in spatial structure, the accentuation of divergence from equilibrium, the conflict situations generated by such fundamentally inegalitarian deviations, and the irregular and nonlinear path of the development process. Conventional equilibrium theories, it is suggested, are applicable primarily at the most advanced stages of development and cannot automatically be extended backward in time to characterize the early, formative stages of modernization. Whatever empirical regularities may exist in the development process are more likely to be reflected in the sequential patterning of deviations from a presumed equilibrium state than from an inexorable, linear, and unidirectional increase in equilibrium conditions.

THE GEOGRAPHICAL CONCENTRATION OF DEVELOPMENT

The analysis of spatial development, as previously noted, is firmly rooted in the assumption that space both affects and reflects the basic processes which operate within society. These processes—migration, urbanization, industrialization, social mobilization, acculturation—assume a spatial dimension in two important ways. First, they are *space-forming* in that they work to shape and structure human interaction in space—in the development of transport and communications networks, in the growth of urban and administrative systems, in the territorial distribution of political authority, in the evolution of a differentiated and integrated space economy. At the same time, they are *space-contingent.* Their space-organizing influence is itself shaped by the existing spatial framework. The decision to locate a modern factory, for example, or an administrative center, both powerful and persistent forces in the organization of space, will depend in large part upon earlier locational decisions.

The interdependence of the space-forming and space-contingent properties of modernization has a profound effect on virtually all patterns of spatial development. The degree to which locational selection, for example, is channeled by past decisions, particularly when derived from similar political and economic objectives (as in a continuing colonial situation), contributes to what is perhaps the most outstanding empirical regularity of spatial development: *the tendency toward geographical concentration of the attributes of modernization.* Once initial decisions are made to locate a particular activity or institution at a specific point, a kind of self-generating momentum is established which continues to attract related enterprises and indeed "multiplies" the impact of a given social, economic, or political investment. The remarkable tenacity of initial locational decisions is evident in all stages of development, from the bulging primate cities in Africa, Asia, and Latin America to the megalopolitan giants in the northeastern United States [Ullman, 1958].

A primary component of any spatial theory of development, therefore, must be an explicit formulation of the dynamics of geographical concentration. In this context, the general and partial equilibrium models of traditional location theory, classical economics, and most of the derived concepts and models of modern spatial organization theory are of only limited use. Progressive development concentration is not simply the result of agglomeration economies, the indivisibility of capital inputs, or a more broadly viewed minimization of distance "costs." Nor does it appear that mechanisms automatically arise through the normal sequence of development to counteract effectively increasing development concentration and the potential diseconomies which may emerge from it. Rather, the tendency toward an increasing geographical concentration of development is part of a general pattern of *deviation-amplification,* which hinges largely upon locational advantages derived during the early stages of growth and appears to be central to the development process throughout the world.

These observations require further elaboration, for they are of critical importance to the spatial analysis of development. Much of contemporary geographical theory rests upon a "locational ideology" geared toward cost minimization or profit maximization. Locational decisionmaking is attuned to the factor of distance (e.g., in the form of transport costs) and derived from an assumption of rational and omniscient economic behavior. From this rubric have emerged interlocked sets of deterministic equilibrium models postulating optimal location patterns for industrial firms, agricultural land uses, and urban settlements. Generated within a developed Western context and essentially morphostatic in that the system being dealt with is presumed to be in an integrated equilibrium state, these models have been combined with development economics into an interpretative framework for analyzing growth and change.

In many ways, most of regional development theory is an extension of the equilibrium models of economic theory and an adjunct to the fundamental tenets of structural-functionalism so pervasive in the social sciences. All concentrate on the mechanics of *deviation-counteracting systems.* Economic processes, for example, are viewed as operating to reduce deviations or inhomogeneities within the system—to pull the system toward a rational and optimal equilibrium state. Regional inequalities in income or nonoptimal ("nonrational") location patterns are presumed to iron themselves out through the normal operations and development of the economic system (e.g., through the free flow of labor and capital).

But the process of development, at least in its early and formative stages, is characterized more by deviation-amplification than by deviation-reduction, by morphogenesis (the evolution of structure) rather than by morphostasis. Any evolutionary process, such as biological growth or economic development, increases structuredness and inhomogeneity through the progressive *differentiation* of the system or organism. The deviation-amplifying character of evolutionary processes severely strains the deviation-counteracting interpretations of classical economic and spatial organization theory. At the later, integrative stages of development, both the economic and political systems may work to reduce internal inequalities, but it cannot be inferred that the same processes hold true throughout all phases of development. In other words, end-state characteristics (i.e., those of fully developed systems) do not necessarily prevail during states of transition and change. Spatial development theory, therefore, must be centrally concerned with the causes and consequences of deviation-amplification within evolving spatial systems and not wholly focused upon the inexorable march toward a deviation-counteracting equilibrium.

Consider the implications for development planning. Under the traditional rubric, which can easily be associated with an economic philosophy of laissez-faire and competitive free enterprise, locational decisionmaking takes place within a deviation-counteracting system. The individual decisionmakers—governments, entrepreneurs—choose locations to minimize costs or maximize profits. Errant choices are presumably ironed out as the economic system tends ultimately toward a state of equilibrium. Initial locational choices and the advantages they may engender are assumed to be as susceptible to adjustment and equalization as any other locational choice. Through a combination of adaption (rational choice based upon locational knowledge) and adoption (a selection of the "lucky" by the economic system), locational patterns tend toward a stable and optimum ideal. It can therefore be presumed that, with a set of idealized models in mind, continued locational decisionmaking based primarily upon attempts at cost minimization and profit maximization will eventually induce a state of economic modernity. Over-concentration, should it occur, will be counterbalanced by emergent diseconomies.

But why do the rich continue to get richer while the poor get relatively poorer? Why does the population in centers of concentrated and often rapid economic growth (e.g., many of the primate cities of developing areas) increase so rapidly as to hinder the spread of development throughout the economy and population, thus delaying the structural transformations necessary for sustained growth? Why is it that apparently successful development programs often result in greater rather than reduced regional inequalities? It is no wonder that contemporary regional development theory has begun to place greater emphasis on deviation-amplifying processes. While the classical models, with their built-in assumptions of deviation-reduction, continue to demonstrate very effectively how a developed system works, a new body of theory is beginning to emerge to answer a fundamentally different question: What are the processes which characterize the evolution of developed systems?

In this latter view, which is probably best known through the works of Gunnar Myrdal, forces that create inequality and polarization are as "natural" and as much to be expected as those that lead toward economic equilibrium. Moreover, Myrdal stresses the tendency, particularly in the earlier stages of development, for the processes of change to be characterized by cycles of cumulative causation which lead to *increasing* (primarily economic) inequalities. Rather than eliciting a countervailing response to reduce deviation, development generally calls forth "supporting changes, which move the system in the same direction but much further" [Myrdal, 1957:26]. Favored regions, once established (whether they are "rational" for the system as a whole or not), tend to generate *backwash effects,* which further amplify existing inequalities. Backwash, or polarization, involves the migration of resources (people, skills, capital) from peripheral areas to the major centers of development, where greater opportunities are perceived to exist. At whatever scale, the rich get richer and economic, social, and political gaps widen—unless the cycle is successfully reoriented toward increasing *spread effects,* the outgrowth of forces which extend the benefits of development to peripheral areas.

The societal transformations necessary for spread to outweigh backwash require, in Myrdal's view, massive government intervention and control. This infuses the development process with a pervasive political flavor. It also underlines, for the development planner, the need to innovate and experiment in an attempt to control or to channel productively the deviation-amplification which dominates the development process. "Natural" forces of adjustment and equalization are not sufficient to prevent the solidification of regional inequalities and the potential social and political turmoil they may generate at some later stages of development. Given the increasing temporal urgency of development, it is foolhardy to expect the spontaneous generation of conditions—or an invisible hand—to promote sustained and balanced growth. . . .

One of the finest recent attempts to incorporate the concept of deviation-amplification into regional (spatial) development theory has been made by Friedmann, particularly in his work, "A General Theory of Polarized Development" [1972]. Development is viewed by Friedmann as a series of transformations in the institutional and organizational structures of society which, if left unchanged, would inevitably limit the system's capacity for expansion and continuous innovative change. His focus is upon the dynamics of authority-dependency relations which emerge from "the unusual capacity of certain areas to serve as cradles of innovation" and dominate less favored areas, i.e., from the relations between *core and periphery.*

. . . From the complex interaction between core and periphery emerge the dynamic processes of spatial development.

> Dominant core and dominated periphery together constitute a relatively stable spatial system in which the latter is successfully "colonized" chiefly to sustain the continued growth of the former. The further growth of core regions, however, is in the final analysis constrained by the tensions that tend to build up from the ever more visible discrepancies in the rates of expansion and modernization between core and periphery. The increasing flow of information from core to periphery, together with an aroused awareness of potentially modernizing elites in the periphery to the conditions of their own dependency, produce conflict with core region authorities over the extent of permissible autonomy.
>
> [Friedmann, 1972:30]

Progressive development concentration in core areas thereby becomes a major driving force for change. The periphery becomes organized into increasingly core-dependent market, supply, and administrative areas enmeshed in a feedback system characterized predominantly by backwash effects. But at the same time there is a limited extension of development impulses from the core, mobilizing the peripheral populations and creating a counter-elite, capable of challenging the social bases of integration in the spatial system. . . .

Conflict resolution between core and periphery consequently represents a major turning point in the development process, leading either toward continued deviation-amplification or to the emergence of deviation-counteracting forces within the spatial system. Both alternatives, at least in the short run, can sustain continued growth, although in the case of increased deviation-amplification this is likely to be "growth without development"—an incremental expansion in one or more components of the societal-spatial system (e.g., regions, population groups, or sectors of the economy) without significant change or transformation of its basic structures. Continuation along this course, even when increasing the total real wealth of the whole system, most probably decreases the potential for peaceful structural transformation and increases the likelihood of greater repression and/or major revolution. . . .

What is being proposed here is a general perspective on the development process which is keyed to the changing patterns of authority-dependency relations between core and periphery. The essential components of this perspective and its implications for the study and planning of development can be summarized as follows:

1. The early course of development is characterized by accentuation of the differences between core and periphery. Peripheral areas increasingly become organizational dependencies of the core as the geographically concentrated impulses of development polarize space through the mechanisms of deviation-amplification. During this phase, regional inequalities are intensified as the transport system expands, the market economy spreads, urbanization and industrialization increase, and new ideas and technology diffuse over space.
2. This tendency toward a more pronounced concentration of modern activities and attributes in only a few areas of an organized spatial system represents a fundamental departure from the spatial equilibrium conditions postulated in conventional geographic and regional development theory. Since deviation-amplification is an inherent part of the early development process, it cannot automatically be viewed as abnormal, nor as an indicator of stagnant growth. Although centrifugal forces of development dispersal must be present in sufficient amounts for growth to continue, development can and does proceed apace under conditions in which

backwash and deviation-amplification outweigh spread and equalization effects.

3. Crucial decisions are made during the early phases of development which establish a framework of locational advantage that tends to embed itself tenaciously within the developing spatial system. Despite the potential predominance of deviation-counteracting forces at later stages of development, the basic structures solidified during the period of deviation-amplification tend to maintain themselves throughout the modernization process. If only for this reason, it is of great importance to understand the causes and consequences of deviations from a presumed stable and integrated equilibrium, rather than dismissing them as temporary and patternless inconsistencies.

4. Deviation-amplifying processes, coupled with expanding social mobilization in the periphery, generate conflict situations as existing patterns of authority-dependency relations are challenged. This crucial threshold can be reached at almost any point in the development process since it typically reflects local social, economic, and political conditions (e.g., degree of cultural homogeneity, locational patterns of core and periphery, dependence upon foreign trade, the nature of the political system, fundamental ideologies). The important influence of specific local conditions makes the outcome, as well as the timing and intensity of these developmental crises, difficult to predict.

5. Expanding patterns of political participation can greatly complicate growth patterns during transitional phases of development. Given a competitive political party system and free elections, an expanding mobilized periphery is provided with a powerful avenue of expression and influence. If sufficiently organized, even at low levels of development (when, indeed, its population may greatly outnumber that of the core), it can press for increased recognition and priority either as a separate political party or (more likely to be successful) through an existing core party which needs its support. This can lead to more extensive government programs favoring balanced regional growth. Attempts at reducing the imbalance between core and periphery, however, are expensive and can limit short-run growth. Moreover, the intense competition generated in a highly compartmentalized pluralistic political system can itself contribute to the political turmoil and instability which the government seeks to avoid. To a great extent, the important role of single-party systems in many developing countries (and within socialist planned economies) reflects the desire of the dominant core to control or at least selectively accommodate the mobilized periphery during a crucial period of development.

6. Development planning during the deviation-amplification phase must seek a productive and flexible middle ground between balanced regional growth (which can be overly expensive) and excessive core concentration (which can rigidify core periphery boundaries and encourage social and

political turmoil). A major problem here is the degree to which an inherited spatial system, structured to serve significantly different goals during a period of colonial control, can or should be reorganized to serve the objectives of politically independent states.

7. Only at more advanced stages of development is there likely to be sufficient surplus capital and geographically extensive organizational structures available for the costly readjustment necessary to break the binds of core domination and move the spatial system toward a stable and integrated equilibrium. This implies that approximation to equilibrium conditions can be used as a direct gauge of development only in the most modernized societies. In the context of early development, the pace and level of modernization must be assessed with respect to the particular nature, rigidity, and pattern of change in the expected deviations from equilibrium conditions. The critical test lies in the ability of the primary decisionmakers to direct effectively the course of modernization, while simultaneously contending with and productively accommodating the pressures and tensions inherent in the development process. Herein lies the essence of what some have viewed as political development. . . .

CHAPTER 14

Colonialism and the Spatial Structure of Underdevelopment: Outlines of an Alternative Approach, with Special Reference to Tanzania

D. SLATER

PRE-COLONIAL SPATIAL STRUCTURE—A GENERALIZED OUTLINE FOR TANZANIA

Prior to systematic European colonization, . . . the vast majority of mainland Tanzania consisted of independent producers still involved in pre-capitalist modes of production. The village community was the dominant socio-economic unit with the primary purpose of production being the maintenance and reproduction of the communal system. The pre-colonial socio-economic formation was self-sufficient and not likely to disintegrate from internally induced conflicts.

At the same time, we must not assume . . . that pre-colonial Tanganyika was a spatially undifferentiated territory. For example . . . although the most common pattern of agriculture was the cultivation of millet and sorghum by means of shifting cultivation, there were other more specialized environments that supported different agricultural systems. So, one had pastoralism in some areas and banana cultivation in other, especially well-watered regions like the Kilimanjaro and Usambara areas. Also, there were particular areas of intensive cultivation such as the integrated livestock and crop husbandry of Ukara Island in Lake Victoria which, according to Iliffe, have been capable of supporting up to 500 people to the square mile through the use of irrigation and animal and vegetable manure.

Furthermore, the pre-colonial period witnessed the gradual emergence of a number of important regional trading networks. Although there is evidence of trading links between the Swahili coast and the Zimbabwe plateau zone before the arrival of the Portuguese, the fifteenth century

---

Reprinted from *Progress in Planning,* Vol. 4, Part II (London: Pergamon Press, 1974), pp. 146–59, by permission of the publisher. This paper has been edited for this publication.

appears to have been characterized more by economic decentralization in the sense that there was no single port which dominated the coast so as to stifle the growth of others. Sheriff [1972] suggests that many of the 37 Swahili ports counted by the Portuguese may have relied on their own catchment areas for ivory, and perhaps shared in the trade of Sofala, an important centre on the Mozambique coast, south of the Zambesi River. At this stage we may assume that the spatial structure was rather amorphous, and that the failure to create an integrated space economy could perhaps be attributed to a combination of difficulties in communication and deficiencies in socio-economic organization.

At the beginning of the sixteenth century East African coastal settlements were still very much part of a different world from the societies of the interior, yet during the course of the next three and a half centuries coast and interior became intimately, if not uniformly, integrated. Basically, Alpers explains this emerging spatial integration in terms of the development of long-distance trade in ivory, and later in slaves, from the interior to the coast. This is not to say, however, that local and regional trade did not play a part in this context. Sutton, for instance, suggests that the long-distance caravan trade of East Africa may be most appropriately viewed as a widening of much older and relatively local hinterland and interior trading networks dealing in foodstuffs and livestock, and especially iron and salt, by which the interior peoples eventually came into contact with the coast and thus with international commerce.

During the sixteenth century two regional trading networks developed and later merged. First, the port of Kilwa, deprived by the Portuguese of its connections with Sofala in the south, by which it had been able to tax the trade in gold, began to trade with the people of the immediate hinterland. Alpers argues that the items of such trade would have included ivory, bees-wax (for candles), rhinoceros horn, and animal skins. Secondly, one also had the establishment of trading links between the people of Yaoland in northern Mozambique and the coast. Again, the coastal centre was Kilwa from where cloth, beads, and other exotic garments were carried inland to Yaoland. At the same time the Yaos traded in iron products (hoe blades, knife blades, axe heads, spear blades, anklets, bracelets, necklaces, etc.), salt, and, later in the nineteenth century, ivory. One of the important features of this trade was that it subsequently became easier to import cloth than to make it locally, and in this way . . . the skills of spinning and weaving cloth were lost in several parts of East Africa.

A much more damaging disruption of East African society was caused by the establishment of slave plantations on the Indian Ocean islands of Mauritius and Réunion in the 1730s. Although there are not many accurate statistics for this time, in 1811 the captain of a visiting British ship estimated that at least 6,000 to 10,000 slaves annually passed through the Zanzibar market. . . . But it was the nineteenth century, with the setting up of clove plantations on Zanzibar, that really saw the intensification of the East African slave trade and at the same time a closer integration of coast and interiors.

As the power of the Omani State in the Persian Gulf increased, its East African base on Zanzibar began to play a very active role in exploiting the wealth of the interior—i.e., ivory and slaves. As the demand for resources of the interior grew apace, many of the long-dormant towns of the coast opposite Zanzibar were resuscitated, and in other cases new settlements sprang up in their place. Their role was to act as coastal entrepôts for the developing trade with the interior.... Here we have an example of a situation where the opening up of new resources through a more intensive use of already-established trade routes brings about a shift in the spatial circulation of surplus and both a creation of new urban centres and the revival of old ones. [For implications, see Harvey, 1973.]

The slave plantation economy which centred on Zanzibar, under the rule of Sayyid Said of the Omani Empire, produced cloves for the world market. Its effect on mainland society may be briefly summarized. Firstly, it necessitated expropriation of labour, either as slaves or as porters for the trading caravans. Secondly, by the second half of the nineteenth century Zanzibar had become a major market for the food of the mainland. The chief suppliers were the coastal areas which produced rice and millet, and also the Muheza region in the north grew rice for Zanzibar. In addition, grain produce moved along the central caravan route from Tabora to Bagamoyo, and Nyamwezi brought herds of goats and cattle to the coast to sell as meat. Thirdly, wherever there were considerable slave populations, such as on Zanzibar and Pemba Islands and in the coastal towns and plantations, many of these new "detribalized town dwellers" became Muslims and were progressively integrated into Swahili society. As a result, the coastal towns became more heterogeneous in the composition of their population. Finally, the dictates of the slave mode of production led to a greater spatial integration of coast and interior—an integration that was greatly facilitated by the operation of Asian merchants who organized and helped to finance the caravan trade in slaves and ivory....

PENETRATION, INTEGRATION, AND DISINTEGRATION—THE SPACE-ECONOMY OF COLONIALISM

It has been suggested that German penetration of East Africa occurred in three main waves. Initially it was in the form of individual incursions by explorers, missionaries, traders and geographers which reached their peak in the 1880s. Secondly, through the activities of the "Society for German Colonisation," which was re-organized in 1887 into the German East Africa Company, and an increasing number of sub-companies such as the East African Plantation Company, occupation of East African territory proceeded apace, so that even by 1888 some tobacco was being exported to Germany—this form of penetration can be described in terms of occupation by trading companies. Finally, after the intensification of coastal resistance to German trading incursions, the Imperial German government took over the administration of Tanganyika in 1891, and in 1895 all land was made the property of the Emperor, thus signifying the culmination of nineteenth-century penetration in formal annexation.

With the introduction of plantations, parts of Tanganyika came under the capitalist mode of production. Tobacco and sugar plantations were established near Pangani in the 1880s, and by the early 1890s a number of coffee plantations were set up in and around East Usambara. Then, after 1898, sisal began to be planted along the newly built Tanga-Mombo railway, and after the central railway reached Morogoro in 1907 a new area of sisal plantations was brought into being around Kilosa and Morogoro. Further, rubber plantations were introduced in the Tanga region, the first plantation rubber being exported in 1907, when there were some 5 million trees in the colony. By 1910 there were 250 rubber plantations, 54 sisal estates, and 17 coffee plantations. Overall, European plantations increased from 180 in 1905 to 758 in 1912, although this area represented only a small portion of the total territory. In 1911 rubber and sisal were the most important exports by value, although the importance of rubber diminished rapidly until after the First World War it no longer had any significance in the trade of Tanganyika. Sisal, on the other hand, increased its share of domestic exports and by 1919 represented 32.8 percent of exports by value.

In addition, European settlement took valuable land in the highlands, especially in the Usambara, Pare, and Kilimanjaro regions, and to a lesser extent in the south-east and north-west. Also, land was alienated around Dar es Salaam and the Ulugurus, and ranches were established at Bukoba, Moshi, Langenburg (Mbeya), Dodoma, and Kondoa-Irangi. Of the land that was alienated not all was productively utilized, so that, for example, around 1910 in the Wilhelmstral district, due west of Tanga district, 61,116 hectares had been alienated to Europeans of which only 13,691 hectares were cultivated, a further 3,373 hectares were used for pasture, and as many as 44,052 hectares were not utilized. Other data for Kilimanjaro and Meru show that in 1913, although 542,124 hectares had been alienated, only 106,292 had been prepared for cultivation and only 56,753 hectares, approximately 10 percent, actually carried crops. Subsequently, the German policy of large-scale land alienation ran into considerable difficulties because it led to a breakdown of the self-sufficiency of African agriculture in the areas that were affected by alienation, and forced the African population on to poorer lands which were now nowhere near sufficient for the basic foodstuff production that was needed.

Not only did land alienation have a detrimental effect on African agricultural development, but also the outbreak of various epidemics in the last decade of the nineteenth century and the first decade of the twentieth century had a much more damaging impact. The widespread occurrence of rinderpest, which Kjekshus suggests "broke the economic backbone of many of the most prosperous and advanced communities," the extension of bush areas and the attendant infectious incursion of the tsetse fly, leading to a growth in sleeping-sickness, the spread through parts of the territory of the jiggers (the sand-flea plague), which had a devastating effect in areas where its treatment was unknown, causing considerable loss of life, and outbreaks of smallpox epidemics combined together to exert a very adverse influence on the indigenous economy. It has been estimated, for instance, that the cattle population in 1902 was about one-tenth the size of the 1890 figure. Also, certain evidence shows that there was depopulation in Tanganyika in the early part of the twentieth century, the total population falling from about 4½ million in 1902–3 to 4 million in 1913. . . .

The establishment of plantations, which became largely concentrated in the Tanga-Usambara, Kilosa-Morogoro, and Lindi regions, naturally necessitated the creation of a labour force. Some idea of the dimensions of this labour force can be acquired from the figures for the 1902–1913 period. . . . These show that the numbers employed in plantation work rose from about 4,000–5,000 at the turn of the century to just over 90,000 by 1912–13. The next two most important employment categories were railroad construction and caravan porterage. Now, in spatial terms there are four main points we should note concerning the creation of plantation enterprises.

In the first place there was the geographical delimitation of zones of capitalist production which were integrated into the international capitalist system by virtue of the fact that their *raison d'être* was to supply this expanding system with certain highly required primary commodities. At the same time the external economic integration of plantation zones was sustained and reproduced by the colonial state, which, in addition to ensuring internal order and security, introduced taxes and/or imposed forced labour so that the plantations would be guaranteed a labour force.

Secondly, then, the measures employed to create this labour force had consequences on those areas that supplied the labour. In 1910 a list of the plantation workers living in the Pangani South native reserve showed that the vast majority were coming from areas that were known to have been deeply involved in the nineteenth-century trading economy. One of these regions—Unyamwezi with its centre at Tabora—showed several signs of the disadvantageous consequences of labour migration. A number of village reports that were written around 1910 . . . reveal a series of adverse changes. One source noted the absence of men throughout the region and the associated rapid decline in agricultural production and cattle-keeping. . . . About the same time, a White Father stressed that the continued sapping of the labour force from central Tanganyika was fraught with great dangers, not only causing the ruin of the indigenous economy, but also exposing the entire province to the very real dangers of depopulation, evidenced by the existence of abandoned or half-empty and decaying villages. Neither was the Unyamwezi unique in being a labour-supplying area as other labour pools were created in areas like Kigoma, Ruvuma, Singida, Mtwara, and Ufipa.

Thirdly, within the restricted zones of capitalist production, a large percentage of the generated surplus value was siphoned off and not utilized to either diversify production or to improve the welfare of the population in these areas. Consequently, within such areas one had a sharp contradiction between the relatively high value of production on the one hand and the prevailing poverty of the population on the other. Finally, attention must be given to the relationship between plantation enterprise and the development of the colonial transport "network." In the Tanganyikan case a brief historical sketch is now in order.

. . . [R]ailways played a pivotal role in colonial spatial organization. In Tanganyika there were two major lines. In the north-eastern section construction of a railway line from Tanga towards the interior was started in 1893, but did not reach Moshi, 220 miles away, until 1911. Alternatively, progress on the Central line was much more rapid, so that by 1914 Kigoma had been linked with Dar es Salaam—a distance of approximately 700 miles, which had been spanned within more or less 9 years. . . .

. . . [A]fter the completion of the Central line the Germans constructed a naval base at Kigoma. Exactly how crucial geopolitical considerations were in this respect is difficult to assess, and certainly there were other factors at work. . . .

One such factor was the interests of German finance capital, represented by the Deutsche Bank, which . . . was the most aggressive financial institution in Germany at this time. In 1912 the Bank obtained prospecting concessions in Katanga and it definitely encouraged the railroad scheme to Kigoma, making a substantial profit from the loans it provided. Copper was exported from Katanga via the Central line to Dar es Salaam, whence it was shipped through the Suez Canal to Europe, and . . . Belgian copper freight accounted for a large share of the total value of freight carried on the Tanganyika Central line, so that in 1930–31, for example, it made up 50 percent of the value of freight carried. . . .

. . . [T]he crucial factor underlying the westward extension was a financial one. . . . [C]olonial finance, including railway finance, was controlled by the Reichstag, and hitherto it had made only small annual grants for railway building so that, for example, it took 14 years to construct the northern railroad from Tanga to Mombo. The colony had to be self-supporting to a considerable degree, owing to the fact that a large sum from the German government would not be forthcoming, and even though the Deutsche Bank would put up a certain amount, this would not be sufficient for the task. However, a smaller loan would be feasible provided that the colonial administration could service the interest on the capital advanced. Hence an East African railway would have to more or less immediately raise revenue. . . .

When we next come to consider roads, it is immediately obvious that in the German period their role, in comparison with the railways, was very limited. There had, however, been some road construction in the late 1890s and early part of the twentieth century, and this was undertaken and maintained through enforced local authority tasks. In the north direct roads were built between Tanga and Arusha, going on to Mbulu, and between Pangani and Kondoa-Irangi. In the central part of the country a road linked Bagamoyo on the coast with Mpwapwa, Tabora, and finally Ujiji on Lake Tanganyika. In the south there was a road connection between Kilwa on the coast and Songea in the interior. . . .

Finally, with regard to transport and communications, telegraph and postal services were gradually introduced over Tanganyikan territory from about 1890 onwards. Three main telegraph systems evolved, two of which radiated from Dar es Salaam, one going towards Mwanza and the other south to Lindi along the coast, and one based on Tanga followed the Usambara Railway. Postal agencies closely followed the opening of military stations, and wireless stations were eventually established at Dar es Salaam, Bukoba, and Mwanza. Also, apart from the sea links between the coastal ports, there was some movement on Lakes Victoria, Tanganyika, and Nyasa. . . .

A final point concerning the general features of the space-economy under German capitalist penetration that ought to be mentioned relates to urban centres, and particularly to coastal settlements. As we noted above, commercial agriculture had existed in Tanzania before the German invasion, especially along the coast, where crops like rice, sugar, and copra were produced for export to Zanzibar and other parts of the Indian Ocean market, both by Arab-owned slave plantations and by Swahili-speaking peasants. One of the results of German penetration was to destroy this complicated coastal economy and to ruin many small ports in a similar way as the import of metal goods and textiles from Europe and Asia destroyed many small industries. In 1907 a German governor remarked with much relevance that:

> One finds only one place that has progressed under German control, viz. Dar es Salaam. . . . Comparatively large places like Pangani, Sangani and Bagamoyo have retrogressed. . . . A whole list of medium-sized places . . . have practically lost all importance. Such places are Mkwaja,

> Winde, Mbweni, Mbwamaji, Kisiju, Kiumangao, Nyamsate, Msindaji, and others. Ruins of many stone houses bear testimony to the former flourishing condition of these places.
>
> [As cited in Iliffe, 1969:134]

By the end of the German period Dar es Salaam had an estimated population of 19,000, which by 1931 had risen to 24,000. This urban centre had become the dominant focus of colonial exploitation in Tanganyika, whilst other coastal ports had greatly diminished in importance. These urban changes within the coastal economy are best understood in terms of a re-direction in the spatial circulation of surplus which took place within Tanganyika consequent upon its integration into the German Empire, and which led to a small part of the extracted surplus being largely concentrated in one particular coastal locality that came to act as a coordinating centre for German and, later, British capital. . . .

[A]fter the First World War Tanganyika was ceded from the Germans and became a League of Nations Trust Territory, which the British subsequently acquired as a mandate. In comparison with, say, either Kenya or Rhodesia it held less attraction for British settlers, partly because of the restrictions that were placed on large-scale alienation of land and partly owing to the inferior nature of colonial infrastructure in relation to other territories. . . .

In terms of spatial structure, one of the most salient features of the 1930s and beyond was the existence of three types of socio-economic zone or region. . . . They were:

(a) Those areas which specialized in production for export: the sisal estates, the main cash crop areas, and the towns.

(b) Surrounding zones which supplied the export-producing regions with food and other services. For instance, Uzaramo, Uluguru, and Rufiji were supplying food to Dar es Salaam; Bonde, Ukaguru, and Ulanga supplied food to the sisal estates; Kondoa supplied cattle for the Tanga and Korogwe markets; and so on.

(c) Finally, extending out beyond the export and food-producing regions were peripheral zones which either supplied migrant labour or involuted in near-isolation from the dominant export-oriented areas. . . .

Whereas those more peripheral zones that functioned essentially as labour-supplying areas covered fairly wide expanses of territory, the zones of capitalist production were very small in areal extent. Luttrell [1972] accounts for such a phenomenon by suggesting that the limited economic and political objectives of colonial control, together with the relatively high costs of rail and road construction in the colonies, led to a situation in which almost inevitably only a small part of the colonial territory was opened up for direct economic exploitation. Therefore in the case of large colonies like Tanganyika the actual zones of capitalist production comprised only a small fraction of the total surface area, and even in the post-independence period capitalist development in breadth has only taken place on a very limited scale.

As we noted in a previous section, foreign ownership of the plantations led to a siphoning off of saving and investment potential which took the form of interest payments to metropolitan financiers and dividends to shareholders, while retained capital tended to be utilized for further expansion of the same export industry rather than for agricultural diversification. On this topic Guillebaud [quoted in Rweyemamu, 1973:26] has collected information on the average profits (net of depreciation) of 10 sisal plantations in the years from 1951 to 1956 and these show very clearly, especially as they do not include interest income, the extent of profit repatriation with as much as 37 percent of total recorded profits going as payments to metropolitan shareholders. This repatriation of potential investment surplus and also the dependence of the plantation system on imported capital goods, metropolitan techniques of production, and skills and technical know-how prevented any multiplier effect from ensuing and aborted any possibilities of capitalist development in other agricultural and industrial sectors.

In contradistinction to the limited areas of export production there were other regions of considerable agricultural and mineral resource potential that remained undeveloped and neglected. There are several such examples: the failure to utilize the food grain resources of the Rufiji and Kilombero valleys—rice production being particularly important here—the neglect of south-central and western Tanzania for maize and wheat, the undeveloped nature of beef production in Masailand and central Tanzania, the untapped fruit and vegetable resources in the Rungwe district of south-west Tanzania, and the lack of interest in the coal and iron ore deposits of the south-west region all testify to the total failure of colonialism to utilize the resources of the colonialized territory for the needs and development of that territory.

We have briefly assessed some key aspects of penetration and integration on Tanganyikan spatial structure. Lastly, then, in this section, it is necessary to make one or two points concerning disintegration, which from the perspective of contemporary policies of spatial restructuring is highly relevant.

Since one had the neglect of a wide range of areas that possessed substantial resources and because of the concentration on export-oriented activities, interdependent regional relationships tended to be whittled away and often broken. Writing generally about the Third World but in the above context, Freyssinet [1966] argues that in pre-capitalist economies, although the absolute value of inter-regional and intra-regional trade was small, these exchanges formed a coherent whole, corresponding to a certain level of economic development. He refers to the flows of agricultural surpluses, regional commerce in artisan products, the manufacture of which was already specialized, and long-distance trade in commodities that had a high scarcity value. The appearance of the capitalist sector, however, broke these circuits, and instead of having an integrated whole in which trade in agricultural and artisan products was multilateral and inter-regional one had the creation of what Freyssinet calls a bilateral system of trading relations which was limited to the unequal exchange of primary commodities for manufactured goods.

An example of the process which Freyssinet describes can be found in Poncet's [1973] essay on Tunisia where he writes that the pre-colonial inter-regional trade which took place across Algeria and Morocco was broken by a colonial conquest that de-urbanized and de-industrialized the society, causing some zones like central and southern Tunisia to become depopulated after outside penetration had expropriated the means of production in these regions and monopolized their internal resources. Also, a similar process took place in Tanzania, and the present-day tenuous nature of many regional linkages can be traced back to the effects of the penetration of the capitalist mode of production.

The contradictory combination of internal disintegration and external integration also helps to explain the manner in which the development of transport and communications took place. Luttrell [1972] has remarked for post-independence Tanzania that basically the country still has a colonially structured transport network with a spatial pattern that is very much externally oriented. . . .

Finally, with respect to Tanzania, one should remember that the diminutive size, peripheral location, and small number of towns, together with an absence of an industrial base, can only be explained in the context of the colonial mode of spatial organization. Coastal centres such as Tanga, Mtwara, and Dar es Salaam played the primary role of organizing nodes for the shipment of surplus extracted from the interior rural areas, and interior centres such as Iringa, Dodoma, Tabora, etc., had an important part to play in maintaining control over their respective hinterlands and policing the countryside.

In the period from 1957 to 1967 the average growth rates for the major towns in Tanzania were consistently above 5 percent per annum, Dar es Salaam recording an annual growth rate of 7.8 percent during these years. More significantly, the 1960s, or our neo-colonial phase, witnessed a noticeable increase in the concentration of industrial activity in the major urban areas and especially in Dar es Salaam. Import-substitution industrialization has predominantly taken place in Dar es Salaam, Tanga, Arusha, and Moshi, and various locational quotients computed by Rweyemamu [1971b] for a range of industries show that in 1966 Dar es Salaam was a particularly attractive location for furniture and fixtures, paper products, printing and publishing, rubber products, non-metallic mineral products, metal products, and the assembly of transport equipment.

IMPLICATIONS FOR CONTEMPORARY SPATIAL PLANNING—A CONCLUDING NOTE

When writing about underdeveloped countries in general, Alonso and Meyer [1972] suggest that a crucial issue in regional planning relates to whether a region, however interpreted, should continue to be closely integrated into the international capitalist economy or alternatively ought to be integrated into the national economy of which it is an essential geographic part. This issue, it seems to me, faces Tanzania at the present. It is difficult to reconcile socialist transformation with a continuing orientation of the society towards the international capitalist system. In fact, it would seem that the reconciliation is impossible unless one has a very warped and erroneous conception of socialist transformation.

For the regions and districts of Tanzania to become self-reliant resources must be utilized so as to meet the needs of the people as a whole.... [I]f Tanzanian-type economy is to be transformed out of its present-day underdevelopment, an organic link must be forged between the structure of production and the structure of needs of the society as a whole. One of the achievements of countries such as Cuba, North Korea, Vietnam, and China is that their previous spatial disintegration has been replaced by a system of spatial organization which utilizes to the full the resources of all the various zones of their territories so as to satisfy internal needs. And as a result, inter-regional interdependencies and exchanges have evolved and matured as an important solidifying agent of national economic integration. So long as a space-economy is internally atomized and externally tied, such development is not possible. To plan is to choose, and to choose in the context of socialist transformation is to reject the contradictory combination of internal disintegration and external integration.

CHAPTER 15

Center and Periphery in the Development Process: The Case of Peru

BRYAN R. ROBERTS

My aim in this paper is to explore those limitations on the effectiveness of central planning and national development policies that result from small-scale political and economic activity in provincial areas. Since I see this local level activity as being an integral part of contemporary development processes in Latin America, I also question government emphasis on centralized organization and planning in producing social and economic development. Such an emphasis contains a centralist bias which overlooks the extent to which national development continues to be influenced by the particular history of local populations and by their relative autonomy in social and economic organization.

I focus my arguments on Peru, which has experienced a recent attempt, on the part of government, to initiate a systematic reform of social and economic structures.... This reform activity has taken place in the context of an explicit official ideology of choosing a path to development that is neither capitalist nor communist. According to this ideology, the primary obstacles to national development are rooted in the lack of social and economic integration of the rural and much of the urban population. This lack of integration is, in part, attributed to entrenched national or foreign economic interests, such as those of the large landowners or mine companies, which have benefited from the dependence of much of the population. In contrast to this official view, I argue that the problems facing Peru and its government are due to the disorganization of its urban and rural population by their integration, from early in this century, into modern forms of capitalism. This disorganization is expressed in a complex pattern of urban migration, in the proliferation of small-scale entrepreneurial activity, and in the absence of any well-defined class groupings in either towns or countryside. In such a context, government reform occurs without the active participation of the mass of the population and

---

Reprinted from *Latin American Urban Research* 5 (W. A. Cornelius and F. M. Trueblood, editors) © 1975, pp. 77–106, by permission of the publisher, Sage Publications, Inc., and the author. This paper has been edited for this publication.

is likely to be overly sensitive to the pressures of centrally located groups in the bureaucracy or in the foreign and domestic business communities. . . . The Peruvian government's reform of structures from the center has not yet benefited much of the population; this outcome of a systematic and seriously undertaken reform program illustrates the centralist bias and the misconceptions of the development process that it produces.

Though views of underdevelopment and of the strategies required to counter this underdevelopment have differed widely in their theoretical approaches and their substantive content, most analyses have stressed central organization as the determinant of development. In the literature on regional planning, there is a stress on the need for a well-integrated urban hierarchy that effectively organizes the social and economic space of a region and articulates it with the national economy. From this perspective the underdevelopment of provincial regions is to be corrected by enabling space to be organized more efficiently through improved communications and other infrastructure and through the planned location of different types of industries. In this organization of space, centrally located groups such as urban entrepreneurs or administrators are described as the crucial agents of change (Friedmann, 1961). Related to the above perspective is that of development specialists who emphasize the importance of introducing rural populations to modern ways of life; rural energies are to be released by placing villages in better communication with the urban centers, through education and through the mass media [Lerner, 1965]. In this approach, as in the first, government officials or technical advisors are given the main responsibility for "developing" the provinces by introducing new techniques of cultivation or industry and by advising on integration with the regional economy.

This emphasis on the central determination of local possibilities is one that is shared by commentators working from the very different theoretical perspective of "dependency theory." In the early accounts of the social and economic dependency of countries and regions in Latin America, underdevelopment is viewed as the product of vertically imposed structures of dependency; economic possibilities at the local level were determined by a chain of exploitation that ran through landowners and merchants, in lower-order centers, to the bourgeoisie of the metropolitan countries [Stavenhagen, 1965; Frank, 1969a]. In subsequent modifications of dependency theory, a greater stress has been placed on variations in the internal structures of class relationships that interpret the external domination; these structures are still, however, centrally determined and their analysis leaves little scope for considering the possibilities of internally generated change [Cardoso and Falleto, 1969].

One assumption implicit in these different approaches is that rural society in underdeveloped countries is relatively static and represents a social and economic configuration that reacts to external stimuli. I argue that these approaches underestimate the dynamic of provincial organization, and that this dynamic enables people not simply to respond to external stimuli, but to interpret them. People on the "periphery" initiate activity that involves them with the national polity and economy and play a part in transforming the pattern of national development. There is a considerable body of material that suggests that in Latin America, at least, even the poorest sectors of the population are active in coping with, and manipulating, their situation. Patterns of migration, "informal" economic and political activities, and the proliferation of "self-help" devices in town and countryside indicate that the less privileged sectors of the population organize their own environments as much as these are organized for them (Leeds, 1971; Peattie, 1968a; . . . Cornelius, 1974). These attempts to organize the environment affect, and even limit, the scope of action of the economically and socially superior classes by creating uncertainty and reducing their possibilities of controlling their environment (Roberts, 1973).

Apart from these suggestions as to the scope and importance of local level activity, there are fundamental reasons for questioning the capacity of central structures to determine development in Latin America. One of these reasons is the limited organizational capacity of the type of centralized capitalist economies that have developed in Latin America. Though primarily urban-based, industrial capitalism now predominates in most of the area, showing little capacity to expand so as to include an increasing proportion of the labor force and to effect a thoroughgoing transformation of local economic structures. This dominant capitalism controls the direction of economic activity; the expansion or contraction of other economic activities is, to a large extent, determined by the behavior of the capital-intensive and technologically sophisticated industrial sector. This sector is unable to impose the degree of discipline with respect to the behavior of entrepreneurs and of the labor force that characterized earlier capitalist industrialization. By discipline, I mean the routinization of tasks, the increasing standardization of working timetables and income levels, and the organized dependence of the work force, as consumers, on the market. This discipline was made possible in developed countries by the large-scale provision of specialized and relatively stable work in the factory environment; it enabled both government and capitalists to plan their investments in economically efficient ways. In Latin America, the development of productive capacity—mainly in capital-intensive industries and through sophisticated technology—has not occurred through a rapid expansion of the urban industrial labor force nor of the labor force in commerce and services associated with this modern sector. Webb (1974) reports that while employment in the modern sector of the economy (businesses with more than five employees) expanded from 18 to 21 percent of the labor force between 1950 and 1966, employment in the urban traditional sector rose from 24 to 32 percent in the same period. These gains in the proportion of each sector in the total labor force were at the expense of the agricultural sector. In practice, these trends mean that the growing urban populations of Peru have been absorbed mainly by small-scale enterprises, personal services and self-employed activities. Indeed, capitalist expansion in Latin America has not only provoked a degree of initial disorganization as customary rural activities were disrupted, but this disorganization has continued as the urban labor force expanded rapidly in the absence of a set of economic activities requiring their services. . . .

THE NATURE OF LOCAL LEVEL ACTIVITY

Though I focus, in this section, on the types of economic and social activity that occur in a provincial region, my argument is intended to apply also to the "marginal" populations of the capital cities of Latin America. Indeed, the interrelationships between these urban populations and those in the provinces are often an important factor in the vitality of small-scale activity in the city. The small-scale activities that I describe are not ephemeral or transitional in character. These activities display consistent patterns of development in different locations and in different economic sectors; they also contain the means to reproduce themselves and constitute relatively enduring features of the kind of economy that many Latin American countries now possess. In this sense, they are as much part of the structure of contemporary capitalism in Latin America as is the dominant form of urban-industrial capitalism.

Despite many decades of capitalist penetration into even the most remote provincial regions, there is often little evidence in the rural areas of Latin America of either the consolidation of landholdings or of the consolidation of industrial and commercial firms into larger enterprises. I shall take my examples from the Mantaro area in the Central Sierra of Peru, which is considered as the most economically and socially developed of the Peruvian Sierra regions and is well-connected to Lima and the coast by rail and a hard-surface road. . . . In agriculture it is the *minifundia* that predominates, and the fragmentation of farmholdings has increased since the turn of the century. . . .

Commercial and industrial activity is also predominantly small-scale. The largest industry in the area (a textile mill) employs some 250 people; the next largest (a knitwear factory) employs about 50 people; the average number of employees for the factories of Huancayo, the largest city, is 7.1. Fifty-two percent of the manufacturing labor force in this city is employed in enterprises where there are five workers or less. The remaining large-scale industry is Cerro de Pasco Company's smelter and associated plants at La Oroya; this labor force, which numbered some 5,300 workers in 1970, has declined slightly in size from its height of 5,600 workers in 1956.

Indeed, the trend in industrial activity has been clearly towards fragmentation and the dominance of the small-scale. Whereas in 1953 there were some 3,000 workers employed mainly in four textile factories in Huancayo, this work force has been replaced by some 1,600 workers in over 400 textile workshops. Commerce is also mainly small-scale and there has been only a very slow development of larger retail outlets; there are over 2,000 grocery stores in the city of Huancayo and three small "supermarkets." This situation is most marked in the transport sector, where the 658 trucks registered in the Mantaro area (Huancayo, Concepción and Jauja provinces) are individually owned and there is no single transport firm. The 81 bus companies servicing this area and linking it to the coast and to the northern and southern Sierras are also mainly individually owned, and a bus company is normally made up of ten or more owner-drivers.

These figures give a sense of the small-scale economic activity that characterizes this one provincial area. It is noteworthy that this small-scale activity pervades all economic sectors, despite the important differences in their modes of operation. The dominance and expansion of the small-scale also appears to be true of Lima's economy where, even in the industrial sector, a large percentage of the manufacturing work force is employed in enterprises of less than five people; and this proportion is likely to have increased from 1960–1972 (Webb, 1974).

The type of small-scale economic activity that predominates in the Mantaro area is similar to that activity which is included in the concept of the informal economy (Hart, 1973; McGee, 1971b; Weeks, 1973). It operates largely outside the regulation of the state and is characterized by a heavy reliance on personal and often kinship relationships and by a tendency toward ad hoc decision-making and acts of exchange. I find the concept of the informal economy to be, however, a misleading one. The issue that is central to understanding the dynamic of small-scale activity is one of scale and not one of the existence of a relatively autonomous or relatively distinctive economic sector.

Small-scale economic activity is closely related to the large-scale sector of the economy. I include within this large-scale sector not only the larger enterprises in industry, commerce or the services, but also government employment. In Lima and in the provincial cities, small enterprises and service workers "export" much of their goods and services to the large-scale sector and to those working within it (Webb, 1974). Likewise, those working in small enterprises or in personal services spend a good part of their income on factory goods, food and on "modern" services. . . .

The significant factor in the economic activities of the majority of the population in city and countryside in Peru is that they are carried out within small-scale units. These units, whether in farming, manufacturing, commerce or services, do not make intensive use of capital; labor costs are likely to form the major part of the costs of the enterprise. . . . [T]here is a relatively free labor market and enterprises are not restricted in their hiring or firing or in their wage bargaining by government regulation or worker organization.

This relative freedom of enterprise explains part of the persistence of small-scale activity; in itself, however, it would simply represent the last vestige of a system doomed to eventual destruction by the greater power of the large-scale modern sector which controls the most lucrative markets and the access to capital. Even the most successful of the small-scale enterprises that we studied did not expand much beyond ten or so employees. . . .

Though incomes in the urban small-scale sector have not been rising nearly as fast as incomes in the large-scale sector, a small rise has occurred despite an increase in the numbers of those working within this sector (Webb, 1974). This capacity to increase per capita earnings indicates an internal dynamic that enables this sector to organize itself to take advantage of opportunities and to regulate competition. To understand the basis of this economic dynamism, it is necessary to consider carefully the logic of operation of the small enterprises.

An important feature of the mode of operation of small-scale enterprises is that they operate within a general context of market uncertainty. In part, this uncertainty is produced by the prevalence of small-scale activity; it also reflects the low level of economic development in Peru. In agriculture, the large number of small farmers, making individual decisions as to crops and the timing of the agricultural cycle, complicates both the coordination of production and marketing and the operations of large-scale agricultural traders. . . . Even the largest of the traders had been unable to secure a constant supply by purchasing or renting land or by arrangements with large producers; the predominance of small landholdings in the area and the uncertainties of climate made such a process of vertical integration impossible.

The structure of trading was similarly affected; the necessity of working with a large number of producers, whether in agriculture or crafts, meant the development of personal relationships and local knowledge which required both time and considerable effort and which limited the expansion of any one enterprise. It was a common practice for a trader to prefer to delegate a new source of business to a relative or long-time associate, on the assumption that close personal attention was required for success. Small industry, which in this area produces mainly textiles, also operates in a situation of uncertain supply and demand; the extreme fragmentation of production is accompanied by a considerable degree of secrecy and lack of coordination among the producers. These small textile enterprises are closely tied to small traders who actively seek out markets in the villages of the Sierra, as well as travel to Lima and the coast. They purchase their wool or synthetic fiber in small lots from retail outlets in Huancayo or travel directly to Lima to obtain it.

Market uncertainties give a special character to the labor policies of these small enterprises: They tend to use labor in place of machines or other capital outlay. This avoidance of capital investments is not due to any reluctance to rationalize or innovate, but to the difficulties these small operators have in obtaining capital. Their credit is not good with the banks, and large-scale enterprises have become increasingly reluctant to grant extended credits to small operators. The capital of the small enterprises was obtained from savings (often pooled with those of a kinsman or friend) or from indemnity payments from previous employment. Several of the largest traders in the wholesale market of Huancayo began their businesses on the basis of savings from mine work; many of the small textile industrialists began on the basis of machinery and payments given to them when the large-scale industry closed in Huancayo. Capital emerged, in general, from a slow process of accumulation and, often, through informal means, as when agricultural credit was used to finance a trucking enterprise.

The intensive use of labor in the small enterprise is possible and profitable because the labor supply is differentiated in ways that enable it to be used flexibly and at low cost. One source of supply is family labor and, often family labor which is given in the spare moments from other activities. In a survey of 400 small businesses in Huancayo (a random sample of the total of some 2,000 businesses in the city), I found that the majority of workers in these businesses were family workers; in bakeries and butcher shops where average employment was over four workers per business, 67 percent of all workers were family workers.

Even where these enterprises use non-family labor, it was common for this labor to be maintained and recruited in such ways as to reduce its costs to the firm. Board and lodging would be provided to a kinsman or fellow villager who had come to town to attend school, for health reasons, or to seek work; in return, the migrant would work part-time for the enterprise. Workers were also recruited from poorer regions and received wages lower than those normal in the city; though they might move on after a time, their place would be taken by other cheap labor. . . .

Part of the mode of operation of the small enterprise is that its labor costs are subsidized by the uneven nature of development in Peru. In part, this subsidy derives directly from the different levels of development present even within a fairly small region. It is the small-scale enterprise that takes greatest advantage of the profits to be derived by the recruitment of a cheap labor force from remote and impoverished areas; in Huancayo, recent migrants from the remote villages are concentrated in the small enterprises. Also, labor costs are subsidized by the health and educational services provided by the government that attract people to the city and often lead migrants to accept low wages in order to be able to stay in the city and make use of its services. . . .

Small enterprise represents the internal logic of capitalist transformation in Peru; it is the legatee of the impact of foreign capitalism and of the slow, locally-based processes of capitalist transformation that antedated the massive penetration of foreign capital. Although the effect of foreign investment has been to limit the development of a centralized, native industrial capitalism, foreign capital did stimulate local differentiation and diversity. This occurred primarily through the wages paid by mines and plantations, and through other local investments that the large enterprises made to improve transport facilities and to obtain foodstuffs.

These processes helped monetize local economies and weakened local jurisdictions and traditional obligations. These changes set in motion an irreversible internal development process. . . .

In this context, the small enterprise becomes closely linked to the development of the modern sector of the economy. The small enterprise expands in response to the opportunities created by modern capitalist development. It is not an historical residue of traditional activities, but a qualitatively different form of activity that emerges out of earlier stages of transformation. The small-scale activities I am discussing are distinct from the customary ways by which villages or small towns obtained their manufactured goods and services. Rather, small-scale activities represent entrepreneurial attempts to take advantage of those needs created by capitalist transformation but not supplied by it: the feeding of urban or mining populations with new types of foodstuffs, whose production cycle is different from that of the more traditional foodstuffs; the provision of low-cost transport services based on a greater job mobility and movement of goods; the provision of specialized repair services and personal services; the production of clothing, made necessary by new patterns of life, which is quite distinct in form and mode of production from that of traditional dress.

The significant aspect about the "urban traditional activities" by which most of Peru's city dwellers earn their living is that hardly any of them are traditional (Webb, 1974). Indeed, they are as "modern" in the historical sense as the technologically-based activities which they complement. The danger of describing these activities as "traditional," "household" or "marginal" lies in giving the impression that they represent outmoded forms of economic activity in process of extinction. It is on this point of interpretation that I am most in disagreement both with Quijano's (. . . 1974) concept of "marginalization" and McGee's (1971b:170) concept of the "bazaar economy" in Southeast Asian cities.

The relation of the small enterprise to the large-scale, capital-intensive sector of the economy is that the small enterprise, for reasons I have outlined above, possesses a flexibility which is a functional component of the current capitalist development process in Peru. This flexibility is most marked in the relative ease with which the small enterprise can change location, alter its labor force, and even change its lines of products or services. . . .

This flexibility ensures that resources are tapped and developed that are too uncertain for the profitable operation of large firms. The increasing intervention of the Peruvian government in the economy and the continuing strength of labor unions have restricted the operation of large firms. Size is crucial in this context, since it is only the smaller operations that can easily escape government regulation or labor organization. . . .

The small enterprise is dynamic not simply because it is economically viable, but also because its operation draws more and more of the population into the orbit of the capitalist economy. The small enterprise, not the large, is the active agent of capitalist penetration in Peru. This dynamic means that the term "marginal" applied to much of the urban and most of the provincial population of Peru is misleading in at least one important respect: These "marginal" populations are not passively reacting to economic changes originating from above, but are engaged in their own internal processes of development and in using changes elsewhere in the system to further that development.

SMALL ENTERPRISE AND PATTERNS OF STRATIFICATION

Despite the degree of complementarity that exists between the small enterprise and the large, the relation is an unequal one. The large enterprise has greater capacity to raise productivity through technological improvement and controls the most accessible and profitable markets—the middle and upper income populations of Lima and of some of the provincial cities. The rate of increase in per capita incomes within the small-scale sector of the economy is low; the increases in sales and services to the rest of the economy have been achieved mainly by expanding the labor force.

In this context, small-scale economic activity has somewhat contradictory implications for social and economic differentiation. There is differentiation within the small-scale sector, since those individuals or enterprises most closely linked to the wealthiest sectors of the economy are likely to receive the most income. In Lima, for example, certain classes of domestic and craft service and certain small industrial activities that directly complement larger firms are likely to generate the most income. On the other hand, this differentiation is not based on any structural characteristics that enable particular sectors or individuals to consolidate their favorable position. The operation of the small enterprise is highly particularistic, depending on the skills and energy of the individual. These skills and energies enable certain people to secure a relatively favorable position; but there is no means of erecting monopolies to ensure continuing access to rewards.

One implication of this is apparent in the distribution of income among the different sectors of economic activity in Huancayo. . . . The shares of income in this city, which is dominated by small-scale activity, are similar for all economic sectors, with the exception of government. There is no evidence that any sector is a privileged domain of activity that ensures a higher income to its members. This is true not only for sector of activity, but also for two other significant indicators of economic structure—size of enterprise and work position. In the absence of any technologically advanced economic activity in Huancayo, bigger enterprises do not obtain higher incomes for their members; similarly, the difference between being an employee and being self-employed is not related to the amount someone is likely to earn. . . .

Though the population is differentiated in terms of income levels and, to a certain extent, by habits of consumption, this differentiation does not occur in ways that are likely to lead to relatively enduring and visible class groupings. This is partly because the spatial mobility and economic flexibility entailed by small-scale activity discourages those commitments that allow differences in access to resources to be reproduced. These commitments relate to housing, educational facilities, local associations, and political activity.

Although income differences do exist in Huancayo, they are most pronounced between those employed by the government and those working in other sectors of the economy. . . . Within the small-scale sector, income differences are diminished by the predominantly family character of these enterprises. . . .

The ambiguous nature of the pattern of stratification in Huancayo is further illustrated by the position of government employees and by that of the employees of the few large concerns (mainly banks). In terms of educational levels and income levels, these employees constitute the dominant social and economic group in present day Huancayo. The commercial and landed families that once had dominated city life have, almost entirely, moved permanently to Lima (Roberts, 1974a). This is not simply an accidental feature of Huancayo's history; it results from the general pattern of economic development in Peru, in which capital has become increasingly centralized in Lima, and provincial centers depend increasingly on government employment or on small-scale activity.

These government and private firm employees are well represented in the activities of the provincial Council and in the prestigious social clubs of the city; but they are people who, in the main, have little long-term commitment to Huancayo. Many of them stay only two or three years; of all the occupational groups, these employees were the least likely, in our survey of Huancayo's population, to say that they would spend their life in the city (Roberts, 1973). Some of the most important officials have, in fact, not taken up permanent residence in Huancayo, but commute to Lima on weekends.

The degree of social fluidity encouraged by the contemporary economic structure is reflected in statistics on passenger transport flow and hotel and lodging house occupancy in Huancayo. There is a considerable flow of professionals, white-collar workers and traders between Huancayo and Lima, numbering, for these categories alone, over 10,000 individual movements in one month. The traders are the largest category of people moving between the two cities; almost twice as many traders are recorded as professionals and white-collar workers. These traders are mainly based in Huancayo and its area; only a small proportion of those registering in hotels or boarding houses are traders, and our own observations of the pattern of trading activity confirms this relative absence of Lima-based traders. In contrast, the professionals and white-collar workers make up almost 60 percent of all registrations in the hotels and boarding houses. The picture is one of an outflow of local traders, moving down to the capital to sell local products and to return with manufactured goods. The reverse flow is that of government bureaucrats or of employees such as teachers and company officials who come to the Sierra to supervise, temporarily, the development of the local economy and polity. . . .

The Central Sierra is increasing its population at approximately the national rate of growth. Population dynamics are affected by several different migration flows: migration out of the region to the coast or other parts of the Sierra; return migration of the locally born, often during the period when their families are growing; and the in-migration of people born elsewhere, usually in the poorer regions of the Sierra. These migration flows contribute to the dynamic of small enterprise and link the area to Lima and to other regions. They also increase the complexity of social stratification patterns by juxtaposing, within the same area, people whose social and economic characteristics are not simply different, but are only loosely integrated.

This situation is different, in several important respects, from that obtaining during the dominance of an agro-mining economy in this area. In the earlier part of the century, social and economic groups with clearly differentiated characteristics were detectable; the commercial and industrial elite was based in Huancayo and was of foreign or Lima origin (Roberts, 1974a). This elite organized the industry and commerce of the whole region. There was also a group of medium-sized farmers in the villages and small towns, whose land produced a large enough surplus to enable them to extend their fortunes through education, trade, and even small industry. The majority of the area's population of nearly 200,000 provided the labor for local businesses engaged in subsistence farming and, eventually, supplied the manpower for mines and plantations.

The interests of these different groups were relatively well defined.

The pursuit of these interests provided the dynamic of political activity in the area, giving rise to labor organization in factories and mines, "peasant" movements, and the extension of political organization at the regional level through the APRA political party, Acción Popular and a number of locally based political organizations. . . . Even in this period, however, the articulation of the various organized interests did not result in clear conflicts of interest or of class position. For example, with the exception of the few resident landowners, the elite had no interest in controlling local resources and made their profits by selling imported goods (mainly to the mines) or through industrial speculation. They were often the allies of the medium-sized farmers, small traders, and small industrialists, to whom they extended credit and whom they helped in their attempts to destroy traditional obligations and jurisdictions (Roberts, 1974b).

The degree of political fragmentation now present in this area is much greater than in the earlier period. The dominance of small-scale economic activity and the complex patterns of stratification provide no consistent basis for the political organization of any sector of the local population. The lack of any basis for such organization does not imply that there are no conflicts of interest present at the local or national levels, or that people are politically inactive. The economic dynamism of small-scale activity means that those involved in it have an inherent conflict of interest with large-scale capitalism and, even, with government. Though there is an important degree of complementarity in the economic system between small-scale and large-scale enterprise, the expansion of individual small enterprises is limited by the monopolies of large-scale enterprise and by government regulation.

The small-scale enterprise depends on securing even small gains by "cutting the corners" of existing regulations or economic practices; thus small-scale entrepreneurs are active in the struggle to control local political and administrative offices. We found that in the villages and towns of the area, there was a considerable degree of political activity, often directed against the implementation of government reforms, that was closely linked to the strategies of small entrepreneurs. . . . My argument is, however, a general one; in the context of contemporary Peru, and possibly of other Latin American countries, the co-existence of two forms of capitalist development—large and small scale—entails a political dynamic based on the different interests of the two sectors. The intervention of the Peruvian government does not, consequently, occur in a political vacuum, but is part of this political process.

THE EXTENSION OF STATE INTERVENTION AND THE DILEMMAS OF PLANNED CHANGE

In this section I want to examine in greater detail the particular role that the Peruvian state has come to play in local and national development. In part, my object is to assess the ways in which government can effectively intervene to secure a more even distribution of national resources among the population, and in a way which best utilizes local capabilities. Such an assessment necessarily involves an examination of the degree of autonomy that central government agencies have in planning and initiating change.

The autonomy of the State is not only an issue in underdeveloped societies which have strongly entrenched propertied classes who permeate the State and orient its policies in their favor . . . ; it is also an issue in nations like Peru which do not have such strongly entrenched groups. In such nations, the State has no class referents—whether a bourgeoisie, an industrial working class or a relatively homogeneous peasantry—by which to assess the effects and implications of its policies (Barratt Brown, 1974: 273–75). Moreover, without the degree of internal economic and social organization presupposed by the existence of such classes, it is difficult for the State to implement any policies without a considerable investment of its own resources. In this situation, public policies can become surreptitiously affected by the centralist bias to which I referred earlier, so that centrally located groups, such as bureaucrats, managers or professionals, are disproportionately advantaged by state intervention.

The rapid expansion of government agencies in the provinces of Peru has coincided with the disruption of the forms of provincial social and economic organization that I described earlier. These agencies have been mainly "developmentalist" in character. In the 1960s their work took the form of community action programs, cooperative organizations, and agricultural extension work; the agencies had limited resources and operated by attempting to persuade local communities to cooperate in development projects. Much of the present military government's intervention in provincial life has been a consolidation of this previous work, especially by bringing together previously separate agencies in a greatly expanded "social mobilization" organization—SINAMOS. The important point, however, is that these government agencies replaced forms of local organization that had themselves been "developmentalist" in orientation and, in a particular way, had been effective in developing local resources.

The small-scale programs of local development that proliferated in the Mantaro area in the early part of the century were sponsored, and mainly paid for, by local communities. . . . These programs included irrigation projects, road and bridge building, public buildings, water and electricity supply systems, schools and health clinics. They originated in the strategies and ambitions of the developing medium-sized farming group (owning 25 acres or more) of the area. . . . With the development of markets in the urban and labor centers, this group became more interested in commercializing their crops; its members held the main offices of the village, and they undertook development projects, using the free labor of the community. This group extended its political and economic contacts at the regional and national level, obtaining the help of their migrant relatives in the different centers. This was the period when Lima politics became penetrated by the lobbying and political strategies of provincial groups, and migrant associations proliferated in the capital and other work centers (Roberts, 1974b).

The local improvement projects were encouraged by provincial and national elites and by the large mining and agricultural enterprises, which often contributed to their financing.... The provision of educational facilities by local communities, the construction of bridges and roads, all directly contributed to the development of the Peruvian economy and to its greater integration. It was also a form of development that more directly exposed provincial people to exploitation from the center.

This occurred because local communities undertook essential works of social and economic infrastructure that would otherwise have had to be financed by the State and, through taxation, by the large-scale enterprises. Also, the type of work undertaken contributed to destroying the basis for local economic autonomy. Community work was heavily concentrated in constructing schools and other social facilities; what was noticeably absent was investment in the development of local productive resources, whether in agriculture or in small-scale industry. This is not surprising since those mainly responsible for initiating the community works had diversified their economic activities and often based them on labor or other resources outside the village; they were less interested in providing local sources of work than they were in those facilities that would secure the comfort, prestige and mobility of their own families. The activities of these rural groups linked the provinces with the capital; but the type of development they sponsored was, ultimately, a disorganizing influence on local society. The community improvement projects did not provide the means to achieve self-sustaining local development; they contributed, however, to making people more aware of supra-local economic and political possibilities. With the shift from an agro-mining economy to one more concentrated in urban-industrial development in Lima, those groups—the Huancayo and village elites—who had maintained an element of control over local processes increasingly moved to the capital. It is in this context of increasing social and political dislocations produced by the decline of the previously dominant forms of provincial social and economic organization that government intervention in the provinces has increased....

The government's efforts at hacienda reform, reform of local peasant communities, and cooperative organization in the Central Sierra illustrate the difficulties of state planning where there is no social or economic organization on which to base it. In such a situation, the intervention of the State follows its own logic of centralization and control; but it does not have the capacity to improve the lives of the majority of the local population. Instead the intervention is often subverted to particularistic ends....

[One] element of agrarian reform in the Central Sierra is government promotion of cooperatives. This policy is an extension of that initiated by previous Peruvian governments; it is aimed at promoting the development of cooperatives in the fields of production, marketing and services. The cooperatives receive tax concessions and are able to obtain government loans at low interest. They can also obtain the assistance of government and foreign technicians in their development efforts. In the Central Sierra, the government has promoted this policy in industry through the cooperative organization of a large textile mill, Manufacturas del Centro; through two agricultural marketing cooperatives that correspond to the left and right banks of the Mantaro valley; and through a series of smaller cooperatives, including one trucking enterprise.

The basic aim of the government policy has been to persuade people to pool resources and, by so doing, to effect economies, rationalize their enterprise, provide better services or products, and raise their own incomes. Our studies of the various cooperatives revealed, however, a different picture; the cooperatives tended to accentuate local economic inequalities. . . .

In the case of one of the cooperatives, the office-holders were mainly traders and commercial farmers who were supporters of the APRA political party, and who used the cooperative to maintain their political contacts. Though there were constant attempts by the government and its agents to ensure that members kept to the spirit of cooperative organization, the interventions were largely ineffectual. The government had invested little material or technical resources in the cooperative and could not constantly monitor the activities of its members.

The textile cooperative has a different history; its industrial organization meant a greater homogeneity of interests and situations among members than was the case in the agricultural cooperatives. The textile workers, however, had interests which conflicted with the spirit of cooperative organization and which reflected the dominant pattern of small-scale enterprise in the area. Many retained land in the nearby villages and returned home daily or on weekends; others had set up their wives or kin in trade and used cooperative products in this activity. Also, the relatively high wages and high educational levels of these workers had enabled many to educate their children to levels at which they wished to take white-collar jobs.

The textile mill came to be seen by many members as a private resource to be exploited to the full in the present, rather than a community resource in which to make long-term investments that would provide work for themselves, their children and their relatives. To improve production and raise profits, the cooperative members, who were all aging after many years in the textile industry, took on surreptitiously contracted labor that was not protected by wage or social security legislation. Though the cooperative was highly successful economically, these practices led to government intervention and to a restructuring of the cooperative's organization and membership that strengthened the control of a professional manager.

These various cases illustrate the difficulty of promoting structural change in the face of an ongoing and vital set of local economic activities. Part of the problem in the Central Sierra was a lack of sufficient investments on the part of government to give members an equal and full-time commitment to the cooperative enterprise and to its social obligations. Apart from setting up the cooperatives, providing some technical aid and tax relief, the government maintained little consistent presence in these cooperatives; the dynamic that was present—that of the small-scale activity of the area—thus easily subverted these cooperatives to its own purposes.

A more fundamental problem was the lack of a suitable social and economic infrastructure for government planning and intervention. The cooperatives did not enclose homogeneous groupings which had previously been structurally dependent on other forms of organization. In the case of the textile workers, their relations with private management had been relatively harmonious in the last years of the factory; management had increasingly acceded to worker wage demands and control in the factory had become increasingly based on cooperation with the powerful textile union. The redistribution of income effected by the cooperativization of the mill was almost non-existent; most cooperative members had been highly paid loom operators, and these took a substantial cut in salary with cooperativization.

THE IMPLICATIONS

Despite the difficulties attending government intervention, the presence of the Peruvian state is increasing in provincial regions and has important implications for the course of local development. In the first place, the State has replaced the foreign and Lima-based commercial and industrial elites as the dominant external force in the provinces. The government is replacing the Cerro de Pasco Corporation as the largest single employer in the Central Sierra, employing some 19 percent of Huancayo's adult male population, as well as teachers and government agents in the smaller towns of the area. The government also pays the highest wages and has become one of the most important economic resources that the area now possesses. The position of the State adds a further dimension, as we have seen, to the complex patterns of stratification in the area.

In earlier periods the income-maximizing strategies of local people were oriented to making use of the opportunities of the agro-mining economy through migration and household diversification; now strategies have changed to tap the resources of the State. Education and white-collar employment have become, for many local families, the most important objectives to be pursued; much small-scale enterprise is geared to making profits for investment in education, or to locating itself so as to maximize educational opportunities. This orientation is another factor preventing the small-scale enterprise from expanding through further internal investments.

State intervention which is not accompanied by a thoroughgoing investment of personnel and capital produces an uneasy situation. Small-scale activity flourishes in the interstices of state regulation of large-scale enterprises; but this small-scale activity is restricted in its expansion and possibilities of rationalization by the governmental presence. This presence also diverts part of the area's entrepreneurial talent into "unproductive" careers in the growing bureaucracy.

There is little in the thrust of government policy that has positive implications for the economic development of the Central Sierra. . . . Very little of the population of the Central Sierra is affected by the government's redistributive measures; indeed, the price controls on basic foodstuffs act to reduce the profit possibilities of the small-farming sector of the population, which is one of the bases of the area's economy. Little government attention has been given to these problems of horizontal distribution of income; yet, through indirect taxation, food price controls, and the scope of government regulation, the State already limits the dynamic of the small-scale sector. The primary beneficiary of the cheapening of food costs is the large-scale sector of the economy, which can most afford to pay them; much of the urban population even of cities like Lima is integrated into small-scale enterprises, which are either directly or indirectly linked to small-scale farming through economic and social exchanges. In the present situation, the state cannot organize economic activity and is an irritant to local development; the form of capitalism which is developing in the region and which articulates it with the national economy is not one that is susceptible to control or direction.

Unless the government in Peru can massively commit resources to a provincial area—and there are good arguments for making such investments—state intervention is simply a form of tampering with local structures that can become counterproductive. This is not meant as a total criticism of current reforms but as a comment on the limited nature of the benefits that they entail for the mass of the Peruvian population. In this respect, it is important to remember the contributions made by small-scale enterprise to generating development, in what are uncertain and highly marginal economic circumstances. The limits on accumulation imposed by the small size of the enterprises, and the personal relationships on which their operation is often based, give the small-scale sector a continuing capacity to absorb labor. More than this, the nature of the accumulation dictates that, to a large extent, surpluses are invested locally. The profits of these enterprises are rarely more than sufficient to build a substantial house, furnish it, and maintain a comfortable, "urban" standard of living. Small-scale activity also impedes economic centralization and reduces the impact of economic monopolies by directly trading with factories and markets in Lima. Gone, for example, are the large merchants and industrialists who once controlled Huancayo's commerce and factories, and who remitted their substantial profits to Lima. The dominance of small-scale activity together with the wages paid locally by the government may have reversed, to a small extent, the outflow of capital from the Central Sierra to Lima that was characteristic of the period of the agro-mining economy.

An area whose economy is based on small-scale activity appears to have a relatively high capacity to retain population. In the last decade, the Central Sierra has increased its population and has not suffered a net loss of population through out-migration. This capacity to retain population in active and productive ways is, itself, a significant contribution to national development, given the rapid increase of Lima's population and the slow expansion of jobs in the large-scale sector of the capital's economy. The State retains an important role in provincial development, even where it is unprepared to invest substantially in that development; the infrastructure on which small-scale activity depends—roads, schools, electricity, water, health services—must be provided by the government. Likewise, facilities for small loans and for obtaining expert advice can usefully be made available by the State. What is unnecessary is the close supervision and direction of local activity by central authorities.

When state employees intervene in local processes to administer, regulate or to mobilize, they are usually outsiders with little understanding of local customs or of local social or economic processes. Though some of the employees may be local people, long absence to obtain education or to pursue a career has often made them unfamiliar with local events. The degree of information that these employees possess about local conditions is often not sufficient to enable them to define public policies to meet local circumstances. It is best left to local people to organize themselves; they are most capable of making the most efficient use of local resources.

MIGRATION FROM RURAL AREAS

INTRODUCTION

Migration is one of the chief means whereby a spatial redistribution of the population occurs, even in the Third World where urban populations may be increasing by an excess of births over deaths even more rapidly than are rural populations. Indeed, the problems of Third World cities have often been attributed to the prevalence within them of large numbers of persons who are only lately acquainted with the demands of urban life. That more than half the residents of Third World cities were not born there should not surprise us, for the cities can only be growing as rapidly as they are by attracting many newcomers. (Third World cities grow at from 5–10 percent per year; since natural increase is at most about 3 percent per year, migration accounts for one-half or even more of the annual growth.)

Why this migration occurs, what the motivations of migrants are in undertaking the move, and what some of the consequences are, not only for the migrants personally but for the development prospects of the society as a whole and for the emerging class structure in urban areas, are basic questions in the field of comparative urban studies. Here, as in other areas of urban studies, the expectations generated out of western experience and sanguine economic development theories have been confounded by the Third World. In the first article in this section, McGee examines the more traditional theories concerning the nature and functions of rural-to-urban migration and finds them inapplicable to the situation which now prevails in South and Southeast Asia. He then suggests a more fruitful approach to explore the impact of the penetration of the world capitalist system upon migration in that part of the world and develops a number of categories of migration from which he builds several models of alternative development. His most useful distinction cross-classifies place with sector, a distinction he uses in another piece included later in this book. He suggests that just as rural areas may contain a peasant sector and a capitalist sector, cities also have both peasant (informal) and capitalist (wage labor) sectors. Geographic movement need not be only from the rural-peasant to the urban-capitalist sector, the kind we conventionally discuss under the heading of migration, but may be in other combinations.

Clearly, the type of movement should have a considerable influence on both the motivations that drive people out of one area and those which attract them to another. In Cornelius' paper the literature on Latin American migration is evaluated and the findings summarized. He too is critical of the theories and assumptions of previous scholarship, noting that the consequences anticipated from migration often are not found, at least in Latin America. But whereas McGee stresses the economic and class consequences of migration, Cornelius' emphasis is on the social and political consequences. These are also central to the last paper in this section, dealing with an African city. Furedi describes some of the changing consequences of migration for Nairobi both before and after independence, and his analysis may help to explain why similar migratory movements should have such different political consequences in Asia, Latin America and Africa. Furedi stresses the importance of young male migrants from the countryside (the so-called "crowd") in the political movements of the pre-independence period and notes their increasing repression and irrelevance once *local* African elites came to power.

CHAPTER 16

Rural-Urban Mobility in South and Southeast Asia. Different Formulations . . . Different Answers?

T. G. MC GEE

> And so the diversion of the population from agriculture is expressed in Russia, in the growth of towns . . . , suburbs, factory and commercial and industrial villages and townships, as well as non-agricultural migration. All these processes which have been and are rapidly developing in breadth and depth in the post-Reform period, are necessary components of capitalist development and are profoundly progressive as compared with the old forms of life.
>
> V. I. Lenin, 1899, p. 580

This quotation occurs at the conclusion of some twenty pages of analysis on a similar theme which, if read out of context, might appear as a panegyric on the beneficial effects of capitalist development in moving population from agricultural to non-agricultural activities. The purpose of the quotation is not to suggest the absurd: that Lenin was a defender of capitalism. Certainly the quotation is not intended to argue that the historical experience of Russia before 1900 is identical to that being experienced by the countries of South and Southeast Asia which are the subject of this paper. Rather the quotation catches the basic theme of the paper which argues that the understanding of the reasons for and features of internal mobility within nations in South and Southeast Asia rests upon the delineation of the broad processes of capitalist penetration of the region.

In the volume entitled *The Development of Capitalism in Russia,* Lenin shows the manner in which the growth of capitalism led to radical changes in the organization and production of agriculture and industry. The development of a commodity market for agricultural produce broke down the old precapitalist relationships and set up an economic differenti-

Condensed and edited from a paper to appear in *Human Migration: Patterns, Implications, Policies* (Bloomington, Ind.: Indiana University Press, 1977). Reprinted by permission of the American Academy of Arts and Sciences and the author.

ation among the peasantry which created a basic dislocation force. At the same time the growth of large scale industry offered opportunities for employment in urban areas, as well as producing goods which began to put the domestic producers of the countryside out of operation. This forced another group from the countryside. These developments were associated with a growth in transportation which greatly aided the circulation of commodities as well as the mobility of the population. The growth of the non-agricultural population dislocated from the rural areas was clearly shown in the growth of towns. . . .

The changes induced by the growth of capitalism which produced this substantial rural-urban mobility are also operating in the region of South and Southeast Asia. This does not mean that the process of capitalist penetration is the same as that experienced in Russia, for in these countries it is *peripheral capitalism* that is operating which means that these changes are sometimes different in form. In addition, the demographic, economic and social components of these Asian societies are different from Russia which means that the penetration of capitalism is not always the same. However, an important thrust of studies in the political economy of the Third World by such writers as Baran (1957), Frank (1967), and Amin (1974) has begun to delineate the major features of peripheral capitalism so that is possible to incorporate their ideas into the earlier formulations of Lenin. In addition, the conceptual framework of the peasant systems suggested by Chayanov (1966) and Franklin (1965, 1969) greatly aid the analysis of internal mobility in the South and Southeast Asian region.

CONVENTIONAL APPROACHES TO THE STUDY OF MOBILITY

Three main theoretical frameworks which have provided the sets of assumptions that are the basis of the most frequent studies of rural mobility may be delineated. These are: (1) the conventional economic approach, (2) the situational approach, and (3) the historical approach. Each of these approaches is discussed in this section.

(1) The Conventional Economic Approach

This approach to the study of rural-urban mobility takes the unequal geographic distribution of the "factors" of production (e.g., labour, capital, natural resources and land) as a given *apriori* and assumes that this will determine the unequal remuneration of these factors. Thus labour will move from a region where it is abundant and capital scarce to a region where labour is scarce and capital abundant, assuming the factors of natural resources and land were equally distributed. The approach, of course, assumes some form of economically "rational" choice on the part of the migrant involved in the labour transfer. A rather more sophisticated formulation is the basis of the influential work of Todaro (1969, 1971). This model assumes that the decision of the migrant to relocate is the function of two variables: namely the gap in income between the city and the countryside and the possibility of being employed in the city. A knowledge of these variables enables likely migratory behaviour to be predicted.

Amin ... offers the following critique of Todaro's model. First, the approach is descriptive, not explanatory. The geographic distribution of factors is primarily the result of the pattern of development which in the case of South and Southeast Asia has been characterised by capitalist penetration within the political framework of colonialism. Thus, for example, the marked differences in the income levels of the population of the West coast states of peninsular Malaysia and the East and Far North is a consequence of the pattern of colonial development. It has not led to major flows of migration from the poorer states to the richer regions as Todaro's model would predict. Indeed, Todaro's model is nothing more than the increasingly criticised push-pull hypothesis of migration dressed-up in economic jargon.

Secondly, the approach assumes economic rationality on the part of the migrant. The decision to migrate is presumably made with full knowledge of the variables of income and potential employment. Surely this is an inaccurate assumption. Migrants from the countryside in the South and Southeast Asian region are moving to cities where the rates of growth are double that of the national population, unemployment (particularly in specific age groups) is high and poverty and squalor a major part of city life. Yet still they continue to move to the cities, being dislocated by processes which are consequences of the pattern of capitalist penetration over time. The introduction of individual land ownership, the development of cash crops for exports, the increased monetisation of the South and Southeast Asian countryside, technological changes and the marketing of the cheap industrial products manufactured in the developed countries have all led to new patterns of social and productive relationships in the countryside. This has marginalised some groups forcing them to migrate, and created affluence among others who have also migrated to the cities in a process of upward mobility. Unfortunately there is no comprehensive study of how this process has occurred throughout the region. Certainly it can be argued that rural-urban migrants are reacting to the impact of these broader processes which are part of the national system and that their decisions are a consequence of the system.

(2) The Situational Approach

The majority of studies of rural-urban mobility are characterised by this approach. Broadly these studies have two theoretical thrusts. The first rests upon the model of rural-urban differences and focuses upon the problems of migrant adaptation as they move from one milieu to the other.

The second thrust of research concentrates upon the decisions of the migrants. I will not run through the familiar litany of the inadequacies of the model of rural-urban differences developed out of Western experience when applied in the region of South and Southeast Asia. Suffice to say that one of the major critiques of the rural-urban continuum is that it does not take into account the movement of people from rural to urban areas. Such movement brings people whose values, attitudes and institutions have been developed in the countryside. These personal attributes do not simply disappear in the urban environment, especially when the rural migrant moves within a network of close personal ties and information which aid his assimilation into the urban environment. In other words, the model of the urban and rural differences is inaccurate. . . .

However, most of the studies of population movement in the South and Southeast Asian region have been carried out within the framework of ideas developed in the model of the rural-urban continuum. In this category I would also place a majority of the many excellent empirical studies of internal migration in the region which have emerged over the last twenty years, despite the fact that they have focused upon questions such as migration differentials and flows.

Of course, the strongest argument in favour of the situation approach is the fact that external determinants such as population density, political and social structures will vary from situation to situation. Thus a particular constellation of these determinants is often an important element inducing migration. For instance, there is ample evidence from my studies of the internal migration of Malays in West Malaysia that a traditional cultural practice of *merantau,* in which men left a matrilineal society for periods of time to travel, is well adapted to the patterns of short-term labour recruitment into government jobs introduced by the British in the nineteenth century, for the men from Minangkabau areas figure prominently in the migration streams. . . . The problem is that there is no way of predicting that two similar situations will not have different rates of out-migration. It is only through an understanding of the manner in which processive forces such as education, technological change, changing forms of production, etc., diffuse into each situation that the real causes of mobility can be investigated. The problem with giving priority to the situation, and within each situation the decision-making process of the individual migrant, is that individual motivations are nothing more than rationalisations of behaviour within a system. This does not take account of the fact that this behaviour is caused by the system of which the individual is part.

(3) The Historical Approach

There have been many historical models of population movement through time, often concentrating upon typologies of population movement, but few have attempted to relate population movements through time to the model of the "demographic transition." An ambitious attempt has been made by Zelinsky (1971) which has had considerable impact on geographic studies of migration in the Third World. In this paper Zelinsky develops an evolutionary model of spatial behaviour which argues that:

> There are definite, patterned regularities in the growth of personal mobility through space-time during recent history, and these regularities comprise an essential component of the modernization process.
>
> Zelinsky, 1971:221–22

He then goes on to postulate unilinear stages of the mobility transition which are related to levels of socio-economic development which are linked to a phase in the demographic transition. Five main stages are delineated. First, there is a preindustrial, traditional phase of low residential mobility and natural increase. The second phase is early transitional, in which a sudden increase in fertility is accompanied by large scale rural-urban migrations and the colonisation of domestic and other areas. The late transitional phase is characterised by a decline in the rates of natural increase, rural-urban and rural-rural migration. Various forms of residential circulation increase, particularly inter- and intra-urban movement. The fourth phase of the advanced industrial society experiences a leveling off of rates of natural increase, a further decline in rural-urban and rural-rural migration. Finally the future super-advanced society of the future will probably have a substantial decline in residential mobility because communication developments will allow place of work to be brought closer to place of residence. In some cases place of work and residence will be the same.

From the point of view of the region under discussion most of the countries are clearly at the phase of early transition at which time, to quote Zelinsky:

> The onset of modernization (or more precisely the onset of major changes in the reproductive budget in Phase B, along with the general rise in material welfare or expectations and improvements in transport and communications) brings with it a great shaking loose of migrants from the countryside.
>
> Zelinsky, 1971:236

This is the process analysed by Lenin which was paraphrased in the introduction.

But the empirical evidence on the mobility patterns being experienced by these societies raises questions as to whether they are experiencing the same mobility transition. First, the theory of the mobility transition assumes that the same processes of modernisation are occurring in this region as occurred in Western Europe; in fact, the processes produced societies in which the majority of people were urban dwellers. While the impact of colonial powers in the region certainly developed a large number of urban settlements, particularly the large port cities such as Bombay, Madras, Calcutta, Singapore, Batavia, etc., the period of colonial rule did not see "a great shaking loose of migrants" from the countryside of the various countries. Rather, selective and often recruited labour migration (much of it international) was the main feature of rural-urban and rural-rural migration. Overall patterns of mobility continued very much at the first phase of the mobility transition. In the period of colonial devolution since the end of the Second World War rural-urban migration has considerably accelerated, particularly in the early part of the period in the case of countries such as India, Pakistan, Burma, Malaya and Indonesia where political disturbance in the countryside was often the major dislocating force. Since 1955 with the exception of the former French colonies of Indo-China political disturbance has not played such a significant role in dislocating the rural populations, and processes connected with the form of economic development have become more important. (See McGee, 1967. . . .) But in the heavily populated countries of the region, India, Indonesia, Sri Lanka and more recently Bangladesh the level of urbanisation has increased only slowly. In other countries such as Malaysia and the Philippines the level of urbanisation has risen during the period since 1955, although leading to an increased concentration of urban population in the largest metropolitan area. Thus the mobility transition appears to have some applicability to these latter countries, while of no great value in the heavily populated countries such as India, Indonesia and Bangladesh.

Zelinsky's model suffers from the assumption that societies will necessarily pass through these various phases and will experience the same set of processes. At least in demographic terms none of these societies in the South and Southeast Asian region has the possibility of emigration as did the societies of Western Europe. Secondly, demographic processes are different: natural increase is higher in the cities and this contributes an important part of urban increase. At the same time the ability of the cities to absorb the rural population into viable occupations is inadequate and the rates of unemployment and underemployment reinforce this observation. Is it possible then that one will see a dual pattern in the region of South and Southeast Asia? One group of countries will bog down in the early transitional phase while the other moves out of it. The analysis of urbanisation in the Asian region which is presented in the next section lends credence to this view.

This brief review of the major conventional approaches to the study of mobility has presented a broad critique of the theoretical underpinnings of the major approaches to the study of rural-urban mobility. The major critique that we have presented is that the approaches either ignore or misinterpret the process of the penetration of capitalism. In the next section an attempt is made to review the empirical facts concerning the urbanisation process in the region as a basis for the structural model of rural-urban movement which is presented in the following part.

PATTERNS OF URBANISATION IN SOUTH AND SOUTHEAST ASIA

Viewed in the context of the underdeveloped world, Table 1 shows that the region of South and Southeast Asia is one of the least urbanised regions. Only Middle and Southern Africa and East Africa had levels of urbanisation that were lower. What is more the proportion of population living in the rural areas in the region has fallen only slowly in the last twenty years. Thus if Kingsley Davis' [1969:I] estimates are correct, the percentage of Southeast Asia's population living in rural areas has fallen from 86.4 percent in 1950 to 80 percent in 1975. The proportion for South Central Asia fell by 2.3 percent in the same period. These figures are, of course, subject to revision but the available census data from the 1970–71 round of cen-

TABLE 1. Trends in Proportion Urban, 1950-1970, Underdeveloped World

	Percentage Urban			Change in Percentage Urban	
	1950	**1960**	**1970**	**1950-70**	**1960-70**
Northern Africa	24.6	29.6	34.6	10.0	5.0
Western Africa	10.6	14.7	19.7	9.1	5.0
Eastern Africa	5.5	7.5	9.9	4.4	2.4
Middle & Southern Africa	6.6	11.6	15.4	8.8	3.8
Middle America	39.2	46.2	53.0	13.8	6.8
Caribbean	35.2	38.5	42.5	7.3	4.0
Tropical South America	35.8	44.7	53.1	17.3	8.4
East Asia (excl. Japan)	12.1	18.0	25.3	13.2	7.3
Southeast Asia	13.6	16.6	20.1	6.5	3.5
Southwest Asia	24.3	29.5	35.5	11.3	6.0
South Central Asia	15.4	16.4	17.7	2.3	1.3

SOURCE: Kingsley Davis, **World Urbanization 1950-70, Volume I, Basic Data for Cities, Countries, and Regions,** Population Monograph Series No. 4 (Berkeley: University of California Press, 1969).

suses in the region support this picture. Thus the level of urbanisation in India, which makes up a major portion of the population of this region, increased by only 1.91 percent in the period between 1961 and 1971 to reach a level of 19.87 percent. . . . In Indonesia, which contains 42 percent of the population of Southeast Asia, the percentage of urban dwellers increased from 14.9 percent in 1961 to 18.8 percent in 1971. Although I have not been able to extract figures for Bangladesh, I suspect that a similar trend would be apparent. Thus, in the four countries that make up almost 82 percent of the population of the region under discussion, three main points emerge. First, urbanisation levels have increased only slowly. Secondly, the countryside has absorbed a major part of the total population increase. . . . Thirdly, despite the fact that the population resident in urban areas grew considerably from rural-urban migration, natural increase and urban reclassification, the actual volume of net rural-urban migration was only a small trickle out of the potential sending population. . . .

In these heavily populated countries we, therefore, have a situation in which the urbanisation process—"that great shaking loose of migrants from the countryside" to use Zelinsky's phrase—has hardly begun. This raises many questions concerning the future of urban centres in this region. Even allowing for the hopeful (perhaps excessively hopeful) possibility that birth control programmes will reduce rates of population increase, it does appear that the long-term prospective is one of increasingly large metropoli. Fairly conservative estimates suggest that Bangkok-Thonburi will reach 13 million by 2000 A.D., Jakarta could be as high as 20 million and the picture of Indian cities approaches science fiction . . . a Calcutta of 66 million. It may be, of course, that as new urban hierarchies develop, the urbanising population will be funneled into new and medium-sized towns. But present trends do not give much credence to this possibility.

The incipient nature of this rural-urban migration raises even more questions concerning the possibilities of providing employment for these urban populations. Already the cities of these countries have high unemployment rates and there are large proportions of the population eking out an existence in the low income "informal sectors" of the cities' economic structure. Despite the growth of industrialisation in such a country as India there is no indication that Indian centres can create viable employment opportunities if the rural-urban movement accelerates and releases the very large volume of rural migrants that would be involved in lifting the urbanisation level to even fifty percent.

Faced with this sort of future, planners have reacted in two ways. . . . One group of planners has a vision of these countries which will see them follow along the capitalist path to high consumer-orientated affluence of fully urbanised Western capitalist societies. A second body of thought does not believe this possible and foresees an inevitable persistence of poverty which will force political and social reorganisation. In the other countries of the region it may be argued that the incipient rural-urban migration picture which characterises the countries already discussed is not so grave a problem. Singapore, a city state, is in the fortunate position of being able to cut off its boundaries to international migration, and with a rapid fall in the rate of natural increase in the seventies should not be faced with severe problems of employment creation or managing its urban environment. Malaysia, while still faced with grave problems of creating sufficient urban employment opportunities for its Malay majority, has already exceeded the 50 percent level of urbanisation and has embarked on ambitious programmes of rural development designed to siphon off rural population from overpopulated areas. . . . Both the Philippines and Thailand will still experience considerable rural-urban migration over the next thirty years, but assuming that political conditions improve, there does seem some possibility that they can manage the urban transformation. Finally, the former French-Indo-Chinese territories are now developing within socialist frameworks; at the moment they are all experiencing deurbanisation (except for North Vietnam) and we can expect that the major spatial focus will be on the socialist transformation of the countryside in the first few years.

To summarise, it does appear that the major problems and potential of rural-urban mobility are likely to occur in the heavily populated countries of Bangladesh, India, Sri Lanka and Indonesia. Despite the fact that these countries have adopted differing mixes of foreign aid, state involvement and priorities in their development plans, their central goals are still that eventually they will move towards industrialised, urbanised, high mass-consumption societies similar to the rich countries. However, the demographic and economic facts appear to be that this can only be accomplished within a structural context of a continuing increase in the numbers of the agricultural workforce at least in the next fifty years. At the same time, there will be some shift in the industrial structure towards industry and services. Thus, this developmental strategy involves the creation of a massive number of employment opportunities in both rural and urban areas. Given the fact that in all these countries there already exist what Bose (1971) labels "pools" of under-employed and unemployed populations this will further exacerbate any effort to increase employment opportunities.

Of course, there is nothing new in emphasising this employment problem in the context of the Third World. From the middle of the nineteen sixties a rapidly proliferating series of papers emerging from individual writers and from such international organisations as the International Labour Office, United Nations' regional organisations, etc. . . . have drawn attention to the fact that an increasing proportion of the labour force of the rapidly growing cities of the Third World is not being absorbed into what has been labeled "full productive employment" (see Friedmann and

Sullivan, 1972:1). In addition to high rates of unemployment among the urban populations, often young and relatively well educated, there is "... a reserve army of low-productivity workers in rural and urban areas, among whom the waste of human potential is massive" (Turnham and Jaeger, 1971:10). What is more, this pattern does not appear to vary greatly between the countries of the Third World which have experienced rapid economic growth and those which have had slower rates of development.

A series of studies has focused upon this problem, in the three continents of the Third World, and the statistics they present, even subject to the problems of the applicability of data, all tell the same story—a picture of a rapidly increasing labour force within both rural and urban areas, increasing rates of unemployment, and absorption into low-productivity informal activities in urban areas which will accelerate in the decades of the nineteen seventies and eighties in most of the Third World countries.... The reasons for this situation are well known: a combination of population explosion, urban growth and limited employment opportunities generated by capital-intensive industrialisation have all combined to create the problem....

It must be obvious from the theme of this section that I am sceptical of the possibilities of the levels of urbanisation increasing to more than 50 percent in the next 50 years in these heavily populated countries of Asia. Assuming there is no structural transformation and various mixtures of state capitalism continue, the scenario I envisage is of increasing labour involution in countryside and cities; that is a continuing absorption of labour into low productivity, labour intensive activities which the I.L.O. has labeled rather inadequately the "informal sector." There is nothing new in this position which was first developed in 1968 (see Armstrong and McGee, 1968) but the prospect of continuing rural-urban involution raises important questions of the types and patterns of rural-urban mobility that will prevail over the next few decades. In the next section a structural model of rural-urban mobility is presented which takes account of the processes that have been delineated in this section.

A STRUCTURAL MODEL OF RURAL-URBAN MOBILITY

In the introductory section of this paper the point was made that many of the conventional approaches to the study of rural-urban mobility were inadequate because they failed to recognise the important role of processes operating at a national level. By far the most important process is the penetration of capitalism into the precapitalistic system of production which I have illustrated by reference to Lenin's study of Russia. In the context of the peripheral economies of South and Southeast Asia this process appears to be best conceptualised within the framework of the well established dualistic model of the economic organisation of these underdeveloped countries. Initially this dualistic model . . . distinguished between the capitalist towns and subsistence countryside, but the work of Geertz drew attention in his work on a Javanese town to the existence of a dual economic structure appearing in urban areas as well. In this work he pointed out that the town was divided between a *firm-centred* economic sector

> . . . where trade and industry occur through a set of impersonally-defined social institutions which organize a variety of specialized occupations with respect to some particular productive or distributive end, and a second sector labelled the bazaar economy based on . . . the independent activities of a set of highly competitive traders who relate to one another mainly by means of an incredible volume of *ad hoc* acts of exchange.
>
> Geertz, 1963a:28

Lately this model has been greatly expanded in the work of Santos (1971).

However, from the point of view of the conceptual framework of rural-urban migration the most fruitful concepts arise out of the work of Chayanov (1966) and Franklin (1965, 1969) which distinguished between three main systems of production in which "the fundamental differentiator is the labour commitment of the enterprise" (Franklin, 1965:148). In this framework the peasant economy is characterised by the commitment of the *chef d'entreprise* to the utilisation of his family (kin). The capitalist and socialist systems of production are different because ". . . labour becomes a commodity to be hired and dismissed by the enterprise" (Franklin, 1965:148). This is a scheme which, as Franklin points out, is not impinged upon by the agricultural-industrial division, or the rural-urban division. This is the conceptual model which I have utilised for much of my research into street vendors in Hong Kong and Southeast Asian cities (see McGee, 1973a . . .). . . .

If this conceptual framework is accepted then it is possible to suggest a model for the analysis of rural-urban and urban-rural mobility as set forth in Figure 1. It should be emphasised that the relative proportion of capitalist sector and peasant sector in the countryside and city is a diagrammatic representation of the number of workers in each system. It should be stressed that this model is primarily one of labour mobility, although some types of mobility such as retirement to the rural peasant sector from the urban capitalist sector may be regarded as a form of social movement. A further feature is that the rural capitalist sector would include plantations, mineral works and smaller towns between, 5,000 and 20,000 in size. There would be some rural peasant workers such as street vendors resident in these towns. In the urban areas there are some difficulties in placing some types of occupations in the correct sector. Thus, daily labourers who earn wages and also work, for instance, as hawkers would be in the capitalist system and the peasant system at the same time. But the majority of the workforce can be clearly placed in one of the four sectors. Government employees are also something of a problem but I have assumed that they are part of the capitalist sector.

Figure 1 suggests that there are five main types of rural-urban mobility. The first, A — B movement, is most frequently circulatory, and seasonal migration involving movement backwards and forwards between the peasant sectors of countryside and city. There will also be permanent migration in this type, in both directions. The second type of rural-urban mobility between rural and urban areas involves a shift from the capitalist sector of the countryside to the capitalist sector of the urban areas. In my work in Malaysia I found that transfer of government servants was a major component of this type of migration. Thirdly, there is the movement from rural peasant sector to urban capitalist sector. This was the major migratory stream occurring during the urban transformation of the Western countries. In the region of South and Southeast Asia at the moment this type of migration appears to be made up of two main types of migrants. The first type consists of migrants who have been educated out of the countryside; the second of unskilled marginalised population who have been forced off the land. Traditionally during the colonial period there was a certain amount of circulatory migration in response to recruited labour and this continues for instance in India. (See Bose, 1971.) There are potential problems of work adjustment in this type of mobility but in general in this Asian region they do not appear to have been severe. Movement from the urban capitalist sector back to the rural peasant sector is largely retirement migration. Finally movement from between the capitalist sector of the countryside and the urban areas is not common except at times when the prices of primary produce fall dramatically. This model, of course, takes no account of the mobility within the urban and rural sectors but this movement may be quite significant. Thus, my research on the occupational mobility of hawkers in Hong Kong [1973b] indicates a prior occupational history of work in the capitalist sector in factory work. There is also ample evidence of peasants in the countryside being recruited to work in plantations, etc.

This model is, of course, simply a static model which allows the description of various types of rural-urban mobility. Figure 2 is an attempt to present a dynamic model of possible migratory results that could occur according to the rates of growth of the workforce in the capitalist and peasant sectors over a period of 50 years. The assumptions upon which these projections are based are very simple and can be easily criticised but they are capable of modification. The assumptions on which the diagram

FIGURE 1. Structural Setting of Rural-Urban/Urban-Rural Mobility

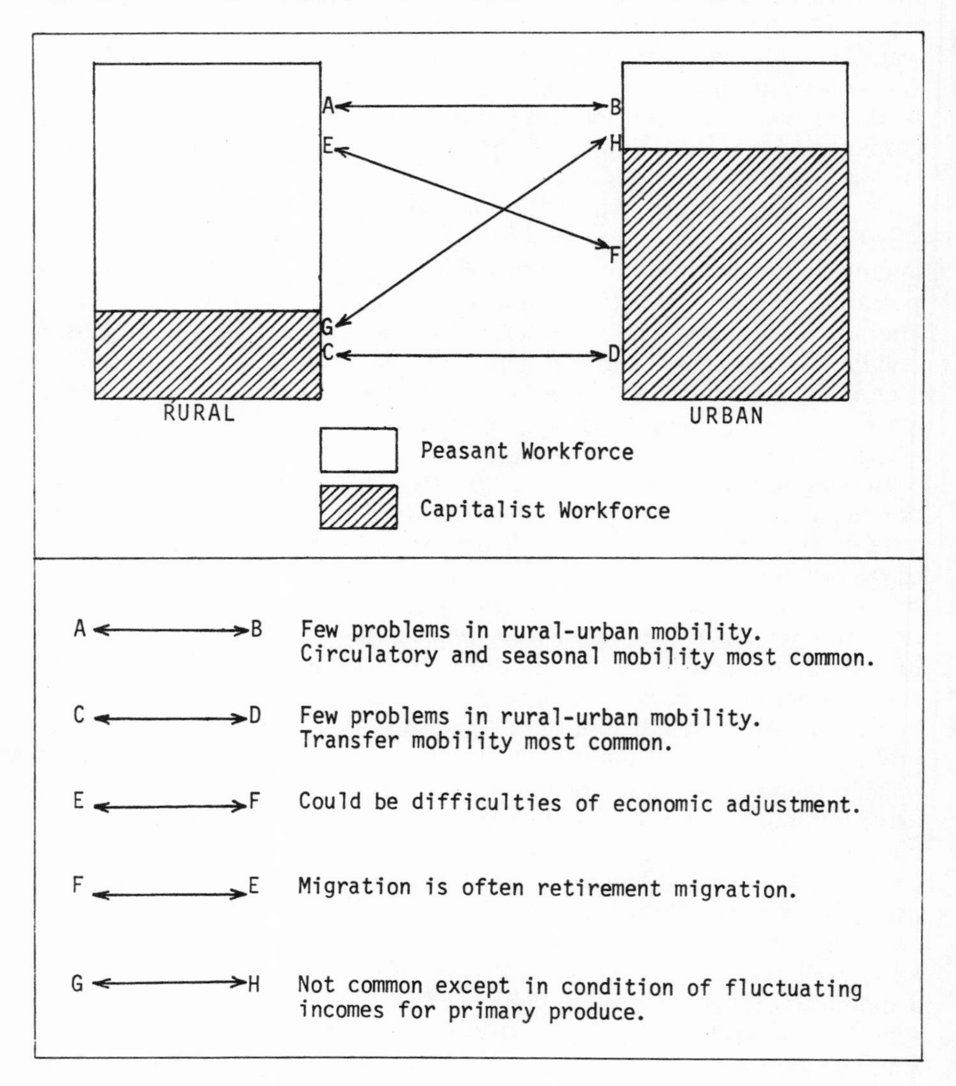

A ⟷ B Few problems in rural-urban mobility. Circulatory and seasonal mobility most common.

C ⟷ D Few problems in rural-urban mobility. Transfer mobility most common.

E ⟷ F Could be difficulties of economic adjustment.

F ⟷ E Migration is often retirement migration.

G ⟷ H Not common except in condition of fluctuating incomes for primary produce.

is constructed are as follows: (1) Equal rate of natural increase of workforce between rural and urban areas for fifty years. (2) A doubling of the workforce every 25 years.

The diagram presents three possible developmental sequences in a hypothetical country of 100 million workers in 1975. Structural setting A assumes that the rural and urban capitalist sectors will absorb a large proportion of the labour force increase as the urbanisation level increases from 20 to 60 percent. This would mean that 248 million additional workers would be added to the capitalist sector in this fifty year period while the peasant rural sector also grew. But the major migratory stream would be from the rural peasant sector to the urban capitalist sector. This developmental path could only be accomplished with massive industrialisation and a very high rate of economic growth.

Structural setting B assumes a much slower increase in the urbanisation level to only 40 percent and consequently the peasant sector in rural and urban areas would be expected to absorb much more of the labour increase. There would still have to be an expansion of jobs in the capitalist sectors both in the rural and urban areas of some 128 million. Thus, there would be a reduction in the volume of migration from peasant-rural to urban-capitalist but peasant-rural to peasant-urban would become more important. This developmental path assumes a slower rate of economic growth than A.

Finally structural setting C represents a truly involuntary situation in which neither the level of urbanisation nor proportion of workers in the various sectors changes over the fifty year period. In this highly unlikely situation the peasant sector would have to absorb a majority of the labour force. Even so the capitalist sector would have to provide 72 million jobs in a situation of very limited economic growth.

Of course, this dynamic model of rural-urban mobility takes no account of positive developments such as a fall in the birthrate and technological developments which allow greater agricultural productivity. Nor does it take into account the contradictions that emerge with the penetration of capitalism. The fact is that new techniques of agricultural productivity can force people from the rural peasant sector. The fact is that there seems little capacity in the expansion of capital-intensive industrial production to absorb more than a small proportion of the future labour force. Of the three developmental sequences it is possible to suggest that the migration pattern described in structural setting B is the one which most closely approaches the situation of the heavily populated countries described in the previous section on urbanisation in Asia.

CONCLUSION

In the introduction to this paper the historical experience of the Western countries was evoked. The description of the urbanisation and rural-urban

FIGURE 2. Dynamic Model of Structural Setting of Rural-Urban Mobility, 1975-2025

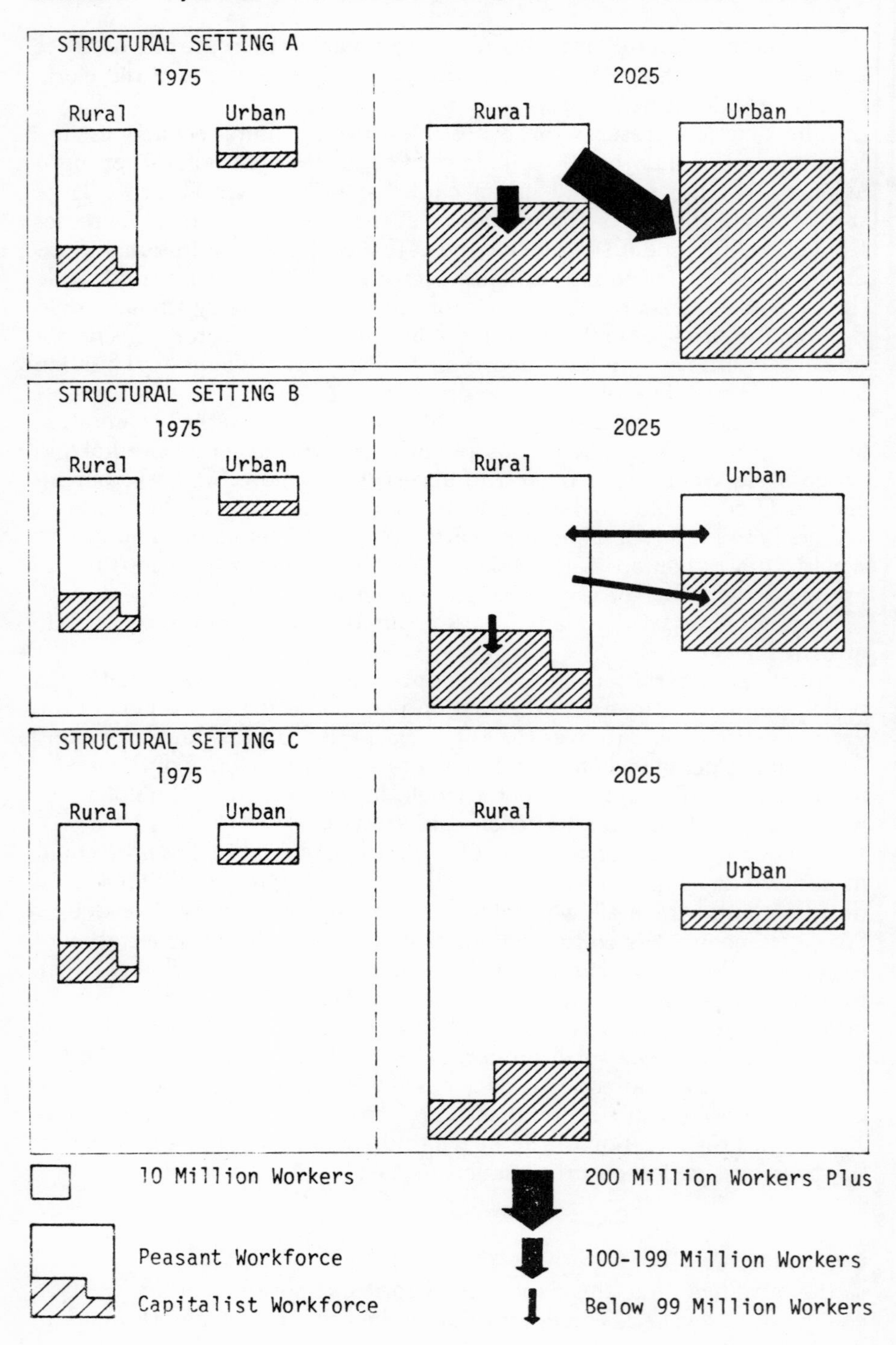

FIGURE 2. Dynamic Model of Structural Setting of Rural-Urban Mobility, 1975-2025

Assumptions

1. Equal rate of natural increase of rural and urban workforce
2. Doubling of population every 25 years

STRUCTURAL SETTING A

1975		2025	
1. Workforce 100 million		1. Workforce 400 million	
2. Urban workforce 20 million		2. Urban workforce 240 million	
Peasant workforce	12 million	Peasant workforce	48 million
Capitalist workforce	8 million	Capitalist workforce	192 million
3. Rural workforce 80 million		3. Rural workforce 160 million	
Peasant workforce	64 million	Peasant workforce	80 million
Capitalist workforce	16 million	Capitalist workforce	80 million

STRUCTURAL SETTING B

1975		2025	
1. Workforce 100 million		1. Workforce 400 million	
2. Urban workforce 20 million		2. Urban workforce 160 million	
Peasant workforce	12 million	Peasant workforce	80 million
Capitalist workforce	8 million	Capitalist workforce	80 million
3. Rural workforce 80 million		3. Rural workforce 240 million	
Peasant workforce	64 million	Peasant workforce	168 million
Capitalist workforce	16 million	Capitalist workforce	72 million

STRUCTURAL SETTING C

1975		2025	
1. Workforce 100 million		1. Workforce 400 million	
2. Urban workforce 20 million		2. Urban workforce 80 million	
Peasant workforce	12 million	Peasant workforce	48 million
Capitalist workforce	8 million	Capitalist workforce	32 million
3. Rural workforce 80 million		3. Rural workforce 320 million	
Peasant workforce	64 million	Peasant workforce	256 million
Capitalist workforce	16 million	Capitalist workforce	64 million

patterns of mobility that now prevail throughout much of the region presents policy-makers with a dilemma. This situation is obviously most dramatic in the heavily populated countries of India, Bangladesh, Sri Lanka and Indonesia. In these countries, if they are to push ahead with development goals formulated on the assumption of the development of capitalism, then the planners will accentuate the processes dislocating population from the countryside. On the other hand, if they try to restrict these processes they will retard economic growth and, while the involutionary capacity of these societies to absorb both rural and urban labour is considerable, it is not infinite.

In the face of this situation it appears highly likely that the experience of socialist Asian countries, notably China, may offer an alternative. For China has been very successful in both limiting the growth of urban population and creating rural employment. But this, of course, has been accomplished within the structural setting of a socialist society. Whether these same measures can be accomplished within a capitalist or even mixed economic system is very doubtful.

Marx was wrong in his prediction that employment-destroying innovations would exceed employment-creating innovations in nineteenth century Europe. But it does not follow that he will be proved wrong about the heavily populated countries of South and Southeast Asia, particularly as the demographic situation is now on his side.

CHAPTER 17

The Political Sociology of Cityward Migration in Latin America: Toward Empirical Theory

WAYNE A. CORNELIUS, JR.

Given the magnitude and ubiquity of the flows of internal migrants to Latin American cities during the past quarter century, it is surprising to find that the number of systematic studies focusing directly upon the attitudinal and behavioral consequences of the migration experience for Latin American political systems has been relatively small. There is a good deal of loosely descriptive, conjectural literature concerning the sociopolitical implications of urban migration, but the empirical research attempting to measure some of these implications, as far as specific types of attitudes and behaviors are concerned, has been quite limited. Granted that systematic social science research in this area is still in its formative stages, what directions can we expect it to take, and what are some of the obstacles to empirical theory-building in the field which will have to be surmounted?

Even a cursory examination of some of the recently published discussions of this topic provides partial answers to these questions. The continuing scarcity of hard data on attitudinal and behavioral correlates of urban migration in Latin America and other developing areas has led both developmental theorists and Latin American specialists to generalize freely from the urbanization experience of advanced Western nations in an attempt to deal in some fashion with the same type of societal change process in non-Western or underdeveloped contexts. Many of the theses and propositions derived from Anglo-European experience have come to be accepted as the working truth, forming a part of the conceptual framework of numerous government officials, politicians, and academicians—both North American and Latin-American—dealing with Latin American affairs. Constant repetition in innumerable books and articles (particularly textbooks) has given these concepts and propositions continuing currency,

---

Reprinted from *Latin American Urban Research* 1 (F. F. Rabinovitz and F. M. Trueblood, editors) © 1971, pp. 95–147, by permission of the publisher, Sage Publications, Inc., and the author. This paper has been edited for this publication.

despite early warnings by some investigators urging caution in the utilization of conventional theory not grounded in the empirics of Latin American experience. Latter-day field researchers working in this area of Latin American urban studies have been fond of isolating and attacking the various elements of "urban mythology" which have held sway over the past several decades (see Mangin, 1967; Leeds and Leeds, 1970). They themselves—and many of their Latin American colleagues writing in this area—seem to have largely rejected the models developed in conventional urban theory. But a glance at the stream of scholarly and popular writings on Latin American social and political development indicates beyond doubt that the confidence of many other Latin Americanists in these formulations remains unshaken. . . .

. . . To what extent is it actually feasible, and desirable, to draw upon conceptual frameworks and techniques developed in more "advanced" areas of social science in the study of migrant political behavior and attitude formation in Latin America?. . . And if the negative consequences of "borrowing-and-testing" do appear to outweigh the positive benefits, what alternative approaches to research and theory construction should be pursued in attempting to deal with problems of migrant assimilation in the Latin American socio-political context? These are the questions of general theoretical and methodological import to be taken up in this paper. With specific reference to the political sociology of migrant assimilation as a substantive research area, we shall attempt to (1) identify the origins of existing propositions and generalizations in this area; (2) assess the validity and explanatory power of such propositions in dealing with Latin American phenomena; and (3) suggest some possible reasons for the inadequacy of our "borrowed" propositions. Finally we will propose some alternative emphases or foci of empirical research and theory-building which might be adopted in developing this area of Latin American urban studies.

SOURCES OF EXISTING THEORY

There is no integrated, formal theory of migrant assimilation and political behavior which informs textbook discussions and other non-data-based treatments of the subject by Latin Americanists. What has been widely diffused through the literature is, rather, an amalgam of generalizations and propositions derived from the work of European social theorists, North American urban specialists and political sociologists, and political scientists in the American politics and political development fields. The borrowed propositions appear to cluster around three basic themes: (1) material deprivation and frustration of mobility expectations; (2) personal and social disorganization; and (3) political radicalization and disruptive behavior. Migrants entering urban centers are assumed to experience instigating conditions (1) and (2), become alienated toward the existing socio-political order, and undergo political radicalization leading to various forms of disruptive activity. Reduced to barest essentials, this is what has served as the most commonly accepted, "working theory" regarding political correlates of migrant assimilation in contemporary Latin America. Intervening variables are sometimes introduced, but these are generally regarded as mediating conditions which may influence the degree of radicalization or intensity of disruptive behavior, which are still regarded as the most probable outcomes of migrant assimilation into Latin American urban environments under existing circumstances. These circumstances, which are said to hold now and in the foreseeable future, include (1) the high rate and volume of in-migration; (2) the limited absorptive capacity of urban centers, capacity being defined primarily in terms of provision of housing, sanitation, educational facilities, public transportation, and other basic urban services; and (3) the large discrepancy between urbanization and industrialization rates, frequently referred to by economists as the condition of "overurbanization" or "hyperurbanization" (Friedmann and Lackington, 1967; Kamerschen, 1969), which ostensibly creates a severe shortage of employment opportunities.

Before attempting to assess the validity and explanatory power of the "conventional wisdom" in this area, it is helpful to retrace its origins in diverse bodies of theory developed largely without reference to Latin American phenomena. The major theoretical input has come from the so-called "Chicago School" of urban sociology and its European background. Within modern urban sociology in the United States, Louis Wirth's conception of "urbanism as a way of life" has been perhaps the most widely cited and influential theoretical orientation. The propagation of Wirth's ideas in largely unmodified form in major urban sociology texts down to the present day testifies to their enormous impact and acceptance as virtual unshakeable truths. In formulating his vision of the city and its life style, Wirth drew heavily upon the work of such European theorists as Durkheim and Simmel, as well as his predecessors at Chicago, notably Park. . . . The resulting description emphasized the anomic, disintegrative, disorienting, depersonalizing features of urban life and entry into the urban environment, manifested in psychological maladjustment, poorly defined social roles, breakdown of traditional value systems and controls on deviant behavior, the weakening of family and kinship ties, and the decline of religious life. The total picture is one of disorganization, normlessness, anonymity, social isolation, and insecurity. Such conditions are assumed necessarily to accompany exposure to the urban environment, and then incidences of various forms of personal and social pathology in areas heavily settled with immigrants (crime, delinquency, prostitution, alcoholism, divorce, suicide, mental illness, drug addiction, etc.) are listed as indicators. . . .

. . . In the late 1950s the models of urban society created by Redfield and Wirth were incorporated and elaborated by political sociologists concerned with the loss of identity, alienation, or anomie in urban-based, "mass societies".... These writers, who, like Wirth and Redfield, conceived of the urban way of life in terms of secularization, atomization of social relations, and the like, implied that the modern city is the quintessence of mass society, with its high potential for political radicalism and totalitarian movements. Kornhauser [1959] places special emphasis on the rate at which urbanization proceeds (the higher the rate, the more likely that the process will generate "available masses"), and upon the disintegration of intermediate social and political structures between cities and masses, making the latter susceptible to manipulation and recruitment into totalitarian movements of the far left or right.

Another type of theoretical input derives from the work of European social theorists and latter-day psychological frustration/aggression theorists, with respect to the attitudinal and behavioral consequences of material deprivation. Both Marx and Durkheim assumed that mass alienation and orientation toward protest would increase as a result of felt or relative deprivation (though not of mere *objective* deprivation, not perceived by the individual actor relative to some external standard of well-being). More recent theorists of revolution and civil violence have also drawn a fundamental distinction between objective and relative deprivation, the latter being defined primarily as the individual's perception of a discrepancy or "frustration gap" between the goods and living conditions to which he believes he is justifiably entitled and the amounts of goods and quality of conditions that he thinks he is able to obtain and retain....

Still another type of theoretical input has been provided by the literature on urbanization and political participation as developed by political scientists working in the American politics and political development fields. The American politics specialists were first to posit a positive association between degree of urbanization and degree of electoral participation and party competition. While the research findings presented to date are far from conclusive, and even contradictory, a number of studies based primarily on aggregate electoral and census data have demonstrated that in some American states urbanization does appear to have facilitated the development of attitudes conducive to higher electoral participation rates and competitive party voting patterns. The notion that urbanization produces an increase in political awareness, electoral participation, and demand-creation by bringing larger numbers of people into close contact with highly salient partisan political activity, a wider range of educational and industrial employment opportunities, and the mass media of communication also proved attractive to comparative politics specialists, who subsequently incorporated it into their multivariate models of social mobilization and political development....

The Adequacy of "Borrowed" Theory

The adequacy of conventional theory in this area can be determined in two ways: (1) by bringing to bear empirical evidence from field studies upon the specific propositions embodied in it, with a view toward verification or falsification; and (2) by examining the overall explanatory power of the theory in accounting for broad patterns of macro-phenomena of urban political behavior, as revealed in over-time data for nations and sub-national units. . . .

Collectively the studies . . . [suggest] that urban migration in Latin America does not necessarily result in severe frustration of expectations for socioeconomic improvement or widespread personal and social disorganization; and that even where these conditions are present, they do not necessarily lead to political alienation. Nor does alienation necessarily lead to political radicalization or disruptive behavior. Indeed, the apparent disjuncture between attitudes and behavioral predispositions among urban dwellers of migrant background is one of the most significant findings of those studies focused primarily upon political behavior (see Bonilla, 1961:7–9; Goldrich, 1964:332–33; Goldrich et al., 1967–68:14–15; Cornelius, 1969; Mangin, 1973). With few exceptions these studies find that urban migrants fail in most respects to conform to the usual conception of a highly politicized, disposable mass. On the contrary, the persistent non-politicization or even de-politicization of these sectors over time appears to be one of their most prominent characteristics as political actors (see Goldrich et al., 1967–68:14–16; . . . Roberts, 1968:201). There is as yet no firm evidence to support frequent expectations of markedly increased political cognition or development of participant orientations among migrants as a consequence of exposure to the urban environment. . . .

Even in those few reported cases in which urban experience has been accompanied by ample awareness of political and governmental activity and perception of a range of possibilities for extracting benefits from local political systems . . . [see Leeds and Leeds, 1970], it has apparently failed to induce a stronger orientation toward protest among the in-migrant population. Prominent among this sector are not large numbers of political extremists threatening to respond to persisting deprivation with "increasingly antisocial activities" . . . , but rather numerous politically moderate-to-conservative individuals seeking small-scale improvements in living and working conditions without recourse to violence or radical restructuring of the social order. Beyond these limited, short-term, highly localized objectives, the mass of migrants to Latin American cities apparently see little or nothing to be gained by direct political action aimed at influencing government decision-makers. Indeed, they have demonstrated relatively little inclination toward collective, broadly-based political activity on any front, let alone for "revolutionary" agitation against incumbent elites. . . . Under certain conditions they may become recruitable for such activity;

but on the whole they exhibit no well-developed sense of class antagonism toward the upper strata of society, do not see politics (especially partisan politics) as instrumental to the attainment of personal aims, and frequently acquiesce in regimes that sustain the status quo. After six years of field work among migrant residents of Lima's burgeoning squatter settlements, one investigator (Mangin, 1967:83; cf. Ray, 1969:157) concluded that

> a paternalistic ideology, combined with a "don't let them take it away" slogan, would be more appealing [to *barriada* dwellers] than a revolutionary "let's rise and kill the oligarchy" approach. Probably not many inhabitants of the squatter settlements would have regrets if someone else took the latter action, but they themselves are too busy.

Such attitudes undoubtedly reflect in-migrant perceptions of improvement in living conditions and life chances (however small by objective standards) usually experienced as a result of cityward migration, as well as their belief in the potential for future socioeconomic betterment within the ongoing system and their low level of tolerance for the risks and uncertainties of anti-system political action, in the face of their considerable investment of energy and resources in the acquisition and improvement of homesites on the urban periphery. . . .

Electoral data from a number of countries reveal that conservative *politicos* (including former military dictators such as Manuel Odría of Peru) have been supported more heavily in working-class districts of some of the large cities than in the hinterland, and that in most cases low-income urban settlement zones containing large proportions of migrants have not voted disproportionately for radical opposition candidates (nor abstained from voting in response to extremist directives), relative to other working-class districts or to all electoral subunits of the cities of which they are a part. . . . Moreover, these patterns have been observed in such rapidly urbanizing countries as Chile and Mexico, where real per capita income among the urban poor as a whole has actually declined in the post-World War II period. . . . The explanation for such phenomena clearly does not lie in prevailing conceptualizations of the attitudinal and behavior concomitants of rapid in-migration.

Some Deficiencies of Conventional Theory

A number of reasons for the inadequacy of conventional theory in this area could be suggested. Part of its inadequacy no doubt arises from the fact that many of the generalizations and hypotheses of which it is constructed were insufficiently grounded in empirical evidence to begin with. In many cases, it would seem, they were never adequately tested even within the

Anglo-European context. The work of the Chicago School of urban theorists is particularly vulnerable to criticism on these grounds. A number of less provincially-minded urban sociologists have pointed out the extremely narrow empirical base of the Park-Wirth *et al.* conceptualizations of city life and cityward migration, derived as they were from a limited number of case studies and impressionistic accounts of urban conditions, mainly in the United States. . . .

. . . The generalizations resulting from this research tend to be extremely time-bound as well as culture-bound; they appear to be no more empirically relevant to the emerging cities of Latin America and the developing world than to the North American metropoli of the 1970s. . . .

Still another type of deficiency afflicting conventional theory in this area stems from inadequate specification of concepts, and imprecision in their use. For example a single concept, "social disorganization," has been used for ordering much of the research and writing on migrant assimilation and urban problems in general. . . . However, . . . the comparative base of "organization" is never clearly specified in the literature. . . . While separate from questions of theoretical validity or explanatory power, such conceptual confusion markedly impairs the utility of conventional theory for research on urban phenomena in Latin America and other developing areas.

But perhaps the most serious deficiency of the conventional wisdom in this area lies in its broad conceptualization of the population flow into urban areas as an undifferentiated mass responding in uniform fashion to a given set of conditions or stimuli, to which all migrants to large cities are presumably exposed. This has led to research which attempts to gauge the potential for political protest, anti-system voting, or anomic agitation in urban centers by examining gross changes in urban population composition attributable to in-migration, often through the technique of ecological correlation. . . . Such generalizations are simply too gross. They leap from macro-societal processes of demographic change to conjecture about individual attitudes and behavior. Such inferential leaps might be justified if the results of macro-analysis could be tested against sample survey data. . . . In the absence of such data, however, there remain substantial doubts as to the long-term promise of research seeking to make statements about personal political predispositions or psychological traits of urban migrants on the basis of aggregate data for macro units such as cities, states, or provinces. . . .

. . . [L]et us return to the notion of the in-migrant population as an undifferentiated, disposable mass reacting uniformly to the urban environment. It turns out that in the Latin American context, relatively few individuals sharing the common experience of "cityward migration" actually go through the same type of migration experience. Some are direct rural-to-urban, or inter-urban, migrants, engaging in a single migratory movement from communities of origin to the city of destination with no extended intermediate steps. Others experience what should properly be described as step-migration, involving sequential movements from village to small town to medium-sized city to metropolis. It is probable that several types of population movement are occurring in varying degree in most Latin American countries. Certainly there is no single, generic pattern of cityward migration for the Latin American region. . . . In actuality the migrant population of a large city contains a large and disparate array of social types, who vary not only in terms of migration and socialization experiences but on a number of indices of pre-migration status and preparation as well. Moreover, there is evidence that in some countries socio-economic differentiation has proceeded to such an extent in the post-migration situation that distinct sub-classes or sub-cultures of in-migrants with different life styles, value orientations, and levels of subjective political competence are now discernible (see Peattie, 1968b; Ray, 1969:51–73, 172). Given such heterogeneity, how could we assume uniformity of response to urban environmental stimuli on the part of in-migrant populations? Yet this assumption clearly underlies both conventional theory on socio-political correlates of urban migration and much recent writing on Latin American urban conditions which has been influenced by it. . . .

Certainly there is little in conventional urban theory that would suggest the persisting importance of rural community orientations within the big city ambience. . . . Nevertheless, field studies have shown clearly that what might be termed "residual ruralism" mediates importantly between pre- and post-migration situations in Latin American countries. Germani (1967:179) has defined this as the "transference from the rural areas of institutions, values, and behavior patterns and their persistence or adaptation to the specific requirements of the urban setting." Thus we may find replicated among in-migrants traditional rural patterns of kinship and *compadrazgo* (godparenthood) relations, of religious observance, health practices, etc. . . . There may even be direct carry-over of attitudinal patterns such as leadership role expectations . . . and concepts of authority (Ray, 1969:50–51, 58–59). Such persistence may result from the effects of residential clustering of migrants from a given locality or region in urban settlement zones, or from the continued importance of small-scale urban production and service enterprises that keep in-migrants in a life style close to that of their rural counterparts. The urban work situation is often comparable to that of the communities of origin, and the social organizational arrangements surrounding it are reinforced by many of the same factors that operated in the rural setting. The same types of mutual aid networks and other arrangements which provided insurance for family survival in the place of origin continue to function. . . . Lewis (1965:495) has found that the *vecindad* or central-city settlement zone within the Mexico City metropolitan area tends to "act as a shock absorber for the rural migrants to the city because of the similarity between its culture and that of rural communities. Indeed, we find no sharp differences in family structure, diet, dress, and belief systems of the *vecindad* tenants, according to their rural-urban origins." Such findings have important implications for any attempt at theorizing about the political consequences of urban migration in Latin America. . . .

The phenomenon of rural carry-over illustrates the need for a theoretical orientation which places more emphasis on the ecological settings from which migrants have emerged and into which they have moved. Available evidence indicates that such factors as proportion of work force employment in large-scale industrial enterprises and other social, economic, and political attributes of the community of origin, the city of destination, and the particular socio-residential environments in which the migrant finds himself at various stages of the assimilation process, are likely to be of greater importance as determinants of behavior and life chances in the urban setting than the personal characteristics of migrants or internal characterological changes stimulated by the experience of migration. . . .

The squatter settlements which have developed in such profusion in and around major metropolitan areas throughout Latin America represent major points of concentration for in-migrant populations, although they may also contain large proportions of native-born urban dwellers . . . [Leeds and Leeds, 1970]. Field studies conducted in recent years have shown that these settlement zones represent a wide range of environmental properties which have important consequences for attitudes and behavior patterns among their inhabitants. The settlements have been found to vary along a number of dimensions, some of which are purely demographic in nature (e.g., size and density of population, stability of residence, and per capita land ownership), others of a physical, structural, or organizational nature (e.g., the type of terrain on which dwellings are built, relative security of land tenure, type of housing construction, presence or absence of formal voluntary associations, and extent of dependence relationships with external political and government agencies). Settlement zones may also be expected to vary importantly according to age, the process by which the settlement is formed (e.g., by organized squatter invasion, gradual, spontaneous accretion, or government initiation as part of an urban renewal program), proximity to employment sources in the central city, and access to basic urban services. Thus we may find a pattern of significant ecological variation among settlement zones within a single city. A squatter settlement may contain several hundred families, constituting practically a small community in itself, with a distinctive sub-culture of its own. John Turner (1965, 1967, 1968:116–20 . . .) has described sharply divergent sets of characteristics and functions particular to the "bridgehead" and "consolidation" settlements of metropolitan Lima. A similar pattern of ecological differentiation was revealed by a recent comparative study of marginal settlement districts in three cities in Colombia, Peru, and Puerto Rico (Rogler, 1967). The investigator found that even geographically contiguous *barrios* or settlement zones may differ sharply in the degree to which they and their inhabitants are integrated socially and economically into urban life. Some settlements are afflicted with recurrent waves of tension, antagonism, and conflict; others show no evidence of such stress. In each type of settlement the entering migrant is subject to a distinctive mix of socioeconomic pressures, social controls, and political influences. . . . The important point to be made here is that the set of environmental conditions or stimuli to which the in-migrant population of a city responds is by no means fixed and uniform, as conventional formulations would lead us to assume. There is clearly no such thing as a generic Latin American squatter settlement, or even a generic Brazilian *favela* or Peruvian *barriada.* Once again, given the diversity of conditions which has been found to exist, both among peripheral squatter settlement types and among the centrally located, high-density, "conventional slums" often inhabited by migrants in the early stages of urban assimilation, expectations of uniformity of response in terms of attitudinal and behavioral patterns seem rather fanciful. Yet the social science literature on contemporary Latin America and other developing areas persists in indiscriminately lumping together all squatter settlements as "tension-ridden zones of transition" . . . , "seed-beds of political unrest" . . . or "lost cities of parasitic misery where promiscuity, vagrance, prostitution, vices, and the worst crimes flourish" . . . ; places where "it takes only a clever and willful demagogue to whip into a frenzy the barely latent discontent" of the residents. . . . Given the importance of the internal differentiation of the Latin American city into distinctive subcommunities, such broad-brush descriptions of squatter settlements and the types of political appeals to which their inhabitants will respond remain mere statements of faith.

Insufficient attention to the squatters settlement as an agent of socialization and integration for migrants in Latin American urban centers, coupled with the traditional emphasis on social disorganization as an inevitable concomitant of "urbanism as a way of life" has caused Latin Americanists to overlook yet another theoretically relevant phenomenon: informal associational activity among in-migrant populations. Apparently many informal networks of social relations do exist that were overlooked or never searched for; networks which may serve to facilitate adjustment to the urban environment as effectively as formal organizational structures. For this reason it seems that, by relying upon participation in formal voluntary associations as a key indicator of "successful assimilation" into urban life (also a carry-over from U.S. urban sociology), the investigator actually engages in "stacking the cards" against the migrant. . . . Surveys in both Latin America and the United States have shown consistently that migrant participation in formal voluntary associations in urban areas is scant, and may even decline with longer residence in the city. By contrast, high levels of migrant involvement in informal networks or non-associational groupings within marginal settlement zones have been noted by anthropologists. This pattern appears to have developed most strongly in the Brazilian *favelas,* where numerous forms of both associational and non-associational group activity have been observed, including mutual aid networks, credit cooperatives, *ad hoc* systems for distribution of water and electricity, cooperative marketing arrangements, group activity to raise money to buy illegally occupied land, and a variety of recreational groupings. . . . Similarly intricate sets of informal organizational arrangements have been observed in the squatter settlements of Venezuela and Guatemala. . . . The existence of such informal networks, however short-lived and ineffectual they may be in terms of actual performance, is pertinent to our main theoretical concerns, for it indicates that disorganization and atomization of social relations as a basis for political radicalism may not be as severe a problem as standard conceptualizations of urban life and migrant assimilation would lead us to believe. . . .

[Political scientists] . . . have neglected a host of matters involving the operation of the political system itself which bear importantly upon the behavior of in-migrant populations. The study of comparative politics begins with the knowledge that different political regimes make different responses to widely experienced but similar problems. Rapid urbanization in contemporary Latin America is such a problem; yet with very few exceptions (e.g., Daland, 1969; Friedmann, 1968, 1969; Leeds, 1968) the extant literature contains little or no recognition of the possibility that differential responses to the problem are in fact being made by political systems throughout the region, and by specific administrations within single nations over time, and that such differences might be important in accounting for variations in migrant political attitudes and behavior. To what extent is political alienation or commitment to the system keyed to the absolute level of basic urban services and physical improvements provided to in-migrants? Or to governmental performance in coping with their needs for housing, education, and medical care? Or to the extent of deliberate efforts by incumbent elites and opposition groups to mobilize and integrate the newly urbanized into the political system in supportive or non-supportive roles? How are rates and directionality of migrant politicization affected by hostile or permissive administrative responses to the illegal land seizures through which most marginal settlements have been formed, or by subsequent government actions in granting or withholding tenure rights? How are internal organizational patterns and propensities for direct political action within the settlements affected by government programs and by attempts of local political party organizations to coöpt *barrio* association leaders and other key activists in such zones? . . .

Unfortunately we can do little more than pose these problems as a string of rhetorical questions, and hope that in future political research on cityward migration and its consequences they will receive adequate attention. It is doubtful that the explanatory power of theory in this area can be raised significantly unless such matters are satisfactorily accommodated in future studies. This will entail a drastic overhaul of conventional theoretical formulations—or a decision to abandon them entirely and concentrate on the generation of new theory from the data of political research. The field studies summarized above strongly suggest that the conventional models of urbanism and urban migration do not stand up to cross-cultural, or even cross-national, comparisons. Indeed, they appear to be of such low relevance to the empirical situation in most Latin American nations today that progress could be made most rapidly through a reorientation of research away from refuting or establishing limits on existing propositions and toward discovering new relationships and formulating new hypotheses. . . .

CHAPTER 18

The African Crowd in Nairobi: Popular Movements and Elite Politics

FRANK FUREDI

The history of Africans in the colonial urban centres is crucial for an understanding of the social dynamics of the nationalist movements. The absence of serious studies by social historians in this area leaves a major lacuna in African historiography. In this inquiry the political relationship among urban African social groups is examined with a view towards making a case for further analyses of the African urban arena.

A recent paper on politics in Nairobi has concluded that "participants in a mass movement, such as the independence movement in Kenya, are drawn from the most disadvantaged groups in society, whereas post-independence participation is a function of special skills, such as education and wealth" [Ross, 1970:23]. The exclusion of the great majority of Nairobi's population from political activity is today more evident than ever before. It reflects the elimination of popular movements as a political force by the ruling African political élite. The difference between the modality and goals of popular and élite politics is striking, and there is constant tension between the politics of these two groups in Nairobi. A historical examination shows that this tension has a long past, and is not just a result of a transformation from "independence" to "post-independence participation styles" [Ross, 1970:22].

The difference in political "styles" has its roots in the matrix of Nairobi's social structure. Two points about the specificity of Nairobi's development are directly relevant to an understanding of the social composition of its African population. Firstly, Nairobi grew up as an administrative centre servicing the needs of the colonial administration and the Uganda Railway, which arrived there in 1899. Industrial enterprises were notably absent, and consequently there was little demand for an industrial working class. (The Uganda railway was an obvious exception, however, though

Reprinted from *Journal of African History* 14, no. 2 (1973):275–90, by permission of Cambridge University Press. This paper has been edited for this publication.

most of the original railway labour force consisted of non-indigenous people.) Thus the Africans who came to work in Nairobi may be characterized as a nonindustrial class of workers, many of whom were casual labourers on short term contracts. In the early part of the 'twenties, they worked mainly in the service sector as domestic help, rickshaw drivers, and manual labourers for various government departments. Secondly, Nairobi, as an administrative centre, had a need for literate Africans to fill various lower level posts in the colonial bureaucracy. It was this group of Africans that formed the nucleus of the future Nairobi political élites.

Like most colonial cities, Nairobi was racially segregated. "Racial segregation was reinforced by a colour bar that was partly legalized by more than 100 ordinances and was partly social (denial of access to hotels, restaurants, and other private establishments, inequality of educational opportunities, pay differentials, etc.)" [Werlin, 1966:187]. Asians, Europeans and Africans lived in strictly defined residential areas. Africans were forced to live in Native Locations, and they were expected to provide housing for themselves. This led to the early growth of African villages such as Pangani and Maskini. Outside Nairobi, peri-urban settlements like Kileleshewa provided homes for workers and petty traders who commuted into town. Before the late 'thirties, housing schemes for African government employees did not exist, and all classes of Africans tended to mix with each other. Africans in domestic service, who made up roughly 30 percent of the African labour force, lived outside the native locations, on the property of their European employer. Owing to the lack of proper accommodation for Africans, there existed tremendous overcrowding in the locations. As a result, illegal squatter villages were established in the 'thirties. In 1938, the Municipal Native Affairs Officer reported that ". . . very large numbers of natives . . . for lack of accommodation are unable to sleep in Pumwani . . . [the official native location]. . . . It is estimated that some 600 are in River Road and Canal Road alone, living in undesirable squalor. About 2,000 . . . employed as casual labourers mainly at the quarries . . . live in huts, sheds, cow houses and goat pens near their work, with few or no sanitary facilities. . . ." [Annual Report, Nairobi District, 1938.]

As a result of this haphazard growth of African settlements in Nairobi, the government could do little to enforce many of the laws designed to control Africans in urban areas. . . . This lack of housing for Africans was symptomatic of the European attitude that Africans were temporary residents of the city, and that their presence would only be tolerated if they served the needs of the European community. The unequal wage structure virtually ensured that only the single African could afford to live in Nairobi. In his annual report for 1943, the District Commissioner clearly stated that Africans could not afford to bring their wives to the city. Social, medical and educational opportunities for Africans were also lacking. In

April 1941, the Nairobi School Area Committee passed the following resolutions:

> This Committee wishes to bring to the notice of the . . . Government the grave inadequacy of the provision being made at present for the social welfare of Africans resident in the Municipal Area, to the very serious overcrowding in the native locations, and to the great increase in prostitution resulting from these conditions.
>
> [*On the Housing of Africans in Nairobi,* 1941:4]

Despite the lack of adequate facilities for Africans in the city, numbers of people continued to migrate from the Reserves. By 1938 there was a considerable number of unemployed in Nairobi—a pattern that has been maintained to this day. Generally, three groups of people migrated to the urban area: the landless, or those with little land, who could not support their families from their small holdings; target workers, who hoped to earn a certain amount of cash and then return home; and divorced, widowed or barren women for whom Nairobi was an escape from the derision of their community. In Nairobi they became brewers of beer or prostitutes, or they took care of the children of European and Asian families. Until the mid–'forties, most domestic jobs were closed to women, and it was not until after 1952 that women came to constitute an important part of Nairobi's African labour force.

Nairobi was not seen as a legitimate arena for African political activity. Hence there were no institutions established for the expression of African political grievances. The need to administer Africans efficiently led to the development of a peculiar system of local government. The native locations were supervised by a Municipal Native Affairs Officer, a European who was appointed by the Municipal Native Affairs Committee of the Nairobi Municipal Council. The District Commissioner of Nairobi was the chairman of this committee. Initially, there were no African representatives on it. In 1939, the strikes and riots in Mombasa led the colonial government to modify its policy. . . . A Native Advisory Council was thus set up. This body of twenty-five members consisted of five Government nominees and twenty representatives of tribal associations and missionary societies. The Council had no legal powers—its main function was to present "native opinion" to the Municipal Native Affairs Committee. In 1945 the African Advisory Council was enlarged, and many of its members were appointed from the village committees that were formed that year. In 1946 the Advisory Council was allowed to appoint two African members to the Nairobi Municipal Council. It was not until 1958 that members were elected to the Advisory Council.

THE BEGINNINGS OF AFRICAN POLITICAL ORGANIZATION

Nevertheless, what becomes striking on hindsight is the speed with which the politicization of Africans in Nairobi took place. In 1921 the European settlers, seriously affected by the fall in world agricultural prices, tried to improve their position by bringing about a uniform reduction of all African wages by one-third. This proposed reduction had a radicalizing effect on Africans in Nairobi. During the month of June, meetings numbering from 3,000 to 9,000 were held to protest against it. At these meetings, Africans expressed their grievances against the colonial system, and threatened to return to the reserves if their salary were reduced. On 3 and 5 June, the rickshaw drivers went on one-day protest strikes. Two days later, some of the speakers at the political meetings formed the Young Kikuyu Association. This was just the first step. The reduction of wages also had the effect of breaking down traditional barriers. All the African workers in Nairobi, whether Kikuyu, Luo or Masai, were equally hurt by the cut in their wages, and the anti-wage reduction campaign brought members of many different ethnic groups together in a common fight. This newly born solidarity found organizational expression: the Young Kikuyu Association transformed itself into a trans-tribal East African Association (EAA). At a Young Kikuyu Association "sub-committee meeting held at Pangani Village on . . . 1st of July, it was unanimously resolved that the style 'Young Kikuyu Association' be changed to 'East African Association' as the latter term is of a more comprehensive nature" [*East African Chronicle,* 9 July 1921].

The EAA was aided by two forces. The Asian community was struggling with the European settlers over a number of political issues. As both Asians and Africans felt threatened by the political power of the settlers, a temporary political alliance was formed between the two groups in Nairobi. Asian organizational and material help benefited African political mobilization in Nairobi. The second important force was the emerging group of educated Africans. The wages issue, which affected them as well, gave the young educated Africans an opportunity to come to the fore as the representative of African opinion. They, more than any other group of Africans, had the skills which were a prerequisite for the establishment of a viable organization. The EAA was run by them. Their point of view was clearly expressed by their leader, Harry Thuku:

> . . . the E.A.A. has a better right to voice native opinion of Kenya than the Kikuyu Association [a group of chiefs and headmen] . . . because the former contains more educated people and also representatives of all tribes . . . than the latter and after all it is the educated portion of the commu-

> nity . . . that can with justification represent the opinion of the bulk of the community.
>
> [*East African Chronicle,* 20 August 1921]

By the end of 1921 the EAA was a force to be reckoned with. Its very success, however, was to lead to its downfall. The organization aroused such a favourable response from Africans living in the Reserves that the colonial government felt that its authority was threatened and decided to intervene. On 14 March 1922, Harry Thuku was arrested.

The arrest of Harry Thuku precipitated one of the greatest political demonstrations of African protest to colonial rule that Kenya ever experienced. On the day of the arrest a general strike was called in Nairobi, which proved 100 percent effective. Europeans found that house servants deserted overnight. Similarly, the staff of hotels, transport companies and government employees went out on strike. A crowd of 4,000 Africans who had gathered around the police station where Harry Thuku was kept was fired upon by the police, and there was a considerable loss of life. The bullets of the police effectively broke up both the demonstration and the strike in Nairobi. Troops reinforced the police and established order throughout the city. Soldiers surrounded the African village of Pangani, ready to quell any potential outbursts.

AFRICAN POLITICS IN NAIROBI, 1939–54

The activities of the EAA were stimulated by three important factors: the general reduction of wages, a temporary alliance with the Asian community, and the important alliance between the African educated and working classes. This unity, made possible by the reduction of wages, which had an impact on all classes of Africans, had only a temporary life and was never revived.

The politicization of Africans in the reserves through the activities of the EAA illustrated the importance of Nairobi within the political system of Kenya. For this reason the Government reacted swiftly to take control of the situation. The EAA was eliminated as a viable political organization, and in the aftermath of arrests and detentions the African population became politically demoralized. Pangani, the former headquarters of the EAA, and an example of uncontrolled indigenous urban settlement, was demolished in 1938. It would have been demolished much sooner had the municipal government earlier been able to find some form of housing for the people affected. From 1937 onwards serious steps were taken to rid the city of unemployed Africans—that year a number of them were returned to the reserves. By 1939 a detention camp in Nairobi held an average of 110 African "vagrants" a month. In the hope of establishing some form of tribal control over the city population, a Nairobi Native Tribunal was established in 1932. Seven years later, as a step designed to increase social control, the jurisdiction of the Native Tribunal was extended to include the hearing of offences against the Native Hut and Poll Tax Ordinance. According to a Government official, ". . . a steady pressure was kept up to enforce the payment of taxes by natives in Nairobi. Not only was this an aid to revenue, but where Nairobi was regarded as an easy refuge for the unwilling tax payer, the knowledge that tax receipts are as likely to be demanded here as elsewhere should serve as a useful deterrent to the idlers and tax evaders who enter the town" [*Central Province Annual Report,* 1939:22].

Another way in which government sought to ensure a degree of stability was by encouraging the growth of tribal associations. The policy of the administration was to encourage "Associations to be founded on administrative districts of origin rather than tribal bases." This materially contributed to the reinforcement of the fragmentation of the urban indigenous population on ethnic lines. In 1939 the government policy of establishing safe channels for African political expression led to the setting up of the Nairobi Native Advisory Council. This move was precipitated by the fear that the example of the Mombasa riots would lead to unrest in Nairobi as well.

The policy of encouraging tribal associations, strengthening the tribunal and establishing a Native Advisory Council led to the integration of a small group of privileged Africans in the colonial system. The Native Advisory Council was composed of leaders of tribal associations (many of whom were close relatives of chiefs and headmen in the reserves), representatives of missionary societies, and government nominees. Some of these men were later appointed to the Legislative Council. This small group (many of whom were in government jobs and enjoyed privileges like government housing schemes for civil servants, etc.) had, by 1940, become conscious of their interests vis-à-vis the colonial regime. While they suffered from various forms of discrimination, their privileged position defined the mode of their politics. An illustration of their political activity may be gleaned from the comments of the Provincial Commissioner of Central Province:

> In Nairobi the Native Advisory Council continues to do its good work. . . . The majority of members are English speaking and well educated. Sub-committees of the Council to deal with business on accepted Local Government lines were formed. One of them was to consider the estimates to the Municipal Native Trust Fund.
>
> It is hoped that this Council will be more representative and suffer less from the sense of ineffectiveness and exas-

> peration of being merely a consultative body. There is a strong desire to be given executive powers and the status of a Local Native Council.
>
> [*Central Province Annual Report,* 1945:10]

Essentially, the political activities of the Nairobi African elite were limited to trying to widen the powers of the African Advisory Council. In 1946, two of their members were appointed to the Municipal Council, a body which hitherto had had only European and Asian members. The inability of this group to obtain further concessions led this rather moderate body to take on a more critical tone. In 1950 the *Central Province Annual Report* stated that the "Members of the Advisory Council themselves feel that it is being used as a means of stifling Africans' ambition to have greater representations and weight in the City Council itself."

The Nairobi African Advisory Council held little significance for the African crowd, which numbered around 50,000 by 1941. Of these, 10,000 worked for the government, railway and Municipal Council. Roughly 22,000 worked for private employers, of whom at least 8,000 were in domestic employment. There were at least 1,500 Africans working on their own as tailors, masons, hawkers and petty traders. Those without any employment numbered between two to three thousand. At least 2,000 African traders (primarily Kikuyu) came daily to trade at the markets of Nairobi. It was estimated that the number of dependents of Africans in employment "was not less than 11,000."

A government sub-committee on African education described the Nairobi crowd as "generally unstable and migrant, predominantly male, badly housed, for the most part on wage levels where it is difficult for a wage earner to support a wife adequately." About 70 percent of the African population lived in the African locations, the rest lived in the residential or commercial areas. The largest single ethnic group was the Kikuyu. Particular tribal groups often came to dominate certain occupations. Most employees of the Public Works Department were Kamba, many of the skilled artisans were Luo, charcoal sellers were predominantly from Nyeri, barbers were from Fort Hall, and all of the municipal street sweepers were from Ndiya location in Embu.

Because of the lack of any social amenities for Africans in Nairobi, there was bitter popular resentment against the colonial authorities. Thus the Municipal Native Affairs Officer complained that police and civilian relations were bad since Africans were so often on the wrong side of the law—petty theft was rampant and the "enforcement of pass laws, registration certificate regulations and laws against congregations of natives is . . . very irksome to the civilian population and embitters its relation with authority" [*Nairobi District Annual Report,* 1941:25].

After the end of the Second World War, the tempo of African political activity greatly accelerated. In 1944 the Kenya African Study Union, later called the Kenya African Union (KAU), was formed. The organization was started by a group of "prominent Africans in Nairobi" who wanted to establish a rather select political body. Initially, this élitist political group had the approval of the colonial government. The party gradually came under the influence of younger and more radical educated Africans, however, and the organization came to articulate a more nationalist position. It is interesting to note that, despite a widely held belief, KAU never attracted widespread popular support. In Nairobi it was the party of younger, educated Africans, who sought reforms within the colonial system.

The most important manifestation of popular political consciousness was the African Workers' Federation. Founded in 1947, the strong appeal of the Federation was its synthesis of trade unionism and nationalist politics. The agitation of the African Workers' Federation had a radicalizing influence on the population. Crowds of up to 5,000 listened to Union leader Chege Kibachia at meetings held at the African locations at the beginning of 1947. This activity led to an increase of the Nairobi minimum wage in March of that year. The threat posed by the Federation's aggressive politics resulted in a vigorous counter-action by the colonial administration. When, later in the year, the Federation announced its intention of holding a general strike, the government deported the leader of the union and was able to divert the strike. It is noteworthy that Jomo Kenyatta, the President of KAU, called a mass meeting and dissociated his party from the strike. According to the Provincial Commissioner, this marked the turning point of the campaign and "from then on the African Workers' Federation steadily lost ground."

Another association of political significance was the *Anake wa 40,* or the 40 Age Group. This organization of Kikuyu came to play a central role in African politics in Nairobi. According to a government official, it "was nothing but a collection of hooligans and thugs ... their main activities were robbery, burglary and the entire senseless boycotting of the Municipal Beershops and Canteen" [*Central Province Annual Report,* 1947:1].

The *Anake wa 40* was founded by Dominic Gatu and a number of young, dissatisfied Nairobi Kikuyu. Its first members were the unemployed, petty traders, thieves, prostitutes and others of the lumpen proletariat of Nairobi. Many of the leaders were ex-soldiers and traders who felt that within the colonial system there was little scope for their skills and ambitions. Many of the leaders saw the 40 Age Group as the vanguard of the African nationalist movement. According to one of its leaders, "... we felt that K.A.U. was going too slow and that the only way to change things was through violence. This was why we started armed robberies. Most of the Africans in Nairobi were behind us and they would not inform the police of our activities" [Interview, 9 June 1972]. The rule of the 40 Age

Group in the African locations was reinforced by the widespread use of oathing and the physical intimidation of political opposition.

The intense bitterness of Africans against their condition in Nairobi protected members of the 40 Age Group from the police. Hence government officials were shocked when, in October 1947, Africans stoned police guarding stores after a fire, and when some stone throwers were arrested, they were rescued by large groups of armed Africans. Similarly, the District Commissioner of Nairobi reported that "The plain clothes squad of Police ... which was employed in the mopping up of spivs was stoned on the 20th July ... (1948) ... in Shauri Moyo Market and had to fight its way clear of the crowd" [*Annual Report,* 1948:10].

The administration tried to control the growing unrest in Nairobi by passing legislation to get rid of so-called "spivs" from Nairobi (the Voluntary Unemployed Persons Ordinance), by carrying out a series of police raids in the locations, and by demolishing the huts of illegal squatters at Kariobangi. Nevertheless Nairobi experienced further unrest when two of the leaders of the East African Trade Union Council were arrested in May 1950. This move led to the declaration of a general strike. The strike lasted for several days, and it took a heavy display of police force to end it. During the first days, the strikers tried to hold meetings in Pumwani; however, the police were successful in breaking up these meetings. Later on, groups of strikers went on demonstration in the residential areas, forcing non-strikers to join them. At this time fights broke out frequently between Kikuyu strikers and Luo non-strikers (many of whom were in government employment). Many Luo joined the Special Constables that were formed in order to quell the strike.

Two points about this strike are worth noting: the exacerbation of ethnic tensions among Africans, and the fact that most (relatively well paid) government employees and railway workers refused to join it. The colonial government, through its close association with the leaders of the Luo Union, was able to form an alliance against the strikers, who were primarily Kikuyu. The strikers, and the people who joined them, were lower paid workers, petty traders and unemployed, many of whom had close links with the 40 Age Group.

The 40 Age Group, through mass oathing, was now in the process of establishing its hegemony over the African locations. In this campaign it came into sharp conflict with the more moderate African élite. The attempted assassination of Councillor Gikonyo (one of the two African members of the city council) in 1950 was symptomatic of this conflict. . . .

The appeal of the 40 Age Group to the less privileged stratum of Africans was strengthened by the deterioration in their standard of living. During 1951 the minimum wage was increased twice to keep up with the steady rise in prices. The city's African Affairs Officer commented that "increases in the prices of necessities added shs. 8/- to the monthly budget of an African living in Nairobi," and felt that, though the minimum wage rate was twice increased, it was "barely enough to maintain a man at subsistence level" [*Central Province Annual Report,* 1951:7]. The rise in building costs led to an increase in rents. This resulted in the establishment of new shanty towns on the boundaries of the city near the African locations. One of these settlements, Buru Buru, became the headquarters of the 40 Age Group.

Between 1947 and 1951, the 40 Age Group had built up an organization which extended into all parts of Central Province and into some areas of the Rift Valley. Many of the leaders of the Group went out to the reserves and established branches in various administrative locations. This organizational drive was particularly successful in the districts of Nyeri and Fort Hall. The Group consolidated its links with the reserves by setting up a series of district committees in Nairobi. Group members from Nyeri, Fort Hall, Kiambu, the Rift Valley, Embu, Meru and Machakos who lived in Nairobi worked within their district's committee in the urban area. This was the basis of the system of informal government which the Group created in the African locations.

The organization also had close ties with a number of trade unions and with the most radical wing of KAU. Through its ability to obtain money and arms, the 40 Group played a central role in the movement which the government called the Mau Mau. It carried out armed robberies, and obtained a regular flow of funds from Asian and African merchants through an arrangement similar to a protection racket. In the eyes of the colonial government, the most notorious activity of the 40 Group was the political assassinations that it carried out. It was responsible for the killing of such important African loyalists as the Nairobi City Councillor, Tom Mbotela, and Senior Chief Waruhiu. The assassination of the latter led directly to the declaration of a State of Emergency in October 1952. This event was the culmination of a lengthy period of political activity by the 40 Group. The Provincial Commissioner of Central Province described the situation thus:

> ... crime reached a climax which was only checked by considerable reinforcements to the existing inadequate Police establishment; the terrorists were also concentrating on obtaining firearms. At this stage one must say that the position was akin to rebellion against the forces of Law and Order.
>
> [*Central Province Annual Report,* 1952:4]

The declaration of emergency, instead of eliminating the 40 Group, strengthened it. During the latter part of 1952 many of the more outspoken leaders of KAU were detained, and in the next year the party was banned. Few of the leaders of the Group were arrested, however, as most of them operated underground. The elimination of KAU left the 40 Group as the only African nationalist body in Nairobi. The banning of KAU had the effect of discrediting constitutional politics among many Africans, and vindicated the Group's policy of limited insurrection. During the early part of 1953 their main activity was to supply ammunition, medical supplies and recruits for the fighters in the forest.

The Colonial Administration endeavoured to check the activities of the "Mau Mau" in Nairobi by eliminating their grass root support. On 19 April 1953, the shanties in the Mathari Valley area were destroyed, and some 7,000 Africans were made homeless. This, and other large scale police operations, proved ineffective in breaking the organization of Mau Mau/40 Group. In fact, the tempo of their political activity increased during the latter part of 1953 and the beginning of 1954. The District Commissioner of Nairobi reported that "... large scale operations proved ineffective against the weight of numbers and the city was permeated by gangsters and other undesirables supposedly in unemployment.... Though extra troops were drafted into the city, crimes of violence persisted and it was clear that even more drastic measures were imperative if the Kikuyu preponderance and virtual control of the African and much of the Asian areas were to be combatted ..." [District Commissioner, Nairobi, *Annual Report*, 1954:1].

In response to this deterioration of security, the government launched one of the biggest military operations in the history of the colony. In April 1954, Operation Anvil was responsible for the removal of 24,000 Kikuyu, Embu and Meru from Nairobi. Pass books were issued to Africans in the city, and African housing was reorganized in the locations on a tribal basis. This operation effectively eliminated the 40 Group as a political force in Nairobi, and the colonial government was able to reestablish its control over the city.

The 40 Group, and other movements associated with it, represent the most successful populist political initiative in Kenya's history to this day. The Group was often looked upon as a Robin Hood type of band, and it could operate within the locations with complete immunity. The basis of its support was mainly among the Kikuyu, but its appeal was not restricted to one tribe, as the government often publicly made it out to be. One of the worries of the government was the fact that many Kamba and even some people from North Nyanza were members. In giving the reasons for Operation Anvil, the District Commissioner of Nairobi indicated that the "dissemination of Mau Mau doctrine amongst loyal tribes" was an important motive for launching the operation. Another important cause for the operation was that the sanctuary of the locations in Nairobi provided the "Mau Mau" movement with organizational flexibility. According to the District Commissioner of Nairobi, "The virtual seizure of the city by an unruly tribe without restraint and capable of the foulest atrocities constituted a threat to the whole Colony, for the city of Nairobi provided a ready and rich supply of recruits, arms, ammunition, money and other comforts for the terrorist gangs. It afforded a secure base for the planning and mounting of operations not only within the city but also in the adjacent settled areas" [District Commissioner, Nairobi, *Annual Report,* 1954:1].

The Group not only challenged the colonial administration, but also attacked loyalist African politicians. The assassinations it carried out obtained widespread support from the African population in Nairobi, illustrating the deep gulf that existed between the loyalist élite and the Nairobi crowd. Men like Jomo Kenyatta and Peter Koinange were hoping to bring about their goal of national independence through the means of constitutional politics. They were members of a class of young, privileged, well-educated Africans, whose frustrations with the colonial system led them to nationalist politics. But, in their roles of recipients of privileges, their interests led them to oppose the politics of popular movements like those of trade unions and the 40 Group. Thus we saw Kenyatta speaking out against the proposed general strike in 1947. Again, at a general meeting of the district officials of KAU held in Nairobi in 1952, a resolution advocating extra-constitutional methods was overwhelmingly defeated.

Nevertheless, the 40 Group was able to exert a radicalizing influence on KAU politicians, and, in Nairobi, the constitutionalists were not able to withstand the popular pressure for direct action. The more moderate politicians were intimidated by the wave of political assassinations and withdrew from politics. Early in 1952, the Nairobi KAU branch was taken over by the most radical wing of the party—and, from this point on, KAU in Nairobi had close links with the 40 Group. The overwhelming support that the radical nationalist leaders had in the city is reflected in the elections to the African Advisory Council. Led by Bildad Kaggia (one of the most respected radical leaders in Nairobi), this hitherto conservative body was taken over by members of the Nairobi branch of KAU, and the council became a political platform for African nationalists. In order not to be discredited and to keep in tune with the mood of the African population, many politicians were forced to take a radical stance. The speech of Kenyatta against European beer in 1952, which led to a boycott of the municipal beerhalls, is a case in point.

The politicization of the Nairobi African crowd may be illustrated by several cases of collective action. Kenyatta's call for the boycott of beer led to a drop in consumption from 13,000 bottles in July, to under 2,000 in September. Even more successful was the boycott of city bus services and

European beer, which was in protest against the emergency regulations. Within the locations, loyalist merchants were boycotted and were often forced out of business. The most significant indicator of the radicalization of the Nairobi crowd was that in the locations they effectively shielded members of the 40 Group from the police. In this context, it is interesting to note that by 1952 more than 90 percent of the Kikuyu population of Nairobi was oathed. At this time probably more than half the African population in Nairobi was Kikuyu.

Unlike the African politicians on the Advisory Council who argued for more trading licences and more representation for themselves on the Municipal Council, the 40 Group addressed itself to the day-to-day question of how to survive in an inhospitable environment. The immediate problems of the dispossessed in Nairobi were concretely linked to events in the reserves through the district committees in the African locations. This gave the African crowd the organizational opportunity to relate its political actions in Nairobi directly to the familiar conflicts within the locations in the reserves.

The leaders of populist movements like "Mau Mau"/40 Group were successful organizers because they lived near and with the African people. Unlike the educated KAU politicians, there was no gulf separating them from the people living in the locations. Most of the leaders had lived in Nairobi for a considerable length of time, and had obtained a certain level of education. Many were petty traders or taxi-drivers, who had a reputation for toughness and intelligence. Others lived by their wits, trading here, obtaining a temporary job there, and demonstrating little respect for legality. Because of their proximity to the African population, they successfully articulated the grievances of the people, and their power was based on their ability to represent the interests of the African crowd. After Operation Anvil, many of them joined the fighters in the forest and others were detained.

The period 1946–54 was one of the most intense periods of radical political activity in Kenya's history. This was due to the fact that an organized popular movement was in existence, led by a group of people who expressed the interests of Nairobi's African people. Often in Africa's colonial history, the nationalist party, led by educated middle class politicians, served as a mediating force between popular rural/urban radicalism and the colonial system. This is reflected in the ready acceptance of electoral politics and other rules of the colonial administration by the nationalist party. Through the use of strikes, boycotts, assassinations and other forms of direct action, the leaders of the 40 Group challenged the moderate politicians and directly confronted the colonial administration. They played a pivotal role in the organizing and financing of the "Mau Mau" throughout Kenya, and thus the role of the Nairobi populist movement in the nationalist struggle cannot be underemphasized.

THE ELIMINATION OF THE NAIROBI CROWD AS A POLITICAL FORCE: 1954–63

Operation Anvil had the effect of eliminating the Nairobi crowd as a significant independent force. For one thing, the complexion of the African population changed as 26,000 Kikuyu, Embu and Meru were removed and replaced by Kamba, Luhya and Luo people. . . .

Most of the people who replaced the expelled Kikuyu were fairly new to urban living and were relatively unpoliticized, as until 1956 all African political organizations were prohibited. In that year the Nairobi District Congress was formed by Argwing-Kodhek, a Luo lawyer. The Nairobi African crowd had lost most of their popular leaders, and to some extent it looked to a new class of politicians to present and defend its interest. The apparent failure of direct action against the superior forces of the colonial government made electoral politics the only feasible alternative. African political leadership in Nairobi was taken over by a well-educated group of younger leaders, many of whom had travelled outside of Kenya. Men like Tom Mboya, Argwing-Kodhek and J. Kiano were among the more prominent members of the new African political élite. Their political style was defined by the interest of their class—independence as soon as possible, with a minimum of structural change. Thus it was necessary to mobilize as many people as possible while limiting participation in the party to a relatively select number.

The most important African political organization in Nairobi during the period 1957–60 was the Nairobi People's Convention Party, led by Tom Mboya. The NPCP successfully mobilized large numbers of Africans during this crucial period of the nationalist struggle. While the party branch was dominated by a small group of politicians, large numbers of the Nairobi crowd were actively involved in the party's youth and women's wings. The political role assigned to the Nairobi African crowd was to be present at demonstrations and mass meetings. Throughout this period public meetings were held regularly and were well attended. The 1958 boycott of buses, beer and tobacco, called by the African elected members, was nearly 100 percent effective. In October 1959, thousands of Africans protested against the arrest of Tom Mboya, and the Riot Act had to be read. The year closed with serious riots between Asians and Africans. This wave of militancy of the Nairobi crowd strengthened the hands of the leading African politicians in their negotiations with the colonial government over the terms of the transfer of power.

In 1960 the Emergency was lifted, and many of the old Kikuyu residents flocked back into the city. There were some attempts to reintroduce populist political organizations. The Kenya Land Freedom Army found considerable support amongst the poorer Kikuyu, the unemployed and the illegal squatters living in the shanty towns. The main focus of the Freedom Army was to give the poor and landless Kikuyu some voice within the main nationalist movement, the Kenya African National Union (KANU). It acted as the pressure group of the poor, with a limited degree of success. One of its most important accomplishments was to obtain the support of KANU against the government decision to demolish the illegal squatter settlement at Mathari Valley in 1962.

After Kenya obtained its independence in 1963, the Nairobi crowd was eliminated as a political force altogether. During the first few months of the "Independence" era, the Nairobi crowd manifested its combustibility by demonstrating in the streets for more jobs and for some tangible benefits from *Uhuru.* The new government was very sensitive to possible unrest. The Triparite Agreement of February 1964, increasing jobs in the private sector by 10 percent and in the government sector by 15 percent, was a direct response to the demonstrations of the unemployed in Nairobi. The government also showed that it would not tolerate active protest—all public meetings were prohibited and a number of demonstrators were detained. This policy of strong urban control over the Nairobi crowd has been maintained to this day. The KANU government, like its colonial predecessor, does not tolerate any challenge to its authority. As a student of Nairobi put it:

> ... Since attaining independence in 1963, the political party and other structures aimed at mobilizing political participation have been relatively unimportant. The leaders of the independence movement are now the governing elite, and are no longer interested in encouraging wide-scale political activity.
>
> [Ross, 1970:10]

During the colonial period, the Nairobi African crowd played a central role in the fight for independence and helped the middle class African élites to obtain political power. In the post-independence period, the Nairobi crowd has not been an important factor in Kenya politics. One wonders when it will again be capable of undertaking independent political action.

SUMMARY

The absence of popular participation in the political process of post-independent Kenya should be seen as the outcome of a political tension, which has its roots in the colonial period. The growth of Nairobi, a colonial urban centre *par excellence,* provided unequal opportunities for its African population. The majority of the Nairobi Africans came to constitute the African crowd—domestic servants, the majority of workers in private and public employment, and petty traders. This group should be distinguished from the Nairobi African middle class which formed the "political élite." The African middle class possessed a fairly high level of education and had remunerative positions with government or were wealthy traders. By the mid-'forties, this group had become well integrated within the colonial system.

The different, and often contradictory, interests of these two groups of people were strikingly manifested on the level of political action. The "popular movements" of the African crowd were direct and often extra-constitutional. Their organizations, e.g., the 40 Group, were characteristically militant, and were often based on mass support. The "élite politics" of the African middle class were strictly constitutional and moderate. Their goal—to consolidate their position within the colonial system—had obviously only limited appeal. The conflict between these two social groups was resolved by the elimination of the African crowd as a political force.

JOBS AND CLASS STRATIFICATION

INTRODUCTION

Underlying the entire process of urbanization is the transformation of the system of production, the division of labor among the participants in that production, and the relative rewards they derive from the product. This is as true today as when Ibn Khaldun wrote it in the fourteenth century, before the Industrial Revolution, and when Karl Marx reiterated it in the nineteenth century, after that fundamental transformation. Therefore, the key issues in any analysis of urban problems, whether in the Third World or elsewhere, revolve around occupation and class. How do people make a living in the city? Within what economic system do they operate? And what are the consequences of their position in that system for their rank in the economic and political pecking order of society?

The first observation—frequently made with reference to cities in the Third World—is that people do not make much of a living. It is claimed that a great many people are either unemployed (idle), underemployed (working too little) or are engaged in activities which appear to be inefficient busy-work (marginally productive). In the eyes of Third World urbanites, too, making a living is their chief problem, but they would be amazed to hear themselves described in economists' terms, as unemployed, underemployed, or marginal. Scarcely idle, they are engaged in a hard daily struggle; but the struggle at best grants them a small and precarious return.

In this section we look more closely at both the intimate details of life of one participant in the "marginal tertiary sector" and the wider issues that create the context for her life. Lea Jellinek describes the work life of Ibu Bud, the proprietress of a mobile cooked-food cart in Jakarta. Certainly, hers is hardly a relaxed or lazy existence. She works eighteen to nineteen hours a day six days a week, is energetic and enterprising, and has built up a "thriving" business by organization, ingenuity and drive. Yet her prospects for rising above the lumpenproletariat are minimal, for everyday she faces the immediate threat of government punitive action against her illegal food stall and the longer term threat of being displaced by a mechanized chain operation. More than energy and enterprise, more than her "Protestant Ethic," are obviously called for.

Ibu Bud operates within a structure of larger economic constraints. Some of these are explored in McGee's paper on the proto-proletariat which tries to classify the types of activities within the so-called "informal" sector (or what he terms the peasant sector) of the Third World city and to place them in the larger setting of a national economy which is involuting. While he has enormous admiration for people like Ibu Bud and a deep appreciation for the real contribution they make to the economy of the city, he is not hopeful about their future nor about the future of their countries.

The analysis by Arrighi takes an even broader view of the question. He asks *why* independent African nations have been unable to provide productive employment for their populations, *why* people have been forced into occupations in the, if not marginally productive, at least marginally paying tertiary sector. He finds his answer not in a failure to develop industrially but in the type of industrial development which has been taking place. Looking at the international picture, he examines the role of the world system, and more specifically, the multinational corporations operating in conjunction with the bureaucratic elite and labor aristocracy of African nations, in fostering capital intensive production for external markets. This he calls "growth without development" for by development he means the raising of the general standard of welfare for all persons, not just those in the labor aristocracy.

It is one thing to discuss abstractly dependent cities and the labor aristocracy of wage earners; it is another reality when the topics are studied empirically through survey techniques and field work. The author of the final piece in this section, Paul Lubeck, lived for two years in the city of Kano, Nigeria, studying employees of the largest factories. As his article indicates, the situation viewed from this vantage point is far more complex. Not only are members of the so-called labor aristocracy drawn from a variety of sources—some urbanites from the traditional walled town, some rural-to-urban migrants, and some commuters who live in the rural area but work in the city—but they do not necessarily view themselves as permanent members of the industrial labor force. Many of them, indeed, have aspirations to go into the very tertiary sector so demeaned in the literature; they see a shift to petty trade as a *step up* in their own mobility trajectory, rather than as a demotion from the "labor aristocracy."

CHAPTER 19

The Life of a Jakarta Street Trader

LEA JELLINEK

Ibu Bud is a Jakarta food seller, a remarkable woman and a friend. Yet two years after I first met her she still remains an enigma. How was it that she thrived when food sellers all around her went bankrupt? Why didn't her helper, Ibu Nanti, run away from her again? Was it only because they were both married to the same man, or because Nanti knew that the police would again be sent after her to bring her back to work on the food stall? Was it determination and ruthlessness, or mere luck, that accounted for Bud's survival?

Jakarta has thousands of small food traders, all of them fighting for survival against a city administration and an officialdom who make their existence precarious. Ibu Bud is but one of these food sellers and in many ways an atypical one. She survived and thrived where others became destitute. This is an account of her life and her trading. I hope it conveys something of the vitality of a remarkable woman and gives some insight into the problems facing the poor people of Jakarta.

Ibu Bud provided an unusual service. No other food seller had the audacity to sell where she set up her stall each night. It was on one of Jakarta's main roads near a complex of army and government offices. And it was there late one evening—for she was a night trader—that I bought a hot lemon juice and subsequently returned again and again until I became a regular customer. I am sure I returned for the same reason as many of her other customers. She sold good food cheaply at a place and time when it was wanted.

To be precise she was breaking the law, for hawkers were not allowed to trade where she did. But in Indonesia, as elsewhere, one must distinguish between what is legal and what is permissible—and her activities were evidently permissible. For Bud that was all that mattered.

As a young, white, English-speaking female I stood apart from her

Reprinted from Working Paper No. 9, Centre for South East Asian Studies, Monash University, Melbourne, Australia, by permission of the publisher and the author.

other customers who were mostly low-ranking public servants and army officers. Within a few weeks of my discovering her food stall she invited me to visit her home, an offer I readily accepted.

Ibu Bud had a river-side house, less than five minutes walk from a road lined by modern buildings, banks and hotels. But that five minutes took one from a street that could have been a part of a modern city anywhere, into a *kampung*[1] that was more like a village in rural Java. One crossed a river to get to her *kampung* by means of a narrow, rickety wooden bridge. Everything that was carried into or out of the *kampung* had to be borne across that bridge or along a tiny network of muddy paths squeezed between houses. The choice depended on whether one wanted to push past children defecating into the river from the bridge or dodge puddles on the paths. I found myself wandering from shanty to shanty asking for Pak Santo, her husband. To my surprise no one seemed to know him. It took me over two hours to find Bud's house and it was only much later that I learned why nobody would show me the way.

Ibu Bud's husband had two other wives and a reputation as a lady killer. When I approached her neighbours asking for Pak Santo's home, they assumed he was up to his old mischief and, in loyalty to Bud, denied all knowledge of him.

Bud's house, like her food stall, was illegally situated. Perhaps the majority of Jakarta's poor live on land which they occupy illegally in the eyes of the city administration. In some parts of the city one could readily believe that the designation of areas as non-residential was determined by the financial interests of the officials concerned. Although Bud's home was centrally located and close to embassies, government and private offices the zoning of her land as non-residential seemed reasonable enough. It was a low-lying area and therefore at risk of flooding from the river that ran through it. Bud and her neighbours were flooded out many times each monsoon season and the water stains on the walls of their houses bore witness to this. But there was no prospect of Ibu Bud moving to legal ground. How could she compete with the businesses, embassies and the rich who all vie for what little land is available?

Bud was planning to build a new house, whatever the law might say. Her existing house measured six by six meters and was barely high enough to enable me, with my five feet six inches, to stand erect inside it. The walls were built of timber and other odds and ends, while the roof was of locally made tiles which could easily be pushed aside for added ventilation. Chicken wire covered the windows and helped keep out the hands of passing strangers. The floor was partly of concrete and partly of trampled down earth. There were two rooms. One meter above the floor level one's eyes encountered the water stains from previous floods. Above her

1. *Kampung* here means "urban village."

bed there was a tiny alcove which was reached by ascending a fragile wooden ladder. Normally the alcove served as a bed for Ibu Nanti and another female helper in the food stall. During floods it served as a retreat for some of Bud's household, including their much treasured cat, and storage for their few valuables. Those who couldn't find room in the house had to seek refuge in a neighbouring place. Fortunately the swirling mud and waters would subside in a matter of hours, but they would leave behind a stench, scattered debris and water-soaked walls. Bud and the others who shared her tiny ramshackle home had to accept this. Jakartans must be grateful for whatever land they can find legal or otherwise. Illegal land (*tanah liar*)[2] is at least not officially taxed. Of course this does not stop a local official demanding a small informal sum to protect against eviction, though it is doubtful whether he in fact has the power to do this.

The composition of the household varied from week to week. Ibu Bud was always there. Pak Santo, her husband, either stayed with Bud in Jakarta or even more often with his youngest wife who lived some thirty kilometers away in a small village. Ibu Nanti who was about thirty, younger, slimmer and much less dominant than Ibu Bud, was Pak Santo's second wife and another regular member of the household. Pak Santo's frequent absence meant that Bud required more reliable helpers for her food stall. They were recruited from the village of Pak Santo's third wife and comprised a teenage male and female. Ibu Bud and Nanti were the only two whose presence I could count on when I visited the house. If both wives were at the food stall I would still be certain of finding someone at home because the house was unlockable and Bud never allowed it to remain unguarded. As her helpers changed frequently I would often find a new face, alone, guarding the house.

Ibu Bud had a bed that could be recognized as such. She slept in a small cubby hole, with the alcove above and a wall to either side. It was suffocatingly hot. It must have been even hotter on those occasions when her husband Pak Santo joined her. He seldom brought his young wife to the city, but as mentioned the house almost always contained some members of the third wife's household or a neighbour from the same village. Though the municipal government had declared the city closed to migrants from the countryside in 1971, Bud's guests continued to arrive. The headman, who was meant to keep a check on people entering the village, was reluctant to forbid so-called relatives from visiting their families in the city. How was he to know who was a relative and who was not? He had his own guests from the village who came in search of work. How could he forbid others in the neighbourhood from having visitors?

Ibu Nanti slept in the alcove above Ibu Bud's bed. She kept all her worldly possessions in a box about the size of a portable typewriter case.

2. *Tanah liar*—literally "wild land."

Perhaps Pak Santo spent an occasional night with her, though I was never sure. Bud's mother was often in the house though she actually lived next door. It was a curious arrangement. Bud had once owned twice the floor space that she now lived on. But when Santo went off and married Nanti, she subdivided her house in fury and sold half to a neighbouring herb dealer (*tukang jamu*).[3] Now her mother sublet a partitioned corner of the herb seller's home.

Thus, Bud's house had two rooms divided by what was in effect a double bunk for her and Nanti. The front room contained the only furniture apart from her bed and cupboard: one table and two chairs. In one corner Pak Santo's work bench lay covered in odds and ends and a few radios he showed no signs of repairing. His radio repair business was far from thriving. It contributed a few broken loud speakers to the room's decorations, alongside the calendars with pictures advertising cars and faraway holiday resorts that were beyond the wildest dreams of the *kampung* dwellers. The village assistants and guests unrolled mats and slept on the floor of this room which was also used as a common sitting room.

The back room was the kitchen, laundry and store room. There were two kerosene pressure lamps, two kerosene stoves, a few pots and pans, some bamboo baskets and some jars of spices and odd foods. Yet, with almost no space and practically no facilities, the kitchen was used to prepare food for over a hundred people a day.

Each day a man came selling water carried in two large tin cans suspended on a pole, slung over his shoulder. This water was relatively clean but costly and used only for cooking and drinking. In the dry season drinking water became scarce and it was no longer brought to the house. Bud had to send one of her village assistants out in search of it. Water for washing was obtained from a privately owned pump which she and her neighbours paid to use. The toilets were strategically situated over the river course but in the dry season the flow slowed to a trickle and was scarcely enough to carry the excrement down the stream. The toilets were make-shift affairs erected by members of the *kampung* appointed by the local headman. They were washed away with each flood. Their low walls enabled the occupant to see out to the surrounding *kampung* and the multi-storey buildings of modern Jakarta beyond.

Each morning at eight from Monday to Saturday Bud set off for market. She was the centre of the household, handled all the money, did all the buying of raw materials and selling of food. Like most Indonesians she shopped daily, for she lacked the money to buy for more than a day at a time and had neither the storage space nor refrigeration required. She visited the same market stalls each day yet had to haggle over the price

3. *Tukang jamu*—herb seller.

of fruit, vegetables, live chickens and fish. Rice, *krupuk*,[4] bean curd, sugar and coffee sold at set prices, but even then Bud had to remain on the look-out for shoddy merchandise and shortweighings. By 11 a.m. she would be on her way back, often using the same *becak*[5] that had brought her to market and another hired to help carry the bunches of bananas, vegetables, rice and squawking chickens.

In those days Bud did her shopping at a nearby market but she may not be able to go on doing so. The municipal government has embarked on a program of modernization in various parts of the city. Old wooden market stalls are being replaced by concrete and wrought iron shops which are more hygienic and better organized. But their rent is much greater and their prices are higher. Before I left Jakarta Bud was already having to go further afield to find ingredients cheap enough for her stall. Of course she is not the only one to suffer from "modernization." Many of the old market traders cannot afford the new rents and so are being forced out of business. The outmoded markets once provided night shelter for thousands of poor people who served as guards for the market stall owners. The new multi-storey markets can be patrolled by a few paid guards and have thus rendered the thousands of former unpaid, unofficial watchmen and their families homeless.

Cooking began when Bud returned with the food from the market. The chickens were slaughtered and plucked, the bananas peeled, the vegetables diced and the rice cooked. Ibu Bud helped with some of the peeling and preparation but soon retired for a midday siesta, leaving Nanti perspiring over steaming and sizzling pots in the stuffy, cramped kitchen. The odour of fried fish and stewed vegetables filled the little rooms and penetrated into neighbouring houses where it met opposition from boiling fat in other kitchens.

By five in the afternoon the food was ready. It was neatly packed into saucepans and bamboo baskets or wrapped in cloth and then carried over the rickety bridge to Bud's mobile stall (*grobak*[6]). Pak Santo was meant to push the stall into position. Indeed Bud paid him a salary to do this. However, to her great annoyance he seldom arrived when required. When she had no male helper staying with her Ibu Bud would ask one of the neighbours to push the stall into position, and he could expect a free meal in return. The stall was set up in a side street, around the corner from the site where I had initially met her. From five to ten in the evening she traded in the side street where she was less likely to encounter difficulties with the authorities. Experience had taught her that by 10 p.m. it was safe

4. *Krupuk*—a delicacy made of shrimp and cassava flour or of fish flakes and rice dough, cut in slices and fried to crisp chips.
5. *Becak*—the pronunciation of this word is readily conveyed by the pre-1972 spelling as betjak; the word refers to a light passenger-carrying (or goods) vehicle pedalled from the rear by its driver.
6. *Grobak*—here refers to a mobile stall, elsewhere to an ox-cart.

to venture into a more important thoroughfare and the stall was moved into a much more prestigious site where no trader but Bud had the audacity to sell. As her stall was pushed to its second site hordes of *becaks* began to appear on the main road from which, under new legislation, they are banned before 10 p.m. and after 6 a.m. each day.

Bud did not arrive at the stall till it had been pushed to its side street location and prepared for selling. Then she came in style, brought by the same *becak* driver—Bambang—who had taken her to market and back in the morning. It would have taken a mere ten minutes to walk, but Bud, as the head of the enterprise, had to set a certain standard. I doubt that she paid Bambang, but he was often to be seen enjoying a meal at her stall, a treat beyond the means of other *becak* drivers. It seemed an informal arrangement between her and Bambang. If a fare had taken him to the other side of Jakarta, Bud would not wait beyond the appointed time but found another *becak* driver. Mostly however Bambang was able to ferry her to and fro. The relationship was not confined to business matters. When Bud visited the village of her husband's third wife to celebrate the young bride's pregnancy, Bambang was among the friends she invited along.

Most of Bud's clientele were regulars. Public servants had a meal after finishing their day's work in the nearby offices and army, police and civilian guards from the various military installations, banks, offices and embassies called by throughout the night. Bud knew what many of them wanted without their asking and they were each served with a serviette, tea and food. They could buy cigarettes of many varieties, hot lemon juice and beer. For an ordinary trader she offered a great range. Her predominantly lower middle class customers could afford to spend twice as much on a meal as most people who lived in *kampungs* like Bud's. She was insistent that food be served politely and elegantly and reprimanded any of her assistants who failed to keep the standards she set. For instance when one of her village helpers forgot to serve tea with a meal, threw a serviette casually across the food stall to a customer or handled food with the wrong hand, Bud commented on how bad mannered or stupid these villagers were. Bud did all the selling and handling of money. She tried to keep her prices stable, for she feared variation would drive her customers away. As the cost of raw materials fluctuated she adjusted the size of the portion she served. She would fill a plate with rice at the expense of the quantity of the vegetables or mix Indonesian rice with a cheaper but less tasteful American variety. She did this so cunningly that few of her customers noticed.

Some of her customers received an extra large helping for which they paid no more. They were a few select army officers whose friendship she cultivated. Some of them would warn her of impending "Trader Clearing Operations" (*pembersihan dagang*) and it was only such a warning or the heaviest thunderstorm which could prevent her from trading. The authorities had a campaign to remove traders from many fashionable streets of Jakarta like the one she was in. But her contacts enabled her to avoid having her goods impounded during such clearing campaigns.

Many traders I knew watched helplessly as army men loaded their stalls and wares on to military trucks, as part of a city clearing operation. The ransoms demanded for the recovery of the traders' possessions were often almost as much as they were worth and almost always well beyond the traders' means. For the average small trader the goods he carried either over his shoulder or in a mobile cart represented his total business investment. The army men were not merely depriving the small traders of their means of earning a living, but confiscating all their capital as well. For the really poor traders it was even worse. They had no capital and the goods they sold at each day's trading and even the push cart or shoulder basket in which these goods were carried belonged to someone else, the trader merely renting them on a daily basis. Thus, the confiscation plunged him into debt from which he might never escape. Some traders frantically turned to an alternative source of credit to pay back their first debt and then had to rapidly find a third source to pay back the second and so on. Others turned to crime. Yet others who were unattached shifted to another part of Jakarta, adopted a new identity and avoided their debt.

Traders had only two means of defending themselves in the face of such trader-clearing campaigns. By far the commonest method was to gather one's goods and run at the first sight of a clearing team. Traders resorting to this method accepted the loss of whatever they dropped whilst fleeing, in preference to having all their goods confiscated. Bud used a less common but more successful means. She relied on her army contacts to forewarn her of a planned campaign and simply did not appear on such days. The army officers who gave advance warning were ensuring they continued to receive cheap, good quality meals at their place of employment. Occasionally these same army officers had just returned from a trader-clearing operation in another part of the city when they sat down at Bud's stall and talked of their day's work. They appreciated that the traders they raided were simply trying to earn a living. The officers however had to protect their own livelihood by carrying out orders.

Bud was reasonably safe from the official campaign so long as the right officers wanted her services. But she still had to contend with the intimidation of individual military men. The latter knew that Bud was trading illegally and tried to use this knowledge to their personal advantage, to gain a free meal or more if they could. Few traders could withstand the threat of having their plates and cups confiscated and so reluctantly gave in to a soldier's blackmail. But not Bud. She would ask for the soldier's papers to confirm that he was on an official campaign. If this failed to deter him she would engage him in an argument and threaten to take him to court. She would assert that she was fulfilling a useful public service, as all her regular customers could testify. The army man would reply that, public service or not, she was breaking the law, and so the argument would go on. Bud knew she was fighting a losing battle, but she was not trying to win. At the first sign of trouble she quietly dispatched one of her assistants to summon support from her sympathetic, more senior army men, so she was arguing simply to delay action until help arrived, as it invariably did. Bud was out to win friends, not enemies, and voiced no bitterness towards the defeated and humiliated soldier. Several who had tried to intimidate her became favoured customers and added to her band of protectors.

Ibu Bud was the prima donna of the enterprise. Each night she appeared well dressed and spotlessly clean. Her large moonshaped shining face, oversized false front teeth and rounded figure dominated the scene. She chatted, joked and argued as the occasion demanded, sometimes describing how she had been bullied by a policeman or how she had witnessed crimes carried out on the main street at night. At other times she would talk about the plight of one of her customers or mention that a baby had died in her *kampung*. She would announce her plan to build a new house or state that she wished to visit her husband's third wife and newly born babe in the village. She always introduced me to her customers, giving them a brief description of who I was, where I had come from and what I was doing in Jakarta. In turn she always told me who each customer was and what he was doing in the city. If Bud thought that one of her clients was particularly interesting or nice, she would insist that I meet him at her food stall one evening and would set a convenient time.

Ibu Bud carefully served meals, giving each customer his portion measured by his influence and prevailing market prices. She unobtrusively collected the money which was carefully stored in a little tin beneath the counter. No one else had access to this tin and she alone knew the amount it contained. Only regular customers were allowed credit. Bud insisted that new-comers always paid for their meals before departing.

Many customers saw Ibu Nanti as an unimportant if not lazy helper of the sparkling and energetic Ibu Bud. They had no idea that Nanti and Bud shared a husband. They had not seen Nanti sweating over the saucepans in the suffocatingly hot kitchen earlier in the day, whilst Bud took her siesta. They did not realize that Nanti had to help the men push the cart into its position each day. They had never heard Bud confess that the business depended on Nanti's ability to prepare tasty meals in that cramped kitchen. All they saw was an exhausted, though attractive and again neatly dressed assistant, falling asleep propped up against one of the poles of Bud's stall. Occasionally she came to life to cook some more rice, heat some water or wash dishes.

The trading usually continued until three or four in the morning, by which time most of the perishable food had been sold. Then the stall had to be cleaned and pushed home and all the valuable pots and pans, kerosene stoves, cigarettes and beer carried back across the rickety bridge into Ibu Bud's house. It would soon be dawn, and at eight Bambang her *becak* driver would call to take her to market for the start of another long day.

Ibu Bud's life is hard and her existence precarious, but she told me that for the first time in years she had a real prospect of improving her lot. It was difficult for me to judge for my only knowledge of her background came from her, her husband, her mother and Nanti. Her story contained its share of the implausible but I doubt she ever consciously deceived me. Furthermore, what seemed implausible to me was to her and her fellow traders often a commonplace.

Ibu Bud had trained as a nurse in Tegal, the coastal town in central Java where she was born. She had received her training during the days of the Independence struggle and I was never sure of her actual qualifications. Her father was said to have been a policeman under the Dutch and after Independence. For a short time Bud worked as a nurse, first in her home town and then in Jakarta. It was here that she met and married her husband Pak Santo, though in fact he also came from Tegal. Pak Santo was stationed in Jakarta after having served with the Indonesian army in the Outer Islands. He had met many Australians during the war and proudly claimed to have sung for them in the harsh jungles of Borneo. While in the army Pak Santo learned what proved so useful in the construction of Ibu Bud's mobile stall. Bud and Santo's fortunes fluctuated markedly. Their one and only baby died in infancy and they returned to Tegal brokenhearted. They soon drifted back to Jakarta in search of work. Santo and Bud both had a variety of jobs through the 1950s and sixties, sometimes prospering, sometimes struggling to make ends meet. For a while Bud and Santo stayed with relations and it was here that Bud first observed the running of a rice stall. Her initial business experience came somewhat later when she was employed in a small Chinese pharmaceutical company where her ability was soon recognized and she was rapidly promoted. However Bud did not like being an employee; she wanted to be her own boss. She had left nursing, she said, because of the awkward hours of work and she soon left the pharmaceutical company when Pak Santo had a lucky break and his *becak* repair business began to earn well. For the first and only time in her life, Bud was a lady of leisure.

Santo's prosperity and Bud's more relaxed life were short lived and both were soon searching around for new ways of making a living. Somehow, despite all these ups and downs, the couple managed to acquire enough money to rent and later buy the house in which they still lived, though they now occupied but half the area they once owned. Pak Santo's final business venture, repairing radios, collapsed with the influx of cheap Japanese transistors when import restrictions were eased in the mid-sixties. His interests turned to women and he began to spend an increasing amount of time away from home. Bud first learned of Nanti's existence after Santo had married her. Bud was furious and demanded a divorce, but Santo would not hear of it. In her anger she sold half the house that she and Santo had struggled so hard to buy. Bud had a way of letting me know if she wished to keep something secret and I never learned how she spent her time whilst Pak Santo was out wooing Nanti. Bud certainly was embarrassed about her childlessness and it was only Santo's suggestion that they could both obtain a child by the existence of a second wife that made her soften to the idea.

Ibu Nanti was part Chinese and part Javanese. She was born in Surabaja and had trained there as a hairdresser. She married in her mid teens but by the time she was 17 her husband had left her, taking their two year old child with him. Nanti felt wretched and desolate. She left Surabaya without any aim in mind and ultimately drifted to Jakarta. By then she was determined to build her life anew. She worked as a hairdresser by day and, in the evenings, sold banana fritters from a rented food stall. It was here that she met Pak Santo. He seemed kind, understanding and sympathetic and, like Nanti, in need of someone he could trust. She married him quite unaware of Bud's existence. She was shattered when she discovered it.

I never understood what it was that made Bud invite Nanti to come and live with her. The two women claimed that they had united in their mutual resentment of Santo's deception. As a second and younger wife, Nanti was considered subordinate to Bud and Bud's dominant personality reinforced the relationship society expected. It was uncommon to find two wives sharing the same roof in the way that Nanti and Bud did. But Bud was pre-eminent in the home as well as in business affairs. I was "her" guest and she became jealous when I was too friendly with Nanti or, indeed anybody else in the *kampung.* Nanti was made to feel inferior in every way. Bud handled all the finances and Nanti claimed that she never saw the share of the takings she was meant to receive. Bud occasionally bought Nanti some batik or shoes, but these were always handed over as gifts rather than payments for her labours. Nanti's eyes troubled her greatly, but Bud never provided money for her to visit a doctor. Even Jakarta's public hospitals require money, for the bewildered *kampung* dweller cannot possibly find his way into the appropriate clinic through a maze of corridors, without paying a bribe to be shown the way by a nurse or hospital attendant. Furthermore, Nanti was needed to do the cooking, and Bud could not spare her the time to wait in hospital queues.

The relationship in the stall, where Bud served the food that Nanti had prepared, happily claiming credit for it, was reproduced in the home. Nanti was generally more gifted with her hands and made clothes for both of them. Yet it was Bud who would parade in her new clothes whilst Nanti sat quietly in the background. Bud would entertain me in the front room whilst Nanti sat cooking next door. If I gave Nanti anything she hid it so that Bud would not become jealous.

Though both women told me they disapproved of polygamy they must have had very different reasons for doing so. The marital roles were quite dissimilar. Nanti seemed more like a servant who had had an affair with her mistress's husband than a second wife. Her quiet dignity and appealing personality were only apparent when the two of us were alone. At other times she was subordinate to Bud and seemed resigned to her fate. She had once tried to run away. Nanti, Bud and their neighbours each independently told me the same story. Somehow the police found Nanti and forced her to return. The neighbours felt that Bud's business would have collapsed without Nanti and suggested Bud had bribed the police to enforce Nanti's return. Nanti never received cash from Bud and had no relatives in Jakarta. The only people she knew were Bud's friends. There was almost no escape. Nanti was destined to remain the unpaid helper of her husband's first wife.

On the other hand, Bud's prosperity was largely dependent on Nanti's labours and skill though she never acknowledged it publicly. Certainly Bud's personality and business acumen were also important but her partnership with Nanti was more successful than any of Bud's many independent ventures had been since she first arrived in Jakarta almost thirty years ago. It was easy to see why Bud may, indeed, have bribed the police to force Nanti to return but much harder to understand why Bud remained loyal to Santo. Bud faithfully paid Santo his salary even though he seldom turned up in time to push the *grobak* to its site. For a long time I thought that Bud's attempt to manipulate and dominate Santo, a technique she practiced so successfully with almost everyone else she encountered, was her most notable failure. In retrospect I am not so sure. True, Santo retained an almost wilful independence but I think Bud had her reasons for accepting his behaviour.

She had never fully recovered from the death of her baby shortly after she married Santo and deeply regretted her inability to have further children. Perhaps she had tolerated Santo's indolence whilst she and Nanti earned a livelihood in the hope that Nanti would conceive. But Nanti did not become pregnant and now Santo had a third wife. Both Bud and Santo told me that before marrying again Santo had promised that any child his third wife might have was to be brought up as Bud's child. Santo had also promised to divorce his young bride if she proved infertile.

Bud took charge of the marriage as of everything else. She paid for Santo's trips to and from the village. She selected the midwife, chose the baby's name and arranged for the appropriate religious ceremonies. Bud fed and clothed the baby. She referred, patronisingly, to its mother as "a silly little girl." Bud felt her training as a nurse and her responsibilities as the baby's provider entitled her to give detailed instructions on the baby's care. Perhaps, I came to think, Santo's role in Bud's world was to provide her with a child and she had come to accept his independence as the price she had to pay.

I often wondered what it was that enabled Bud to prosper when so many other traders I knew were either struggling to remain solvent or actually going bankrupt. Undoubtedly Pak Santo's *grobak* was essential to her success. The authorities would not have tolerated a permanent stall where Bud traded yet her customers were seeking greater variety and quality than the usual mobile store could offer. So ingeniously had Pak Santo constructed his *grobak* out of the left-overs of his old *becak* repair business that Bud could compete with the permanent stalls and restaurants a few hundred meters away. Bud was again fortunate in that, unlike so many of her fellow traders, she owned rather than rented her stall. Charges for even a modest *grobak* were high and would have greatly diminished her profits.

Bud's childlessness was a major misfortune both in her own eyes and in the opinion of those who knew her. But it was an important factor in her ability to save. Bud's household was unusual in having so few dependents. Her mother was self supporting. Her adopted daughter had been kidnapped. Whilst the business was being established Bud and Nanti's earnings only had to support one additional person—Santo. His expenditure was limited to a fixed salary. Nanti and in later years the stream of village assistants were paid in kind. Bud provided adequately though not generously and her personal expenditure was modest. Now of course Santo's new bride, the baby and the new bride's family in the village will make demands on the earnings of Bud's stall and financially things will not be quite so easy.

I was intrigued by the success of Bud's enterprise. Her secret seemed to be in her ability to manipulate to her advantage any opportunity that offered. She had organizational abilities admittedly, as had been recognized when she worked in the Chinese pharmaceutical firm. Bud's organizational flair however could not account for her ability to pitch her stall where no one else dared, to evade raids on traders, to withstand and even befriend soldiers who came to intimidate her, to exploit to the full the priority society conferred on the first wife and, perhaps her most audacious feat, to have a child by proxy as it were.

I used to speculate whether Bud's self-confidence could have been related to her alleged *prijaji*[7] background. It was a formidable asset whatever its origins. I think, however, it was her ability to manipulate people without antagonizing them that made her so effective. Whilst other traders over-extended their credit and went hopelessly into debt, Bud's customers knew they would get nothing without paying for it but Bud always managed to make them pay without any heated words being

7. *Prijaji* means aristocratic or upper class Javanese.

exchanged. Bud also managed to stand firm against her would-be intimidators. Few of them held it against her afterwards yet I doubt if those same soldiers would have tolerated similar behaviour from many other small traders. I was saddened by Nanti's subordination and unable to tolerate Bud's refusal to pay for the long overdue attention to Nanti's painful, reddened eyes. I knew Bud never intended to pay back the money she borrowed from me to buy a sewing machine. She may have been dominating, jealous and exploitative but she was also immensely likeable and vital and I came to regard her as a friend. I believe it was Bud's ability to remain likeable and interesting even though she was domineering, manipulative and, at times, ruthless that lay at the basis of her success.

For all her good fortune Bud's life was in many ways typical of the Jakarta poor. She was simply one of the hundreds of thousands of migrants who had crowded into Jakarta since the Independence struggle of the late forties. Bud seldom, if ever, returned to her home town but she still referred to it as "home." Perhaps her life there had been more stable and secure than in Jakarta where she lived on illegal land and had worked in so many jobs over the years. But a high proportion, if not a majority, of Jakartans live on land that has been declared illegal and it is exceptional for them to have regular employment over an extended period of time. Bud's stall has now been established for several years and, if all goes well, she will soon have a very good house by the standards of Jakarta's poor.

Her success came only with great effort. She worked eighteen or nineteen hours a day six days a week. She did little hard physical labour but had to cope with all the worries of running her business. She knew her survival depended as much on good luck as hard work. Perhaps one day someone will forget to warn her and she will lose everything to the army's insatiable trucks. Even if she avoids such a catastrophe, I think, her prospects look grim. She will find it increasingly hard to compete with shops as her cheap source of supply, the "poor man's markets," are progressively eliminated. Her patrons are fickle. If, through illness, Bud fails to appear for more than a few days another trader could rapidly take her place and be at least as eager to please Bud's former patrons. If Bud's home was demolished she may not find alternative accommodation sufficiently close to enable her to push her *grobak* to its selling point and back each day. If present municipal policies are pursued and *becaks* eliminated, Bud could neither transport her purchases home from the market nor her produce to the *grobak* even if she manages to keep it in the center of the city.

Yet for all this Bud feels more prosperous than she has ever been and, at last, has a baby which she refers to as her own. Her food stall is—and her new house will be—the envy of her *kampung*.

CHAPTER 20

The Persistence of the Proto-Proletariat: Occupational Structures and Planning of the Future of Third World Cities

T. G. MC GEE

> The lower strata of the middle class—the small tradespeople, shopkeepers, and retired tradesmen generally, the handicraftsmen and peasants—all these sink gradually into the proletariat, partly because their diminutive capital does not suffice for the scale on which Modern Industry is carried on and is swamped in the competition with the large capitalists, partly because their specialised skill is rendered worthless by new methods of production. Thus the proletariat is recruited from all classes of the population.
>
> Karl Marx, *The Communist Manifesto,* 1848

INTRODUCTION

Six in the morning is the correct time to experience Jakarta. A haze rises from the streets in the early morning light, softening the harsh rectangles of the large hotels and office buildings and giving a rich hue to the ubiquitous orange tiled roofs of most of the houses. At this time the city is just beginning to move. A few bicycles and *becaks* (tri-cycles) are on the streets. Pedestrians are few. *Pedagang keliling* (hawkers) are already bouncing through the streets with their peculiar hopping rhythm accentuated by the wildly bobbing baskets at either end of their carrying poles. The shrouded forms of the *gelandongan* (the homeless) on the pavements are awakening; other people emerge from the shelter of plastic covers which they have erected over bamboo frames. This is the time when the inhabitants of the *kaki-lima* (the pavement) own the city. By seven the

Condensed from a paper presented to the Comparative Urban Studies and Planning Program, School of Architecture and Urban Planning, University of California, Los Angeles, April 1974. Reprinted by permission of the author.

city is no longer theirs. The motorcar, bus, lorry and motorbike command the main streets and the pedestrians compete for the pavement. The police begin to chase the hawkers and the battle for the city's streets has begun. The inhabitants of the *kaki-lima* are only one of the many groups within the city's population who inhabit specific ecological niches. There are those who dig platforms barely wide enough for a sleeping human being into the banks of canals, erect a tiny shelter, and cling precariously to the almost vertical sides. There are others who erect housing alongside railways, reaching almost to the tracks and there are those, more fortunate, who live in bungalow housing in the suburban tracts which have begun to sprout on the fringes of the city. There is also a prosperous Chinatown with architectural styles that come straight from traditional China. Most of the population live in a mass of sprawling *kampongs* (villages) characterized by housing of varied standards and legality, and served by a minimal infrastructure of streets, sewerage and water.

This picture of Jakarta is not conjured up to evoke nostalgic memories of similar ecological and human dimensions in other Third World cities, but rather to set the scene for the theme of this paper which seeks to describe and analyze the features of the very substantial portion of these cities' populations who are engaged in individual or family enterprises and who are not regular salary or wage earners—a group neither proletarian nor peasant whom I have elsewhere labelled the proto-proletariat (see McGee, 1973a). This definition is, of course, too simple to describe this fluctuating group of people who range from skilled artisans to bootblacks, for their delineation is greatly complicated by conflicting concepts of class, work, and residence. Therefore one purpose of this paper is to attempt to disentangle these diverse concepts and to present what is hoped to be a stronger definition of this population.

Historically, of course, there is nothing unique about the existence of the proto-proletariat for they formed important sectors of the occupational structures in the cities of the developed world as the industrial and urban revolution blossomed in the nineteenth and early twentieth centuries. For instance, Charles Booth's description of Petticoat Lane, London, on a Sunday in the eighteen nineties, is repeated in the crowded street markets of many contemporary Third World cities. . . .

Assumptions concerning the features of the proto-proletariat changed little in the last one hundred years. The nineteenth century writers recorded the all-pervading poverty of this population, their low productivity and the degrading conditions of their low standards of living. Similar viewpoints are expressed concerning the proto-proletariat of the contemporary Third World cities to such a point that there is no need to quote these writers. Implicit in the assumptions of the nineteenth century writers was the view that the proto-proletariat would eventually largely disappear, absorbed either into the bourgeoisie or into the proletariat—hence the quotation from Marx which heads this paper. . . .

. . . While this is certainly happening in some Third World countries . . . it has been the point of much of my previously published work to argue that the urbanization process occurring in many of the Third World countries, while still leading to the growth of cities, is substantially different in terms of the mix of its demographic, social and economic components (see McGee, 1967, 1971a, 1971b . . .). Because of this difference I would suggest that there are good reasons to presume that the proto-proletariat may persist for very much longer periods as a substantial proportion of the occupational structures of the majority of Third World cities than was the case in the developed world. This is not a comforting picture either for the proto-proletariat, whom it would seem must persist in conditions which are generally characterized by poverty . . . or for the planners of these countries, many of whom are dedicated to hastening this redistribution, or finally for the capitalists who surely would wish to see a growing consumer market for their products.

It is perhaps surprising that it is only recently that attention has begun to be focused upon the features of the proto-proletariat. . . . One might suggest at least three reasons for this lack of research attention. First, among many economists, it seems that their models of development tended to concentrate upon the economic growth features of the modern sector at the expense of the activities of the proto-proletariat. Part of this attitude was the consequence of assuming that the only way for poor countries to "develop" was to repeat the historical experience of the developed countries. . . . Thus only the economic activities of the modern sector could be conceived as growth-promoting, while the . . . "informal sector" was construed . . . as an inhibiting sector. To the Marxists and Neo-Marxists its existence was a symptom of the poverty-creating capacity of international capitalism. To the "similar path writers" who favored rapid growth within the capitalist structure, its existence was a blockage. Neither polar position, it seems to me, was willing to explore the question of whether the "informal sector" contributed to economic growth.

A second reason is that the modern-traditional dichotomy, which has been a central conceptual model for the analysis of social and economic change, has encouraged a one-sided emphasis on the process of development as a penetration of traditional systems by the "elements" of "modernization" which are thought to be equated with development. The major limitation of this approach . . . has been its portrayal of traditional society as a static, homogeneous culture with a consistent body of norms and values which are opposed and antithetical to modernization. . . . The activities of the proto-proletariat, despite the fact that they often involve highly ingenious adaptations and utilization of traditional technology or practices, have been ignored because they have been portrayed as traditional and thus of little interest to the process of social and economic development.

Finally, one may point to the fact that the systems of statistical enumeration of occupations have been most inadequate for recording the activities of the proto-proletariat. Most existing systems of enumeration tend to record information pertinent to the economic and service activity of the wage-earning population of the modern or formal sector.... This technique was generally transferred to the various countries of the Third World. Its limitations for the underdeveloped countries are obvious since it is largely based upon the assumption of employment or work in which the majority of the labor force are "employees," and does not account for those forms of "work" which are performed in household enterprises in both agricultural and non-agricultural situations. Thus Hauser, Myrdal and others have argued strenuously for a new system of occupational/labor force data collection in the Third World (see Hauser, 1971, 1972; Myrdal, 1968:II:994 et seq...).

These statistical data collection systems have been further rendered inefficient by the fact that they often collect data only from enterprises which have some form of legal recognition.... Therefore the activities of the proto-proletariat have never been enumerated satisfactorily, further contributing to our ignorance of their role and contribution to the economy.

In the late nineteen sixties, however, there appeared to be a growing realization of the importance of the size and activities of the proto-proletariat. Again, while it is not always easy to identify the causes of this growing realization, one may suggest at least several themes. First, there has been, in some quarters, a growing disillusionment with development as being conceived solely as "economic growth...." [T]here has been a growing interest in the features of the labor-intensive systems of the peasantry and proto-proletariat which do not rely on the high energy consumption of "fossil fuels."

A second and perhaps more important factor in the change of attitude has been the growing focus on what has been labelled the "employment problem" of the less developed countries.... In addition to high rates of unemployment among the urban populations, often young and relatively well educated, there is "... a reserve army of low-productivity workers in rural and urban areas, among whom the waste of human potential is massive" (Turnham and Jaeger, 1971:10). What is more, this pattern does not appear to vary greatly between the countries of the Third World which have experienced rapid economic growth and those which have had slower rates of development.... A series of studies has focused upon this problem in the three continents of the Third World, and the statistics they present, even subject to the problems of the applicability of data, all tell the same story—a picture of a rapidly increasing labor force within both rural and urban areas, increasing rates of unemployment, and absorption into low-productivity informal activities in urban areas which will accelerate in the decades of the nineteen seventies and eighties in most of the Third World countries. . . . The reasons for this situation are well known; a combination of population explosion, urban growth and limited employment opportunities generated by capital-intensive industrialization have all combined to create the problem. . . . Whatever the reasons, the end-product is that the "employment crisis" has resulted in a growing concern with the development of employment strategies for the countries of the less-developed world which has increasingly had to accept the fact that certain activities of the proto-proletariat are likely to play an increasingly important role in the cities of the Third World, and therefore there is a need to focus research upon them.

Finally, one may suggest that a very important reason why the proto-proletariat's activities have come to be recognized has been the radical revision of concepts concerning the process of class formation in the less-developed countries. Hardly surprisingly this has emanated primarily from a group of writers, principally in Latin America and Africa, who represent a stream of broad Neo-Marxist thought, perhaps most forcefully expressed in the writings of Fanon (1963) but analyzed more structurally in the writings of Frank (1969b) and Worsley (1972). In particular, the writings of the latter author have focused upon the significance of the emergence of the semi-proletariat (proto-proletariat) in urban centers as a major element in the class structure of urban areas. . . . While there has been much debate over the role of this group, the importance of the analysis has been to break down the rigid Marxist categories of bourgeoisie, proletariat, and lumpenproletariat by stressing the emergence of this new class—victims of "urbanization without industrialization" (Worsley, 1972:209)—as well as a questioning of the repeated emergence of a large middle class which was associated with the urban revolution in Western countries.

Basic to all these factors which have led to a growing awareness of the role of the proto-proletariat are two sets of polarized assumptions which rest upon an assessment of the demographic underpinnings of the current urbanization process in the Third World. At one pole is a group of writers who see the process of occupational formation occurring in Third World cities as basically involutionary; that is to say, a combination of very high rates of natural increase in both countryside and city and a large volume of actual and potential rural-urban migration involved in the urban transformation place grave demographic constraints on the possibility of labor absorption in high-productive capital-intensive sectors, thus forcing people into the low-productive activities which proliferate and involute. . . . The second group of assumptions may be described as evolutionary; that is to say, while they accept certain of the demographic underpinnings which have earlier been enumerated, they are more optimistic about the progress of occupational formation as one in which there will be sectoral lags in the rates of labor absorption in certain sectors (for instance, industry), which will cause a growth in the "informal sector." This process will be temporary and, eventually, occupational mobility and economic growth will correct these sectoral imbalances in a manner which will produce occupational structures not dissimilar to those of the developed capitalist world (see Kilby, 1971 and Garlick, 1971).

In fact the gross figures of predicted urban and labor force increase in the less-developed countries strongly support the first of these positions.... [F]or the majority of countries, particularly those with large, dense populations such as India or Indonesia, the problem is a grave one for the future, unless successful strategies of labor absorption can be developed. Thus, if the proto-proletariat and their rural counterparts, the peasantry, are going to persist, it is imperative to learn more about the features of this population....

THE DELINEATION OF THE PROTO-PROLETARIAT

... Broadly one may suggest that there are three economic dimensions by which one can seek to delineate the proto-proletarian population of the city. First, structurally it is possible to identify the proto-proletariat as carrying out their activities primarily within ... one sector of a dualistic model of the economic structure of a Third World city.... Geertz ... in his work on a Javanese town ... drew attention to the dual economic structure of the town divided between a *firm-centered* economic sector ... and a second sector labelled the *bazaar-economy*, based on "... the independent activities of a set of highly competitive traders who relate to one another mainly by means of an incredible volume of *ad hoc* acts of exchange" (Geertz, 1963a:28). Lately this model has been greatly broadened by the work of Milton Santos (1971, 1972) who has stressed the relationships between these two sectors in the form of two interacting and interlocking circuits of economic activity summarized in the accompanying Table 1. The types of economic activities found in the upper circuit are banking, export trade, modern large industry, modern services, wholesaling, and some forms of transport (airlines, etc.). The lower sector consists of non-capital-intensive industry, services and trades, etc. There are, of course, intermediate types of activity but these do not invalidate the model. Independently, apparently, economists associated with the International Labour Office have also suggested a model of the dualistic structures of the economies of the less-developed countries which, while it is applied to the total economy, can be equally fitted into the urban economy as well. Thus they distinguish between an "informal sector" of the economy characterized by:

TABLE 1. The Two-Circuit Economy Characteristics

Characteristics	Upper Circuit	Lower Circuit
Technology	capital-intensive	labor-intensive
Organization	bureaucratic	generally family-oriented
Capital	abundant	scarce
Hours of work	regular	irregular
Regular wages	normal	not required
Inventories	large quantities and/or high quality	small quantities; poor quality
Prices	generally fixed	generally negotiable betwen buyer and seller
Credit	banks and other institutions	personal noninstitutional
Benefits	reduced to unity, but important due to the volume of business (except luxury items)	raised to unity, but small in relation to the volume of business
Relations with clientele	impersonal and/or through documents	direct; personal
Fixed costs	important	negligible
Publicity	necessary	none
Reuse of goods	none; wasted	frequent
Overhead capital	indispensable	not indispensable
Government aid	important	none or almost none
Direct dependence on foreign countries	great; outward-oriented activity	small or none

This table was adapted from Santos (1972).

(a) Ease of entry
(b) Reliance on indigenous resources
(c) Family ownership of enterprises
(d) Small scale of operation
(e) Labor-intensive and adapted technology
(f) Skills acquired outside the formal school system, and
(g) Unregulated and competitive markets

and a "formal sector" of the economy of which the principal features are:

(a) Difficult entry
(b) Frequent reliance on overseas resources

(c) Corporate ownership
(d) Large scale of operation
(e) Capital-intensive and often imported technology
(f) Formally acquired skills often expatriate, and
(g) Protected markets (through tariffs, quotas and trade licenses).
(International Labour Organization, 1972:6)

While this division is broad, and is surprisingly little developed as an analytical framework in the I.L.O. publication, from the point of view of our previous arguments concerning the proto-proletariat, they are clearly identified as participants in the "informal sector" or lower circuit of the Third World cities' economies.

A second approach to defining the proto-proletariat is an institutional approach which classifies the proto-proletariat on the basis of the mode of production in which they are engaged. This is an approach which I have favored, following Chayanov (1966) and Franklin (1965, 1969), which distinguishes among three main systems of production—capitalist, socialist and peasant—in which:

> . . . the fundamental differentiator is the labour commitment of the enterprise. In the peasant economy the individual entrepreneur is committed to the utilisation of his total labour supply—that of his family, who may and do often find alternative or additional sources of employment. This accounts for the diversities of historical peasant societies, but if these sources are not available the *chef d'entreprise* must employ his kin. . . .
>
> In the capitalist and socialist systems of production labour becomes a commodity to be hired and dismissed by the enterprise according to changes in the scale of organization, degree of mechanisation, the level of market demand for products. . . .
>
> (Franklin, 1965:148–49)

The advantage of such a distinction between systems of production is that while it recognizes the fact that the majority of peasant enterprises may be family firms, it does not exclude the fact that peasant enterprises, such as family-organized industry or retailing, may exist in cities. This is what Franklin means when he says:

> The scheme ignores the agricultural-industrial division. At this level of generalisation the division has little importance. Agriculture is carried out under all three systems of

> production and so is manufacturing. Admittedly, in peasant-folk-traditional societies, agricultural activities predominate without excluding the appearance of a capitalistic system of production—but it is the systems, not the societies, which are our concern.
>
> (Franklin, 1965:148)

Following this definition most of the proto-proletariat of the Third World cities fit clearly into a peasant system of production. There are, however, problems with this definition in the city situation, because members of a household may often be operating in both modes of production. For instance, among the hawkers in Hong Kong it was common to find that members of a hawker's household might be engaged in factory work but were still working in the hawker establishment in their spare time. Another variation of this model has been devised by Friedmann and Sullivan (1972) combining a labor force and institutional model of Third World cities that presents a tri-sectoral model of the urban economy divided between individual operators, which includes the "open unemployed" and occupations such as street vendors, casual laborers, etc; the family enterprise sector which includes household enterprises; and the corporate production sector which includes employees of government, large firms, etc. While I think this model presents very useful possibilities for analytical insights, it also is not without its problems of fitting populations into the various categories.

The third approach is that suggested by Hart which I now think may offer the most possibilities for a fruitful research method of delineating the proto-proletariat. Hart's work in Ghana has shown that there is a wide range in the size and scale of activities within the informal sector which a delineation on the basis of the Franklin or the Friedmann and Sullivan model could not encompass. Therefore he has developed classifications based on the *income opportunities* available to people in the city. The accompanying list in Table 2 is not exclusive but it is not hard to see how income figures collected within this framework might lead to a more precise definition of the participants in the informal sector. For instance, working at a household level, it may be possible to determine, in situations where households earn incomes from two sectors, which sector is the larger contributor. Another aspect of this framework which offers fruitful possibilities is that some idea of the contribution of the informal sector to total income generated in a city might be produced, in addition to figures on the flows of income between the two sectors.

Thus the three approaches coalesce to enable the proto-proletariat to be more effectively delineated. There is no need to point out that with each of these conceptual categories there is a whole bundle of operational

TABLE 2. Income Opportunities in a Third World City

FORMAL INCOME OPPORTUNITIES

- ***(a) Public sector wages.***
- ***(b) Private sector wages.***
- ***(c) Transfer payments*** — pensions, unemployment benefits.

INFORMAL INCOME OPPORTUNITIES: LEGITIMATE

- ***(a) Primary and secondary activities*** — farming, market gardening, building contractors and associated activities, self-employed artisans, shoemakers, tailors, manufacturers of beers and spirits.
- ***(b) Tertiary enterprises with relatively large capital inputs*** — housing, transport, utilities, commodity speculation, rentier activities.
- ***(c) Small-scale distribution*** — market operatives, petty traders, street hawkers, caterers in food and drink, bar attendants, carriers ***(kayakaya),*** commission agents, and dealers.
- ***(d) Other services*** — musicians, launderers, shoeshiners, barbers, night-soil removers, photographers, vehicle repair and other maintenance workers; brokerage and middlemanship (the ***maigida*** system in markets, law courts, etc); ritual services, magic, and medicine.
- ***(e) Private transfer payments*** — gifts and similar flows of money and goods between persons; borrowing; begging.

INFORMAL INCOME OPPORTUNITIES: ILLEGITIMATE

- ***(a) Services*** — hustlers and spivs in general; receivers of stolen goods; usery, and pawnbroking (at illegal interest rates); drug-pushing, prostitution, poncing ("pilot boy"), smuggling, bribery, political corruption Tammany Hall-style, protection rackets.
- ***(b) Transfers*** — petty theft (e.g., pickpockets), larceny (e.g., burglary and armed robbery), speculation and embezzlement, confidence tricksters (e.g., money doublers), gambling.

SOURCE: Keith Hart, "Informal Income Opportunities and Urban Employment in Ghana," **The Journal of Modern African Studies** II, no. 1 (1973): 69. Reprinted by permission of Cambridge University Press.

problems in collecting data, as I have found in my efforts to collect information on hawkers in Hong Kong, and I am not particularly sanguine about collecting data on income, particularly the illegal forms of income that can only be collated through intensive participant observation using the approach of economic anthropology....

Conceptually then I believe it is possible to delineate the proto-proletariat in Third World cities as being broadly engaged within a peasant system of production which we call the informal sector and gaining the major part of their income from informal income opportunities both legal and illegal....

THE ECONOMIC FEATURES OF THE PROTO-PROLETARIAT

. . . In outlining the economic activities of the proto-proletariat three main aspects will be discussed: first, the forms of economic organization; secondly, the types of economic activity; and thirdly, the relationships that exist between the informal sector and formal sector of the city and the countryside. Turning first to the form of economic organization, it can be stressed that the prevailing forms are either self-employed or socially-based forms of organization in which the household prevails, but other forms such as age groupings or birthplace groupings are not excluded. . . . [T]his social basis of organization has important consequences for the patterns of work, investment, credit availability, etc., which make the activities of this sector very different from that of the formal sector. . . .

Secondly, turning to the types of activities which prevail among the proto-proletariat, there is an immense range, but they may be grouped under four main heads. First, there are those activities concerned with the broad field of distribution. Here there is a wide variety of activities concerned particularly with trade and transportation which range from middlemen in the commodity distributive chains, who break the bulk of produce, to the variety of small retail outlets such as small stores, market vendors and street traders. . . . All sorts of commodities flow through these marketing chains—vegetables, meat, charcoal, water, petrol, textiles—the whole gamut of provisions for human needs, and while reliable figures on the proportion of total consumption of cities which is distributed through this informal sector do not exist, to my knowledge, there are indications from our studies in Asian cities that a major part of the volume of sales (value is more difficult to calculate) is released from this sector. . . . An interesting sidelight is the manner in which some of the commodities produced in the formal sector are packaged in very small quantities and distributed through the informal sector to enable the low-income populations of these cities to purchase them. . . .

Secondly, there are the groups of activities concerned with the provision of services. These again are well known; a simple catalogue would include car washers, car minders, tire repairers, professional queuers, self-appointed house guards, prostitutes, number runners, self-employed mechanics or carpenters and other skilled and semi-skilled artisans, magicians, cult leaders, etc. There is a thin line between legality and illegality in the activities which have fascinated anthropologists . . . but virtually no research has been carried out as to the economic contribution they make to the urban economy.

A third type of activity is industrial in character. Food preparation and processing industries are very significant in this sector. . . . [A]nybody familiar with the superb cooked food stalls of the Chinese in Southeast Asian cities will be aware of the immense diversity and specialization of food offered. Other types of industry such as furniture making, carving, pottery, etc., also abound, often proving surprisingly resilient in the face of plastic replicas from the developed world, but sometimes, unfortunately, collapsing completely. A final category of activities is that concerned with various forms of financial manipulation—money lenders, rentiers, etc., who abound in the informal sector but, once again, our knowledge of their activities is very limited. . . .

Finally, we turn to the third aspect of the economic features of the proto-proletariat which up to the present has been the least analyzed. This concerns the relationships and differences between the formal sector and the informal sector in cities, and the relationship of the latter sector primarily to the informal sector of the countryside, although in the plantation economies of the Third World delineated by Beckford (1972) there are also important relationships between the informal sector of the cities and the formal sector of the countryside. Broadly these relationships may be grouped under the heads of people, money, and commodities and services. With respect to people the first important point is the distinction in flexibility of time and occupational commitment that characterizes the two sectors. People employed in the formal sector are tied to regular hours of work and regular vacations for which they generally earn a contracted regular income. People working in the informal sector often have more flexibility in their hours of work (particularly if they are self-employed), in their commitment to work (insofar as it is social commitment), and far less regular income. This distinction has important consequences for the flow of people between the countryside and city. For instance, recent work among rural migrants to urban areas in West Java suggests that there is much more circular migration among migrants in the informal sector than the formal sector. . . .

A second aspect relating to the flow of people between these two sectors is concerned with the question of the volume of movement between the informal and formal sectors of the urban economy and the capacity of each of these sectors to absorb more labor. Here one must contrast the difference between the capital-intensive, high-productivity characteristics of the formal sector with the labor-intensive, low-productivity features of the informal sector. The formal sector tends towards a rigidity in employment of labor; that is to say, employment bears a close relationship to the degree of capital-intensity and profit. On the other hand, the informal sector is exceptionally elastic with a capacity to absorb very large numbers of people. . . .

A third point relating to the labor mobility between these city sectors relates to the shifts between formal sector and informal sector. In the Hong Kong situation, which I have already indicated is a very special case in the Third World situation because of the substantial increase in formal sector employment opportunities in industry during the last ten years, I found that just over one-third of the hawkers had an occupational history of employment in the formal sector. They had chosen, or had been forced to seek, occupational opportunities in the informal sector where, although individual income was generally slightly lower, they felt the social and personal advantages were more favorable than wage employment....

With respect to the flows of money between the sectors, Hart's model of income opportunities raises some interesting questions. First, the informal sector participants, while extracting money from the formal sector by legal means, such as purchases of commodities offered by informal sector retail outlets, also extract a considerable amount of money from the formal sector by illegal means. I would hypothesize that in general such illegal extraction enables many more people to be supported in the informal sector than it would in the formal sector. Let me give an example of this—an individual from the informal sector steals a hub-cap from the Mercedes of the manager of an import-export agency. The manager goes to the local thieves' market to repurchase the hub-cap for, let us say, $10 and the $10 is then taken by the thief to keep five members of his household for a day. The manager had intended to use the $10 to purchase a box of imported chocolates for his wife. By using such an example, it is quite possible to see how such a redistributive activity as theft may have great income-generating possibilities and certainly enables the informal sector to absorb more population. In general, as Hart points out, such redistributive activities seem to be more common within the informal sector as, for instance, with protection rackets, etc. But I would argue that even a fairly small flow of money from the formal to informal sector will considerably increase the number of people who can exist in the informal sector. In this respect the role of highly paid communities of expatriates living in the cities of the Third World may be of some significance for they are often easy prey for the redistributive activities of the informal sector participants. A second aspect of the income flows between sectors is the important flow of transfers that occurs from the informal sector to the countryside. There is again little research on this subject but the effect of transfers on the rural economy must in some cases be of considerable importance.

Finally, I must turn to the contentious question of the income generated in the two sectors.... The assumption that prevails is that because the activities of the informal sector are considered to be characterized by low-productivity, "inefficient" activity, they generate little income. Certainly as a per capita or per unit of operation statement this is true, but the total income generated by the informal sector may be much larger than economists realize.... I am also concerned about the question of efficiency. Much of the research that we have carried out among the hawkers in Southeast Asia and Hong Kong suggests that they manage an extremely efficient system of commodity distribution which provides labor opportunities and cheap goods. I can see no great strength behind the proposition that would argue that such systems should be replaced by highly capital-intensive chain store distribution because this form is more efficient or productive. The fact is that in many Third World cities informal sector activities are operating efficiently. To dislocate them by legislation or capitalist penetration would only increase the already pressing labor employment problems of these cities. . . .

CHAPTER 21

International Corporations, Labor Aristocracies, and Economic Development in Tropical Africa

GIOVANNI ARRIGHI

The emergence of the large-scale corporation as the typical unit of production in advanced capitalist economies has had momentous implications for the process of development in the still underdeveloped lands. Implicitly or explicitly, this is generally acknowledged by all but those who continue to base their theories on the competitive model. . . . It is also agreed that such implications are, on balance, negative. . . .

All theories that emphasize the size of the market and its growth and/or technological discontinuities as important factors in hampering development are, to some extent, implying the relevance of the increased scale of capitalist production and of oligopolistic behavior. However, this relationship between oligopoly and underdevelopment is often seen in purely technological terms, that is, as having little to do with the political-economic systems obtaining in the advanced and underdeveloped economies. . . .

An emphasis on oligopolistic behavior, rather than on technological factors, can be traced in Prebisch's [1959] argument that the terms of trade between the "industrial centers" and the "periphery" of the world economic system have behaved in the opposite way . . . one would expect from the competitive model. In that model the faster technical progress in the industrial centers, relative to the periphery, ought to result in falling prices of industrial products relative to primary products. However, the market power of workers . . . and oligopolists . . . in the industrial centers is considerably greater than the market power of capitalists, workers, and peasants in the periphery. As a consequence, in the centers the incomes of entrepreneurs and of productive factors increase relatively more than productivity, whereas in the periphery the increase in income is less than that in productivity. . . .

Reprinted from Richard I. Rhodes, ed., *Imperialism and Underdevelopment* (New York and London: Monthly Review Press, 1970), pp. 220–67, by permission of Monthly Review Press. This material has been edited for this publication.

With regard to Africa, Nkrumah has emphasized another aspect of the problem by pointing out that the balkanization of Africa has created a super-structure that makes it impossible for individual nations to cope with the bargaining power of the international corporations which, by means of interlocking directorships, cross-shareholding, and other devices, effectively act on a pan-African scale.

The purpose of this essay is to analyze the relationship between capitalist centers and periphery in order to assess the validity of these assumptions. The analysis will be limited in two ways. In the first place, it will be concerned with tropical Africa and, within that region, with East and Central Africa in particular.... The discussion is limited in another direction. The pattern of relationships between centers and periphery is changing and considerable confusion ... stems from the fact that different conclusions are drawn according to whether the "old" or the "new" pattern is emphasized.... [W]e shall focus our attention on the new pattern, i.e., that emerging from the relative decline in importance ... of foreign capital attracted to tropical Africa by the combination of rich natural resources and cheap labor ... and the growing relative importance of direct investment by large-scale oligopolies.... By limiting the study in these two ways we shall be in a position to gain an insight into the development potential of the emerging pattern of center-periphery relations under conditions of embryonic class formation in the periphery itself....

We shall proceed as follows. In Section I we shall analyze the emerging pattern of foreign investments in tropical Africa with particular reference to the choice of techniques and sectors implicit in that pattern.... In Section 2 we shall analyze the changes in the class structure of tropical African societies associated with that pattern. In Section 3 we shall examine the implications for growth and development of the conclusions reached in the previous two sections. And finally, in Section 4 we shall discuss the limitations of state action in the light of the political economy of tropical Africa.

1

The growth of oligopoly as the dominant structure in the advanced capitalist countries has been accompanied by a relative decline in importance of rentier capital ... and of competitive capitalism as a dynamic factor of growth.... These changes in the competitive structure of the industrial centers have, since World War II, been reflected in the pattern of investment in the periphery.

The declining relative importance of rentier capital has been matched by a decline in portfolio investment in the periphery relative to direct investment on the part of the corporations.... [T]he vast financial resources available to the corporations favored further vertical integration, while oligopolistic behavior encouraged the formation of consortia.... These tendencies were strengthened by the process of "decolonization...." More important still was the outflow of small-scale, competitive capital that accompanied independence.... The upshot of these changes has been the emergence of a new pattern of foreign investment in which financial and merchanting interests and small-scale capital ... have declined in importance relative to large-scale manufacturing and vertically-integrated mining concerns. The typical expatriate firm operating in tropical Africa is more and more what has been called the "multinational corporation," ... i.e., an organized ensemble of means of production subject to a single policy-making center which controls establishments situated in several different national territories.

An analysis of the factors determining the investment policies in the periphery of such multinational corporations is therefore necessary ... to assess the impact that foreign investment is likely to have on the process of development.... It is useful to break down the analysis into two problems: (1) the sectoral distribution of investment; and (2) the type of techniques adopted in each sector....

There is a lack of basic quantitative evidence on the *sectoral distribution of foreign investment* in tropical African countries.... There is, however, considerable agreement on a few broad generalizations:

1. The colonial pattern of capital investment in production for export has basically remained unaltered.... What has changed ... is that complementary investment in the infrastructure, which used to be undertaken by private interests, is now the responsibility of the public sector....

2. Industrial investment other than in mining has been almost entirely concentrated either in primary products processing for the export market or in import substitution in the light branches of manufacturing such as food, beverages ... and other consumer goods....

3. ... [H]eavy industry ... is either nonexistent or, being export-oriented, is totally unrelated to the structure of the national and supranational African economies....

This sectoral pattern of foreign investment is likely to change slowly or not at all for reasons that are partly technological and partly political-economic....

With regard to *choice of techniques*, it seems fairly well established that foreign investment in tropical Africa has a capital intensive bias. This bias is sometimes due to technological constraints ... [but] technological constraints are only one factor ... and, in the case of many industries (e.g., light industries) in which foreign investment shows an equally strong bias toward capital intensity, they are rather unimportant.... Techniques of management, organization and control have evolved in the ... industrial centers and cannot easily be adapted to the conditions in the periphery....

There is another reason, probably more important than management constraints, why labor intensive techniques may be disregarded. . . . [The multinational firms] not only have access to the capital markets in the industrial centers, but . . . are in a position . . . to build up large accumulated reserves of capital for their investment programs. Financial strength makes the large firm adopt capital intensive techniques, not only in the industrial centers but also in the periphery.

In a way, capital intensity is favored also by the qualitative characteristics of the labor force in tropical Africa. . . .

Capital intensive techniques . . . not only require less labor . . . but they . . . require a different composition of the labor force . . . as they make possible the division of complex operations, which would need skilled labor, into simple operations that can be performed by semi-skilled labor. . . .

These two biases of the pattern of investment . . . (i.e., in favor of capital intensive techniques and against the capital goods sector) reinforce each other. . . . This double bias has many implications for growth, development and class formation in tropical Africa that will be examined in the following sections. . . . We must now analyze the impact of the emerging pattern of foreign investment on the class structure of tropical African societies.

2

The analysis in this section will be focused on wage and salary workers and their direct and indirect relationships with other classes and interests. *Wage employment in tropical Africa* is at a low stage of development. . . . Equally important is the fact that wage employment has been relatively static over the last ten to fifteen years. . . . This relatively static wage employment has been accompanied by rising wages. . . . In general, wages are not merely chasing prices but are running ahead of them. . . . Thus the main characteristics of the wage working class are relatively static numbers and rising incomes. . . .

. . . [T]he public sector, as a rule, employs a substantial proportion of wage workers and . . . nonagricultural employment is heavily concentrated in the service sector. The underdevelopment of industry is an obvious determinant of that structure. Another important factor is that the colonial powers superimposed a complex administrative structure on near-subsistence economies. . . . After independence, African governments have taken over these functions and expanded them in their attempts to step up economic growth and to enlarge social services. . . .

. . . The rapid growth of the African elite and sub-elite in the last decade can be traced to the expansion of educational facilities and of job opportunities for Africans in highly paid employment that accompanied and followed the accession to independence . . . due mainly to the Africanization of the complex administrative structure inherited from colonial rule. . . . Another factor favoring this expansion, the Africanization policy of expatriate firms, is of lesser but growing importance . . . [as] firms . . . have quickly Africanized their office staffs, middle commercial posts, and some managerial posts. . . .

. . . [Well below this elite and sub-elite are] the "masses of wage workers. . . ." The mass of English migrants in the early nineteenth century were landless agricultural laborers. In tropical Africa the mass of migrants are peasants with rights to the use of land. While the former were proletarians, the latter are peasants at different stages of proletarianization and therefore present a much greater heterogeneity. . . . The two main characteristics of the labor force under the [African] system of labor migration are low wages and high turnover. The wage rate is customarily based on subsistence for bachelor workers. . . . Low wages strengthen the tendency for the participation in the labor market to be of a temporary nature, which in turn accounts for the persistently unskilled character of the labor force. These factors interact, favoring the development of a poorly paid and unskilled labor force. . . .

Under these conditions the complete *proletarianization of the wage workers,* i.e., the severance of the ties with the traditional sector, is largely optional. It occurs when the incomes derived from wage employment are high enough to . . . allow . . . [a worker] to support his family in town. . . . The nature of the typical enterprise in colonial times militated against a breakthrough in the vicious cycle "high turnover—low productivity—low wages—high turnover" and therefore against the development of a semi-skilled, relatively highly paid, stabilized labor force. . . . Thus traditional colonial employers relied on African migrant labor for their requirements of unskilled labor and on racial minorities (Europeans, Asians, Levantines) for . . . skilled labor. . . . In the 1950s important changes took place. . . . [T]he importance of small-scale colonial enterprises declined and that of the multinational concerns increased. This change was accompanied by the slackening of the influence . . . of the racial minorities on government policies, and by the correspondingly greater influence of the international corporations and of the African elite, sub-elite, and working class. These two changes can be assumed to have been instrumental in bringing about the breakthrough in the vicious circle "low wages—high turnover—low productivity—low wages. . . ." Thus the Africanization of high-level manpower and greater influence of the working class on government policies favored a gradual rise in wages at the lower levels.

Another important factor was the *emerging pattern of foreign investment. . . .* As we saw, the greater capital intensity of production associated with that pattern requires a labor force in which semi-skilled labor predominates. . . . Capital intensity of production (which makes wages a small proportion of total costs and requires labor stability), the ability to pass on to the consumer increased labor costs . . . , and the ability to take a long time horizon in employment and investment decisions, make the multinational companies willing and able to pay sufficiently high wages to stabilize a section of the labor force. . . .

. . . [The result is a] differential between the remuneration of stabilized skilled and semi-skilled labor and that of the semi-proletarianized unskilled labor whose market power is bound to remain negligible.

. . . The conclusion that emerges from the foregoing analysis is that the changes in the pattern of capital investment and in government policies in tropical Africa that have occurred in the last decade have resulted in a breakthrough in the vicious circle . . . ; such a breakthrough, however, concerns only the small section of the working class that is being rapidly proletarianized . . . [and] is . . . achieved at the cost of a relative reduction of the overall degree of participation of the labor force in wage employment. . . . From . . . [the] standpoint [of structure] it is correct to assume that the spiral process of rising wages and mechanization tends to produce a situation of rising productivity and living standards in a limited and shrinking modern sector, while the wage-employment opportunities in that sector for the unskilled, semi-proletarianized peasantry . . . are reduced. To find out to what extent this tendency is a special aspect of a more general trend toward a growing cleavage between the modern capital intensive sector and the rest of the economy we must analyze the impact of the emerging pattern of foreign investment on the other classes of tropical African society.

Let us begin by examining the implications . . . for the rural sector. The first point . . . is that the sectoral distribution of such investment enhances the dependence of agriculture on world markets for its expansion. The bias against the capital goods sector not only restrains . . . the growth of the internal market, but also increases the dependence on foreign sources for the supply of the capital goods necessary for the transformation of traditional agriculture. This transformation comes, therefore, to be subject to balance-of-payments constraints which, as it will be argued in the next section, are likely to become increasingly severe. . . .

It would seem, therefore, that the emerging pattern of foreign investment tends to reduce both the complementary links between urban and rural sectors (i.e., to increase further the lack of structure of the economies of tropical Africa), and the spreading of development stimuli from the modern to the traditional sector. . . . [O]ther things being equal, it tends to produce an impoverished peasantry without fostering its absorption in capitalist agriculture.

The last point we have to discuss in this section is the implication of the emerging pattern of investment for the national bourgeoisie in non-agricultural sectors. In the colonial period most commerce (not directly in the hands of expatriate companies), and small-scale industrial enterprise was in alien hands—the Levantines of West Africa, the Indians of East Africa, the Europeans, and, to a lesser extent, the Indians of Central Africa. . . . This pattern started to change with the approach of independence. In East and Central Africa the Africans, with official support, began to challenge Asian dominance in the commercial sphere. . . . In industry, on the other hand, locally based capitalist enterprises are still largely in the hands of racial minorities. . . . Africanization in this sphere is proceeding more slowly than in petty trade. . . .

. . . [While capital intensive large-scale enterprises tend to drive out smaller more labor intensive firms, making it difficult for an independent African bourgeoisie to rise], there are *operations* which can be profitably subcontracted to smaller labor intensive enterprises by the large-scale expatriate firms. It is not inconceivable, therefore, that investment by multinational corporations in tropical Africa will encourage the growth of a satellite, small-scale national bourgeoisie. Such a subordinate role is all that this national bourgeoisie will, at best, play in the area. . . .

3

In this section we will analyze the implications for growth and development in tropical Africa of the main assumptions that have emerged in the previous discussion. A brief summary of the main conclusions so far reached is in order. In Section 1 we argued that the financial strength and managerial characteristics of the multinational concerns are reflected in the choice of capital intensive techniques. . . . In addition, the oligopolistic behavior and greater calculating rationality of the multinational concerns are reflected in a sectoral pattern of investment which is biased against the capital goods sector. Both biases . . . contribute to the low demand-generating potential of investment. . . . We concluded that this pattern of investment tends to promote the integration of the modern sector in the periphery and of these with the industrial centers but does not contribute to the reduction of the lack of structure of the national and supranational economies of tropical Africa. In Section 2, where attention was focused on the changes in the class structure . . . that can be associated with the emerging pattern of investment, we saw that the multinational corporation contributes to the reproduction of an environment in the modern sector of the periphery that suits its operations: a semi-skilled proletariat, a white-collar elite and sub-elite, a dependent "lumpenbourgeoisie." This tendency deepens the cleavages between modern and traditional sectors. . . .

Growing internal cleavages and greater external integration tend, of course, to reinforce each other in a process of circular causation. . . .

[*Editors' note:* A rather complex economic argument, requiring considerable background, intervenes here, from which Arrighi concludes:]

. . . In consequence, the bias of the emerging pattern of investment in favor of capital intensity and against the capital goods industry cannot be expected to lead, in the long run, to a faster growth of wage and salary employment; it will simply allow a larger outflow of surplus from the area and growing incomes for a small, and, in relative terms, constant or contracting section of the working population. This type of growth . . . we shall call growth without development. In the last section of this essay we must turn to discuss the reasons for the stability of this pattern of growth.

4

The analysis in the previous sections has been carried out in some detail in order to show the complexity of the relationship between the integration of tropical Africa with the international capitalist system and the obstacles to African development. The assumption of a connection between the persistence of underdevelopment and the evolution of oligopolistic structures in the advanced capitalist countries seems to be valid; we need, however, to qualify it in many ways to take into account various technological and behavioral factors that act independently of the form of ownership of the means of production in the periphery and in the industrial centers with which the former is integrated.

It should be clear that the mere participation of the state in stimulating or undertaking major industrial and marketing functions (a phenomenon that can be observed in many countries of tropical Africa), or even the nationalization of foreign enterprises, does not necessarily alter the nature of the relations between periphery and industrial centers and among sectors and classes within the periphery itself. . . . [E]ven if state ownership increases the share of the surplus retained in the periphery, the bias of investment in favor of capital intensive techniques may remain unaffected . . . [and] the other obstacles discussed in Section 1 are not removed by the mere public ownership of the means of production. . . .

Our analysis also implies that a disengagement from the international capitalist system and greater integration with the socialist economies of Eastern Europe and China may not in itself alter the pattern of growth without development . . . [since] the problems connected with the balkanization of Africa which make the individual national economies insufficient planning units would persist. More important still is the fact that many technological and managerial constraints are independent of the mode of production . . . in the industrial centers.

In other words, there is no panacea for African economic development, and African unity is no such panacea either. The fact that international capitalism acts on a pan-African—indeed on a world—scale undoubtedly reduces the bargaining strength and ability to plan of the small African nations. However, . . . the lack of development . . . is only partly due to the balkanization of the area. . . . [No single factor] can be singled out as the crucial variable. Institutional changes alone cannot be expected to change that pattern.

African governments will have to face up to the problems of primary accumulation, a process which has not gone very far in tropical Africa. Broadly speaking, this process has two related aspects: the mobilization of the saving potential implicit in the underutilized productive resources of the pre-capitalist economies; and the reallocation of the surplus from export of investment income and from conspicuous or nonessential consumption to serve the requirements of that mobilization. . . . The existing pattern is characterized by a high capital intensity of production within each sector and by a sectoral distribution of investment implying a low "implicit capital intensity." We have seen that this pattern has a very low development potential because it restrains the growth of the internal market and . . . it creates balance-of-payments problems which frustrate the further expansion of productive capacity. An alternative pattern of growth . . . would have greater developmental potential because it would foster the autonomous growth . . . of the internal market and would reduce the dependence of steady increases in productivity upon the availability of foreign exchange.

The importance of this last point warrants some detailed discussion. Increases in productivity involve a "learning process. . . ." The existing pattern of growth not only restrains the spreading of the learning process over large sections of the population; in addition, even in the state-owned enterprises, it limits considerably the range of experiences. . . . Furthermore, the bureaucratization and narrow specialization that capital-intensive large-scale enterprises entail limit the number of . . . situations in which learning can take place. The use of labor intensive techniques would not only spread the learning process to larger segments of the African population but also make it more complete and varied. The use of labor intensive techniques is also more likely to make possible the mobilization of the underemployed labor of the African pre-capitalist system. Disguised unemployment in Africa is typically seasonal and periodic since no general population pressure on the land exists. The labor migration system (an adaptation to African conditions of the "putting-out" system that has characterized primary accumulation in the now advanced economies), . . . performed the function of mobilizing this type of disguised unemployment for productive purposes. As we have seen, the emerging pattern of investment is displacing the system but no alternative way of mobilizing underemployed labor has emerged. . . . Labor intensive techniques and the development of a capital goods industry would . . . make possible the mobilization of the disguised saving potential of tropical Africa and therefore the internal generation of the surplus necessary for long term growth and development.

It is, however, important to bear in mind that the question of a shift toward more labor intensive techniques within sectors and toward a different allocation of the investable surplus among sectors cannot be divorced from the second question mentioned above, namely, that of the distribution of the surplus among classes. . . .

. . . [If strategies for changing from international investment patterns to locally-generated labor intensive patterns have not emerged yet in tropical Africa] we must attempt to find out why. . . .

The emphasis is usually put on external obstacles. By not dealing with such obstacles it is not our intention to belittle them . . . ; it is more important to understand the causes of the failure to evolve a valid strategy of development, which are rooted in the political economy of tropical Africa itself, namely in the power base of the African governments. As pointed out in the introductory section, in most countries of tropical Africa feudal elements, landowning classes, and national bourgeoisies are either nonexistent or not sufficiently significant, politically and/or economically, to constitute a power base of the state. The implication is that the stability of the existing system of internal and external relationships must be sought in a consistency between the interests of international capitalism and some classes other than the abovementioned. Our analysis has suggested that such classes are, in all likelihood, the African elite, sub-elite, and *proletariat proper* (e.g., excluding migrant labor), which we shall collectively refer to as the "labor aristocracy" of tropical Africa.

The labor aristocracy . . . owes its very emergence and consolidation to a pattern of investment in which the international corporations play a leading role. The displacement costs involved in the disengagement from international capitalism therefore have to be borne mainly by the labor aristocracy itself. The most important consideration, however, concerns the reallocation of the surplus that is necessary for the mobilization of the disguised saving potential of tropical Africa. Such a reallocation directly hits the labor aristocracy, which has most benefited from the present pattern of growth without development, and whose consumption therefore has to be significantly curtailed. State ownership and management of the means of production is not sufficient to prevent the present unequal distribution of incentives. As we saw in Section 2, the steady rise in wages and salaries of the last ten to fifteen years is only partly due to . . . large-scale foreign corporations. Governments' wage and salary policies have also played a leading role. It follows that even though the labor aristocracy may not be opposed to state ownership and management of the means of production, it can be expected to resist that reallocation of the surplus on the part of the state which must be an essential component of the strategy for the transformation in the total situation of the societies of tropical Africa. . . .

CHAPTER 22

Contrasts and Continuity in a Dependent City: Kano, Nigeria

PAUL M. LUBECK

During the last decade, scholars concerned with cities in the "developing" or "peripheral" areas of the world system have turned increasingly to a dependency or a Neo-Marxian framework for their analysis of structural and social change within urban centers. While I agree with this reorientation from modernization theory, generalizations about cities at the periphery are often deduced mechanistically from a venerated formula rather than investigated empirically in each historical instance.

For example, does the specific nature of the world system, i.e., ideologies, technical communications and degree of interdependence—at the historical moment when a society is integrated into the system—affect the development of urban centers in the periphery? It would appear that Spanish urban policy of the sixteenth century differed radically from British notions of "indirect rule" during the early twentieth century. Does the decision to allow European settlers into a peripheral region (i.e., Algeria, Palestine, Kenya, and South Africa) determine to a significant degree the dependent city's spatial form, ethnic segregation pattern, and relations with the countryside? How much independent variation among cities in the periphery is attributable to indigenous institutional development prior to conquest, i.e., a strong, bureaucratic state, a complex economic and class system or adherence to a universal religion such as Islam? Explanations for variability between cities in the world system, and most importantly, how the dependent city's autonomy may be enhanced, are critical questions for analysis and for policy formation.

With these questions in mind, we will examine the experience of a centuries-old urban center in northern Nigeria, metropolitan Kano, in order to compare Kano's urban patterns with those traits associated with dependent urbanization. Clearly, Nigeria and Kano are locked into a pattern of dependent capitalist development for the immediate future. But the question of whether the state can mediate and correct some aspects of this dependent urban syndrome remains unanswered. Since the state is never separate from society it can be argued that Kano's unique features endure and that the existence of an urban tradition, developed prior to integration into the world system, may serve as a basis for state action against the more destructive and fragmenting features of peripheral urban development. After reviewing Kano's historical evolution I shall focus on patterns of labor migration, the integration of labor into Kano's economy and the nature of urban working-class formation in metropolitan Kano.

Kano's growth and development cannot be separated from the nation of Nigeria. As the most populous state in Africa (70 million) and as the richest (annual petroleum revenues of $8 billion), Nigeria has moved from an agricultural commodity exporter (peanuts, cocoa, and palm oil) to the world's fifth largest exporter of petroleum. With the negative example of Latin America in mind, the Nigerian state is attempting to control its own development by the massive infusion of oil revenues into basic industries (iron and steel, refineries and petrochemicals) as well as irrigation projects in the northern states which are designed to make Nigeria self-sufficient in food. If these irrigation projects are completed, by 1981 there will be approximately 1.7 million acres under irrigation and Kano should develop as the food and agricultural processing center of West Africa. These irrigation schemes surround metropolitan Kano and are consistent with Kano's natural resources of flat plains and wet season rivers. Because peasant techniques are comparatively advanced (i.e., manures, fallowing, crop rotation and leguminous plants), population densities in the region have increased to over 800 per square mile. Accordingly, little land is left fallow in the close-settled-zone surrounding the city. In fact, much of Kano's urban population, estimated to be well over 1 million, originates from the dense population surrounding the city.

To emphasize a contrast with the model of the dependent city, Kano's urban origins date to the seventh century and were based upon agriculture and iron smelting, not on export trade with Europe. The city expanded by building adobe walls around an area sufficient to withstand attack and to grow food if under siege. Islamic traders and scholars from North Africa and West Africa exerted influence over Kano's kings until, after the time of Rumfa (1463–99), all kings were at least nominally Muslim. Trade within West Africa and across the desert to North Africa developed parallel to Islamization and in 1804 a Muslim cleric, declaring a holy war or *Jihad* against backsliding Muslim kings, instituted a Muslim empire of which Kano became the economic center.

With the conquest of Kano by the Fulani *Jihad* the contemporary social structure of Kano was formed. The Fulani are an ethnic group originating from the Senegalese valley who, after conquest, formed an office holding and fief holding aristocracy—the *sarakuna.* Their subjects became the Hausa or the *talakawa,* who engaged in farming, commerce and crafts. The Fulani have intermarried with the Hausa, lost their linguistic identity and have evolved into the hereditary ruling class that has dominated modern administrative positions in the northern states of Nigeria. Note that the Hausa are not a tribe but a nation: that is to say, anyone who speaks Hausa, professes Islam, and acts like a Hausa becomes integrated into Hausa society in a single generation. Sub-ethnic identities based upon origin, region and social rank are important.

With relative peace established by the Fulani empire Kano expanded in population, wealth, and industry. The transsaharan trade to North Africa, shifting eastward during the nineteenth century from Timbuktu, allowed Kano to emerge as the entrepôt not only for the northern trade but also for West African trade to the southern regions of Nigeria, the kola nut center of modern Ghana and the other Sudanic states of West Africa. Although the transsaharan trade was important, African interregional trade was equally significant. A handicraft industry, based upon indigo-dyed cotton cloth, leather products and some metal working, produced both for export and for local consumption. Most crafts were organized into guild-like associations. European travelers visiting Kano during the nineteenth century emphasized the rich agricultural base of the society, noting that it exported grain to the Sahara in exchange for salt. An elaborate taxation system administered by a patrimonial and literate (in Arabic script) bureaucracy gained influence and increased the rationality of administration. For example, land suitable for wet season irrigation was taxed by the assessed productivity of the land even if the holder failed to farm the parcel. Through tax measures such as this and the protection of the textile industry from taxation it is certain that a fiscal policy designed to extend the wealth and economic development of Kano was in existence prior to the British conquest of 1903.

When Kano was formally integrated into the world system, the British favored indirect rule whereby the Muslim patrimonial bureaucracy was centralized and rationalized. However, surplus extraction from the peasant population by the aristocrats, *cum* colonial bureaucrats, continued through traditional methods of taxation. Originally, the British intended Kano to be the center of a cotton-growing region and built a cotton gin and extended the railway for cotton exports from Lagos to Kano in 1912. However, instead of cotton becoming the export crop as colonial administrators and British textile interests deemed necessary, the indigenous entrepreneurs, led by Alhassan Dantata who had earlier established himself in the kola nut trade between pre-colonial Ghana and Kano, disseminated seeds and established a network for the cultivation of groundnuts (peanuts). Since then, groundnuts and not cotton became the major export crop of northern Nigeria.

Here we should emphasize that, unlike other cities engaged in the transsaharan trade (Timbuktu, Gao and Jenne), Kano was able to shift its orientation from the transsaharan to the Atlantic trade. Furthermore, the indigenous entrepreneurial class remained in control of significant proportions of trade both before and immediately after integration into the world system. Finally, these factors enabled Kano's old city to remain intact. Today, just as was the case prior to 1903, the Emir's palace and the central mosque form the cultural center of the old city. Wards, organized by occupational specialization and regional and ethnic origins of the earliest inhabitants, form the sub-units of the city. Each ward has a head and an Imam, that is the religious leader of the ward, as well as mosques that are particular to it. Again in contrast to the characterization of Latin American cities whose rural areas are depicted as internal colonies, rural-urban relations in Kano remain intense and are constantly reinforced by trading and Islamic scholarly links between urban and rural communities. Moreover, the formal political structure remains centralized with distinct heads appointed by the center, and village heads, in turn, appointed by the district heads. While Kano is certainly the primate city of the region, it is also true that an early colonial census (1917) ranked Kano's population (59,670) as nearly four times that of the next largest city, Katsina (16,513); and that even then rural densities within 30 miles of the city averaged 350 per square mile. Hence, Kano's high rural densities and primate city features predate integration into the world system.

Yet despite the resilient features of Kano's urban structure within the walled city, the areas of the new city outside the walls have taken on the social class and ethnic segregation patterns associated with the dependent city. For example, adjacent to the walls lies *Fagge*, a transsaharan camel camp during the nineteenth century; it contains the Syrian quarters whose inhabitants performed intermediary trading functions between the large trading companies (i.e., Unilever) and the Hausa population. *Sabon Gari* is a residential area for southern Nigerians, usually Christian migrants to Kano, who were brought by the British to perform clerical roles in the colonial bureaucracy and economy. An early settlement for the migrant, usually Muslim, working class is *Tudun Wada*. It was designed by the British for colonial state employees such as sanitation workers, porters and ex-soldiers of the colonial army. Together with *Gwagarawa* and *Gama*, *Tudun Wada* forms the twentieth century urban area for northern Nigerian migrants. Occupationally it is composed of urban wage laborers, petty traders and craftsmen. The colonial elites—administrators and representatives of trading companies—resided in *Nassarawa*, a government residential area, in which the Nigerian civil servants and managerial classes now reside. Thus, after integration into the world system, Kano's residential patterns changed from distinct ethnic and occupational communities located within the walled city to rather distinct and rigid ethnic and class segregation patterns.

If we compare Kano to Nigeria's capital city of Lagos, the uniqueness of Kano and the uneven development within Nigeria become clearer. Originating as a slave port, Lagos developed with the Atlantic trade and later as the administrative and commercial center of Nigeria. From Lagos the agricultural products were and continue to be exported. It has expanded to a city of approximately 3 million people, with slums, traffic and port congestion, and enormous inequality of wealth and status fully apparent. Multinational corporations have developed their import substitution industries and agricultural processing centers in Lagos. The industrial labor force is better educated (secondary school leavers), and rationalized labor-management relations are the rule. Trade unions are centered in Lagos and have organized a major proportion of the modern sector workers.

In contrast, though Kano is second only to Lagos in industrial development, multinational capital there is insignificant relative to the entrepreneurial capital in Asian, Levantine, and local Hausa ownership. Production processes are less capital intensive and more labor intensive in Kano than Lagos. Entrepreneurial management techniques are personalistic and labor unions are repressed with impunity. Because little modern education was provided to Kano's population, except for the elites, less than 10 percent of primary school-age children are enrolled and few Hausa workers have attended school. Hence most skilled workers originate from the south. While Kano's industrial and urban development is less integrated into multinational capital and while the local entrepreneurs are visibly making the transition from commerce to industrial development, there is no evidence of "risk" or innovative investment. Instead, the state, in alliance with local industrialists, has stimulated industrialization and is expanding its involvement within the economy.

Any discussion of rural-urban relations and labor migration into Kano must acknowledge the high degree of African population mobility prior to incorporation into the world system. Indeed, African and West African history is demarcated by migrations such as those of the Fulani and the Bantu. In Kano and in Hausa-speaking Africa, rural to rural migration was common as was migration of religious communities who dissented from established policy. During the nineteenth century Kano merchants expand their networks throughout West Africa by establishing migrant Hausa communities along trading routes until Hausa became a *lingua franca*. Similarly, temporary dry season migration of rural populations into Kano city did not originate with the British conquest. Barth, the German traveler who visited Kano in 1851, estimated that the urban population doubled during the dry season (i.e., October through April). Then as now, Koranic students came to Kano to study, traders came to trade and resell to rural markets, and craftsmen came to produce handicrafts. Recent research on the textile industry suggests that much of the actual production took place as household industry in the rural villages surrounding Kano. Similarly, traders often camped in the villages where food, water, and animal fodder were less expensive. The point is that during the nineteenth century, the city and villages were well integrated not only by a centralized administration but also by Islamic scholarship and trading networks. The pattern of dry season migration is not a creation of dependent capitalist development, but rather, with only four months of rainfall, the ecology of the area encourages temporary migration to urban centers.

When I studied the villages surrounding Kano in 1970–72, I still found close integration between city and immediate countryside. Not only did villagers continue to farm their plots and to provide raw materials (mud, straw, wood and stone by the donkey load) to the city, but perhaps one-half of the adult males in one village were engaged in wage labor within the urban modern sector. Although I had expected villagers to supply raw materials and to engage in petty trade, the discovery that twenty to thirty percent of the urban factory labor force was made up of rural villagers was an unexpected finding.

After some investigation, the pattern became clearer. Additions to the urban area—i.e., an international airport, military bases, industrial estates, a new university and housing for an expanding population of nine percent per year—required state intervention. When a peasant's land was absorbed, he was paid compensation but not market price for the land. The expansion of the city increased population pressure on the land and forced villagers to seek urban employment or urban trading opportunities. Hence, there were several forces operating against villagers in the periurban area which were creating "proleterianized" villages: the expansion at the urban periphery, the villagers' attraction to modern wage labor, and the high rural density. Gradually, either the villages are absorbed into the city or the peasant families lose sons to urban labor, often followed by permanent emigration to Kano. For those villages too distant to be incorporated into the city, a division of labor between sons develops whereby, typically, the eldest may be engaged in trading with the father, the second son engaged in factory labor, and the remaining sons may continue to farm the small plots and provide raw materials to the urban population.

Although most permanent migrants to Kano left rural life for economic reasons (i.e., "hunger and poverty at home," "a chance to get ahead in life" or "to save enough to become a trader"), many factory workers I interviewed had come to Kano originally for Koranic education. Like trading, Islamic scholarship is a major route of upward mobility for commoners (*talakawa*) in Hausa society. In the rural areas parents place their sons at ages 8–12 with wandering Islamic scholars, or *mallams,* for either a dry season or for several years without returning home. In explaining the practice, informants argue that if a child stays at home, discipline will be lax and the child's parents will spoil him with kindness at the expense of Islamic education. At the same time, it is clear that dry season migration of sons reduces the family's grain consumption during the dry season. Hence, dry season migration functions to reduce the strain on a peasant family because Koranic students are fed through the charitable alms of the more affluent, whether in cities or in the countryside.

When a student arrives in a city like Kano, he often has a letter of introduction, the name of a *mallam* or a student of their *mallam*, or in some cases the hospitality of a local Koranic student whom they met in a rural school. Younger student-migrants usually study in the early morning and evening and work at casual labor during the late morning and afternoon. If there are no employment opportunities, their subsistence is provided by the Islamic obligation of alms giving. If they possess no other lodging, they are allowed to sleep in the *zaure* or entry-room which nearly all Hausa compounds possess. This pattern is an unusual example of the way in which Islamic institutions mediate and eventually recruit labor into the labor market. Normally, the Koranic student decides to marry, and in order to acccumulate enough capital to pay the expenses of his marriage, he takes a factory job whose shift schedule often interferes with Islamic education. Yet in other cases the wealth and consumption goods offered by the city lead the student away from Koranic education and toward factory labor. Furthermore, once in the factory, workers who are integrated into Islamic organizations have a higher tendency to support working class industrial actions and favor workers' organizations than those with little or low integration into Islamic organizations. Thus, Islamic integration buttresses rather than dilutes class integration.

It is important to stress that Koranic students are to some degree a floating population in Kano. Often they are implicated in protest against government policy and once formed a source of support for the opposition political party during the 1950s and 1960s. The degree to which Islamic patterns of education and migration become integrated, or serve as ports of entry into the labor market, is an interesting example of the complementarity of tradition and modernity, rather than the conflict that is often alleged.

Recently, much scholarly interest has focused on the role of the informal sector, that is, the petty service, craft, and small scale entrepreneurial sector of African urban economies. Unlike the modern sector, this sector receives no aid from the state yet employs a majority of the labor force in Kano. Rather than being isolated from the modern sector, the informal sector is integrated into the modern sector; it provides services for both the factory and factory employees at exceedingly low wage rates. In analyzing the career histories and occupational aspirations of Kano factory workers, I found that the service-entrepreneurial sector is an important career experience and future reference. For example, nearly all factory workers of rural origin worked in the informal sector prior to obtaining factory work. Their jobs included unloading trucks, pushing carts stacked with empty oil drums, practicing a traditional craft such as shoe repairing, hawking used clothing, peddling snacks on foot in the market, and literally hundreds of other hawking, handicraft, and repairing industries such as repairing bicycle tires.

If the migrant was fortunate to be placed by a friend, or had saved enough capital to bribe the factory supervisor with approximately two weeks pay, he was able to obtain an unskilled factory job. Once engaged in factory labor, the migrant had the advantage of regular payment and a higher wage rate. However, despite higher pay, approximately eighty percent of factory workers interviewed aspired to become independent traders, citing the insecurity, harsh treatment by supervisors, and rigid discipline of the factory as reason for preferring to return to trading in the informal sector. From the perspective of the migrant, as well as the urban worker, factory labor is a closed mobility system. Modern education is the avenue for advancement within the modern sector. Hence, because they never possessed an opportunity to attend primary school, Hausa workers look to the entrepreneurial sector as the path to independence and economic security. A constant interpretation of factory work described it as "the work of strength and youth." Therefore, once a worker becomes aged, workers fear that they will be dismissed, to be replaced by a younger (and lower paid) worker. Again, the lack of job security is created by the inability of unions to survive, until recently, the repressive tactics of entrepreneurial management. If four-fifths of factory workers aspire to return to the informal sector, albeit as small scale capitalists rather than casual workers, the question must be posed as to whether workers located in the formal sector form a labor aristocracy as dependency theory suggests.

Perhaps the most misleading element in theories of urban dependency and class relations at the periphery is found in the labor aristocracy thesis. It has been argued by Arrighi and others that, because workers located in the state capitalist or international capitalist sector of the economy are better paid, more secure, and often organized into trade unions, they constitute an aristocracy of labor and therefore are alienated from the "semi-proletarianized" workers in the informal sector. And further, because their higher wages constitute a cost of manufactured products consumed by peasants and the semi-proletarianized workers in the informal sector, there is a conflict of interest between formal and informal sector workers. Accordingly, it is alleged that the labor aristocracy is politically conservative because of their high wages and their integration into international capital.

This view is incorrect for a number of reasons. Formal sector workers are more productive, more disciplined, and more alienated than informal sector workers; from a Marxian perspective they are even more exploited than informal sector workers because they receive a smaller proportion of the surplus value which they create with their labor power than is the case with informal sector workers. Moreover, if they were paid less, nothing would be redistributed to the peasantry and the informal sector workers. But, because the urban workers in Kano emerge from the common inequality status of being a *talakawa* before they become urban laborers, and because marriage patterns, mutual aid, Islamic institutions, household and community relationships integrate the informal and formal sector workers, there is no objective or subjective cleavage between them. Rather, as Adrian Peace argues for Lagos, members of the informal sector, by servicing the formal sector workers, participate in a redistribution of high income such that informal workers regard the formal sector workers as a political elite pursuing the class interests of the laboring population as a whole. (For a fuller discussion, see Lubeck, 1975.)

Within Kano society the basic cleavage is between the *sarakuna* who have become a civil service elite in alliance with wealthy businessmen and industrialists and the rest of the population; the latter is composed of petty traders, modern sector wage laborers, informal sector workers, craftsmen, and casual laborers. A few shillings difference in daily pay between formal and informal sector workers makes little difference when contrasted with the generous salaries, benefits and investments of the wealthier classes.

To return to our original question regarding the degree of variability exercised by the peripheral urban areas within the world system, several conclusions can be drawn from Kano's experience. The prior historical development of a city and the manner in which an urban area is incorporated into the world system provide important sources of variation. Because Kano is only the second city of Nigeria, local entrepreneurial capital appears to have survived the initial phase. Furthermore, the tradition of a strong Islamic state as well as Nigeria's economic nationalism may insure some degree of insulation from wholesale absorption into international capitalist development. At the human level, while class formation and ethnic segregation are occurring, the resilience of precolonial institutions, modes of integration and rural-urban relations have not been seriously altered. Whether the influx of oil revenue and the hasty arrival in Kano of multinational enterprises, anxious to recycle petro dollars, will radically alter the situation which we have described depends upon the strength of the Nigerian state. Although the present government is committed to resisting multinational capital, as well as returning to civilian politics by 1979, the present policies can only result in "negotiated dependency" rather than autonomous and self-sufficient patterns of economic development. Hopefully, Kano will continue to remain a variable and unique urban center in the world system.

HOUSING: SQUATTING AND THE SELF-HELP PRINCIPLE

INTRODUCTION

The most *visible* problem Third World cities face is the need to provide decent, sanitary, attractive housing for their growing populations. Certainly, this is the issue on which many writers focus and concerning which there seem to be the most active public programs. The association between housing problems and Third World cities is very strong. The topic of urban centers in Asia, Africa or Latin America conjures up immediate images of agglomerations of make-shift shacks, bulging with families, unpaved and puddled paths, people cooking, doing laundry, etc. out-of-doors, and everywhere salvage and garbage. Indeed, if the tenements and dark row houses described by Engels in his *The Condition of the Working-Class in England,* symbolized urban problems in the West during the early industrial revolution, so the squatter shack settlement in the Third World city symbolizes urban problems in the developing world.

That the problem is widespread cannot be denied. The brief selection from Abrams' *Man's Struggle for Shelter in an Urbanizing World,* while written some time ago, still conveys some of the magnitude and some of the reactions of planners to the problems associated with squatting. For Abrams, the chief problem is a land policy that denies potential settlers in the city access to adequate sites for the construction of housing and which fails to assist them adequately in their attempts to constructively relieve their housing shortage.

The legal status of land and the unwillingness of planners to work with people in solving their housing needs are seen by Mary Hollnsteiner, in her powerfully argued and perceptive article, as the major impediments to successful amelioration of the housing problem in Manila. Both she and Abrams recognize that it is economic necessity that drives poor people to build houses on land that doesn't belong to them, but the policies they recommend work within the system of the present distribution of resources, advocating, at best, transfer payments to assure minimal adequate standards of housing.

So strong is the association between squatter shacktowns and Third World cities that Americans forget that, at a time of extreme economic distress, even in the midst of a much better stock of land and housing, the United States too generated its own squatter settlements, the so-called Hoovervilles that proliferated in large American cities during the Great Depression of the 1930s. In a rare study, made at the time through participant observation, sociologist Donald Roy demonstrates with barely disguised sarcasm that, when an economic system fails to provide jobs and income to its people, they will take their fate into their own hands and make shelters. One need not look beyond for more exotic causes, such as overpopulation, migration, etc. It is easy, in the immediate response to a crisis of visible needs, such as housing, to overlook the fact that poor housing is a symptom of the level and distribution of wealth and power in a society. Poor housing exists because poor people exist.

Rivas, in his article on Peruvian squatters, takes a dim view of the self-help activities of the poor in providing themselves with housing, not because he does not respect and admire the organization and ingenuity of Lima's squatters in planning invasions, in cooperating to design and build their settlements, and in then negotiating with the government for services and a regularization of title, but because he believes that, in the last analysis, the interests of the elite are better served by this hidden subsidy from the labor and scarce capital of the very poor than are the interests of the poor themselves, who really need jobs and a fairer distribution of income.

This position is also taken, in somewhat different form, by anthropologist Anthony Leeds in his article dealing with forms of proletarian housing, chiefly in Rio de Janeiro but in other parts of Latin America as well. He reminds us that although squatter settlements are the most widely publicized form of proletarian housing in Third World cities, there are many other types of residential areas occupied by poor persons, each inadequate in its own particular way. He sees the true causes of such inadequate housing in an economic system which condemns an estimated 70 percent of the population in major Latin American cities to incomes which fall in the poverty/proletarian bracket. And like Arrighi, who earlier discussed the class system in African cities, Leeds suggests that since urban national elites (including the planners) benefit from the proletarianization of the masses, they cannot, therefore, be expected to move to the roots and basic causes of the problem.

Thus, the housing problem is linked to other problems: of economic development and class formation, of migration, of regional distribution—in short, to all the problems dealt with in earlier sections of this book. If this is so, then policies designed to deal with only one of the problems in isolation are unlikely to succeed. If the causes are all interwoven, the solutions will have to address the larger fabric in which they are enmeshed.

CHAPTER 23

Squatting and Squatters

CHARLES ABRAMS

Human history has been an endless struggle for control of the earth's surface; and conquest, or the acquisition of property by force, has been one of its more ruthless expedients. With the surge of population from the rural lands to the cities, a new type of conquest has been manifesting itself in the cities of the developing world. Its form is squatting, and it is evidencing itself in the forcible preemption of land by landless and homeless people in search of a haven. Unlike other forms of conquest that were propelled by the pursuit of glory, trade routes, or revenues, squatting is part of the desperate contest for shelter and land. Of all forms of illegal seizure, squatting is the most condonable.

The old frontier areas of the more developed nations were once also the scenes of squatting, but in time titles were established, the land was often granted or sold to the squatters, and the law of force was supplanted by the force of law. Squatting, however, was rarely carried over into the cities of America or Europe, because law and property rights in cities were too firmly rooted. Members of the British privileged classes who had acquiesced in rural squatting until the time of the enclosures would not long allow the same indulgences for urban property. The urban slum, not the squatter's shack, became the mark of industrialization in Europe and later in America.

Squatting in the cities of the underdeveloped world today is usually open and defiant, tempting more squatting by its successes. It has affected not only government-owned land but private land as well, including tracts provided with costly facilities. When squatting is prevalent, orderly development and expansion are impeded, investment in greatly needed urban enterprises may be discouraged, and the political stabilization of governments may be delayed.

The squatting problem exists in many parts of Asia, Latin America, and Africa—in fact, wherever there has been a mass movement of people to cities and insufficient shelter. There are now [1964] some 240,000 squat-

Reprinted from Charles Abrams, *Man's Struggle for Shelter in an Urbanizing World* (Cambridge, Mass.: M.I.T. Press, 1964), pp. 12–24, by permission of the publisher. This material has been edited for this publication.

ter units (*gececondu*) in Turkey. Squatters make up about 45 percent of the population of Ankara, where some land has had to be turned over to them. They are 21 percent of Istanbul's population and 18 percent of Izmir's. In 1951, they numbered sixty thousand in Baghdad and twenty thousand in Basra, Iraq; in Karachi, squatters represented about a third of the population. Squatters account for at least 20 percent of Manila's population, and in Davao squatters have taken possession of the whole parkway area running from the city hall to the retail center. Urban centers in South America are also experiencing a flood of migrant squatters. In Venezuela the proportion of squatters (rural and urban) is more than 65 percent of the total population, with a 35 percent rate for Caracas and 50 percent for Maracaibo. Cali, Colombia, has a squatter population that makes up about 30 percent of the total figure. In Santiago, Chile, squatters represent an estimated 25 percent of the population. They constitute 15 percent in Singapore and 12 percent in Kingston, Jamaica.

Though usually primitive, the appearance of squatting colonies varies somewhat according to the availability of building materials, the financial status of the squatters, and the prospects of continued possession. Little one-room shacks built of adobe and scrap are cropping up in Medellín, Barranquilla, and Cali, Colombia, and in fact throughout Latin America. The colonies lack paved streets, a sewerage system, and a water supply. Havana has a profusion of rude huts without sanitary facilities. In Algiers, tin-can towns, or *bidonvilles,* stand just five minutes away from the center of the city in almost any direction. The tightly packed shanties with only narrow alleys for passage are built of old oil drums, scrap metal, tin cans, and odd boards. Each hut, about 10 by 10 feet, houses an average of four or more persons and often a goat. In Tunis, the squatters live in caves dug out of hillsides. Around the edges of Johannesburg, South Africa, sprawl squatter colonies that are a chaos of shacks and hovels pieced together by the homeless and destitute. In India's larger cities, squatters can be found hanging on to their precarious hovels in old forts or wherever they can acquire a foothold. They include not only the unemployed but also construction workers, some 250,000 of whom move from zone to zone as they finish one job and start another. Their tin and rag shanties remain long after they have left for other places. Almost 150,000 squatters live in Delhi, about 90,000 of whom are on public land.

Squatting is triggered by many factors—enforced migration of refugees because of fear, hunger, or rural depression, the quest for subsistence in the burgeoning urban areas, and simple opportunism. Usually it is the by-product of urban landlessness and housing famine. Surplus rural labor and the need for labor in the towns combine to speed migrations. When there is no housing for the migrants, they do the only thing they can—they appropriate land, more often publicly owned land, from which there is less fear of being dislodged. Sympathy with the squatters' movements or lack of a consistent official policy encourages further squatting. Existing settlements spread, and new settlements mushroom. Many of Delhi's squatters put up shacks during the 1962 political campaign when they thought that politicians had assured them they would not be harassed. Their shacks were demolished after the election.

SQUATTING IN PAKISTAN

In Pakistan the pattern has varied from major aggregations in well-defined areas (as in Karachi, Lyallpur, Hyderabad, and Dacca) to a few scattered colonies (as in Lahore, Rawalpindi, Peshawar, and Chittagong). Some squatters' settlements stand on costly land ripe for improvement. In Lyallpur, for example, squatters held land valued in 1957 at $1,000 an acre. Squatters line the public avenues in Karachi and have moved into cemeteries and behind sacred old mosques. When construction workers erected temporary barracks abutting on the land where apartment houses were being built, the workers' colony remained long after the apartments were completed. Then the barracks were sold or extended until the area became a bustling shacktown.

The squatter movement began in Pakistan with the partition of India, which caused millions of Moslems to cross the new border in quest of a haven. By the end of 1948 some 6.6 million Moslems had entered Pakistan, while some 5.6 million non-Moslems had moved to India. The influx was resumed when violence broke out again. Fleeing for their lives, the Moslems settled where they could....

SQUATTING IN THE PHILIPPINES

In the Philippines, before the smoke and dust of World War II devastation had cleared, thousands of Philippine families had already moved into the ruins. They put makeshift roofs over the naked columns and raised partitions of ragged cloth or old tin to separate family groups. Others settled on private land and have stayed there. Tondo, the so-called Casbah, and the Intramuros section are typical of squatter concentrations in Manila. The long rows of shacks near Cebu's piers and the thousands of families entrenched on the park-site area of Davao are examples in other provinces. There are also squatters on the shores of fishponds who are polluting the water, squatters in market areas, street squatters who have been evicted from their homes, shop squatters, and even wagon squatters who curl up for the night in their Coca-Cola carts. The variety is never-ending—a somber tribute to human ingenuity in the face of privation.

The sites marked for bombing and those chosen for squatting in Manila were often identical. They were at the city's core, the waterfront, and the centers of work. An anchorage downtown affords the squatter access to the city's hub and shortens his journey to work. . . .

At first, the townspeople felt sympathy for the squatters in their ruins. But as time wore on and the justification for further indulgence faded, the squatters dug in deeper. Other squatters began to arrive en masse. Private property owners, fearing for their lives if they took summary action, often compromised by accepting a nominal rent, or actually paid a ransom to regain possession. The squatters could then move to another piece of property, hoping to repeat the profitable experience. Thus squatting became a business as well as a way of getting shelter in the Philippines. Some squatters have sublet their quarters; others have even sold them.

Seeing the squatters firm in their footholds, others have come, enlarging the colonies or creating new ones. As in a military campaign, some would bivouac during the night with their stock of materials behind a newly placed billboard. Next day, the horizon would be dotted with new rows of hovels, to which others would be added shack by shack, until the expansion was checked by a road, by a canal, or by an owner prepared to spill blood. . . .

SQUATTING IN VENEZUELA

In chronically troubled Caracas, Venezuela's capital city, squatters' colonies dominate the scene both inside the city and on its outskirts. The "ranchos" are perched on the mountain ranges. They are close to the shopping areas and in the city center. They also adjoin the housing superblocks built by the government.

Traveling by automobile from Caracas along the sea to the Macuto area, one never loses sight of growing rancho colonies carved out of niches or built on the mountainside. When the crags become sharp and steep, the ranchos fade, only to reappear on the first gentler incline.

The population of metropolitan Caracas has spurted from 694,000 in 1950 to 1.3 million in 1961, and now public services cannot keep pace with rancho expansion. When the rains come, a lava of human excrement is washed down on the roads below.

In contrast to other cities, whose suburban sprawl is the spread of middle-class families to the outskirts, the suburbs of Caracas blend mansions with ranchos. The mansions are on the flatlands and are provided with ample water and utilities. The ranchos are high up, depending on water hauled in oil cans. In the Caracas suburbs, squatter shacks have appeared wherever there is a new private development, as for example the stylish La Carabelleda Golf and Yacht Club, with its $75,000 mansions. The shacks also abut on the Puerto Azau, which is one of the larger beach apartment developments. The availability of water and utilities for a new development becomes the signal for a new rancho settlement close by. It then grows rapidly until the mountainside no longer yields usable space for settlers.

The type of rancho construction varies from house to house. Earth, cardboard, old boxes, tin, scrap, stucco, and brick tile are common materials. Most are one-story, but there is also a well-built four-story rancho for tenants. . . .

The ranchos exist in other Venezuelan cities and are numerous in rural areas as well. Some twenty thousand squatters in 1962 lived in the projected steel city of Ciudad Guayana, even while it was being planned and when the steel mill was providing only a thousand jobs. The 1950 census of agriculture and livestock placed the number of rural squatters at 35.8 percent of the total of 248,738 agricultural workers. Squatting is the second most widespread form of tenure. One estimate is that 65 percent of the country's population have no legal title to their land. . . .

TYPES OF SQUATTERS

The types of buildings erected by squatters vary with the materials available, but most are one-story makeshifts made of mud, scrap lumber, or tin. Sometimes there are substantial houses. In Lima, Peru, squatter groups have even been said to hire surveyors to lay out sites. Some houses are built so that the owners can take them to other sites if officials or private owners persevere in harassing them.

The types of squatter tenure are not uniform and may generally be classified as follows:

The *owner squatter* owns his shack, though not the land; he erects the shack on any vacant plot he can find. Public lands and those of absentee owners are the most prized. The owner squatter is the most common variety.

The *squatter tenant* is in the poorest class, does not own or build a shack, but pays rent to another squatter. Many new in-migrants start as squatter tenants, hoping to advance to squatter ownership.

The *squatter holdover* is a former tenant who has ceased paying rent and whom the landlord fears to evict.

The *squatter landlord* is usually a squatter of long standing who has rooms or huts to rent, often at exorbitant profit.

The *speculator squatter* is usually a professional to whom squatting is a sound business venture. He squats for the tribute he expects the government or the private owner to grant him sooner or later. He is often the most eloquent in his protests and the most stubborn in resisting eviction.

The *store squatter or occupational squatter* establishes his small lockup store on land he does not own, and he may do a thriving business without paying rent or taxes. Sometimes his family sleeps in the shop. A citizen of Davao, in the Philippines, can get a dental cavity filled by a squatter dentist, his appendix removed by a squatter surgeon, or his soul sent on to a more enduring tenure by a squatter clergyman.

The *semi-squatter* has surreptitiously built his hut on private land and subsequently comes to terms with the owner. The semi-squatter, strictly speaking, has ceased to be a squatter and has become a tenant. In constructing his house, he usually flouts the building codes.

The *floating squatter* lives in an old hulk or junk which is floated or sailed into the city's harbor. It serves as the family home and often the workshop. It may be owned or rented, and the stay may be temporary or permanent. In Hong Kong, there are so many thousands of junks and sampans in one area that one is no longer aware of the water on which they rest.

The *squatter "cooperator"* is part of the group that shares the common foothold and protects it against intruders, public and private. The members may be from the same village, family, or tribe or may share a common trade, as in the case of groups of weavers on evacuee land in Pakistan.

Although there have been admirable attempts by squatters to build clean houses and to rear their children properly, squatter colonies are generally safety and health hazards. Many of the foundations are unsafe; most of the shacks are overcrowded. The criminal, the prostitute, and the derelict make the colonies their retreats. Most often fires are accidental, but there has also been suspicion of arson committed by an owner, the value of whose land would soar with the elimination of the squatter occupants. . . .

In Singapore, 130,000 people live in squalid and insanitary *attap kampongs* throughout the municipal area. They have standpipe water and the most primitive sanitation. "It is a physical impossibility to eject these people; they have nowhere else to go. Although the municipality does excellent work in trying to keep these areas properly drained and free from disease, nevertheless they constitute a menace to the general health of the whole city" [Fraser, 1952]. Singapore squatters demand fantastic prices for possession; a parcel of land free of squatters is three times as expensive as land that is squatter-occupied. When a fire ousted 16,000 persons from a squatter area, the government acquired the land for a housing project. Because it would have had to pay the value of the land as a cleared site, it passed a law fixing the price at one third of the value. When acquiring squatter-occupied land, it often compensates the squatter for his "rights."

Part of the blame for the intense squatter traffic in Singapore is ascribed to speculators who are aware of the physical difficulties obstructing demolition. "The Land Inspectors are intimidated in the execution of their duties and enforcement of instructions becomes a dangerous process." The effort to enforce orders by constables has proved insufficient to prevent "a disturbance of the peace. . . ."

To look upon all the squatters, or even the majority of them, as lawbreakers, is to misjudge the problem completely. Had land been made available to him, the squatter would not have appropriated it. A land policy that would have granted him a site, however small and humble, might have prevented a critical challenge to social and political equilibrium in the underdeveloped areas. But this obligation has never been accepted by the more developed nations of the world as a charge upon their consciences.

CHAPTER 24

Hooverville—A Community of Homeless Men

DONALD FRANCIS ROY

In January, 1934, the writer was hired by the Washington Emergency Relief Administration to investigate Hooverville, one of the newer and increasingly popular residential districts of Seattle. During a widespread and protracted slump in real estate and the building trades, this area had been favored with an extraordinary "boom"—an expansion in open, noisy disregard of carefully drafted graphs and diagrams which showed clearly the critical state of a bed-ridden economic system. From the sandy waste of an abandoned shipyard site, almost in the shadow of the multi-story brick and steel sanitaria of indisposed business, was swiftly hammered and wired into flower a conglomerate of grotesque dwellings, a Christmas-mix assortment of American junk that stuck together in congested disarray like sea-soaked jetsam spewed on the beach. To honor a distinguished engineer and designer this unblueprinted, tincanesque architecturaloid was named Hooverville.

The objective of the W.E.R.A. was to understand the structural and functional aspects of the phenomenon as it had come to exist. Under the assumption that one who participates in the "native" domestic life can make more accurate observations than an outsider who snoops and quizzes, the writer accepted $15.00 from state relief funds to cover the cost of a furnished dwelling, convinced an easily influenced Hooverite that in three five-dollar bills there inhered values very favorable in comparison to those of home and fireside, closed the deal, and moved in. . . .

There are 500 tiny shanties huddling in the rain and steaming in the sun on a former shipyard site where World War rivets were once slung. Old wooden pilings still thrust scaly heads through the sand, and here and there big blocks of cement lie embedded in the earth like ruins of an ancient city. Sometimes these blocks provide flooring for the flimsy dwellings of the Hooverites. There are no streets or boulevards, but paths weav-

Reprinted from *Studies in Sociology* 4, nos. 1 and 2 (Winter, 1939-Summer, 1940): 37–45, by permission of the Department of Sociology, Southern Methodist University, Dallas, Texas. This paper has been edited for this publication.

ing in and out like animal trails. Some of the better worn "runs" offer transit for two-wheeled junk carts. . . .

On first impression Hooverville appears to be an odd assortment of junk painfully assembled to form a conglomerate of shacks all more or less of a uniform type, constructed from materials picked up along docks, railroads, alleys, and dumps. On closer inspection, however, differences in the qualities and proportions of lumber, tin, and paper used, stand out; but the most striking differences lie in construction. The shanties range from small bungalows to semi-dugouts that might be described as "lairs." Two dwellings may be made entirely of lumber; one will be a shed, the other a neatly and compactly built little cottage with double floor, weather-boarded or ship-lapped walls, tar-papered roof, a window on every side, and a latticed porch. But the shed type is predominant. Tin is universally popular as roofing material. Some builders nail the tin down tightly, others lay it on loosely at a gentle angle and weight it down with rocks or scraps of iron. Painted walls are by no means rare, though in general exteriors are left unfinished. Paper is used chiefly for interior covering. Many walls and ceilings are equipped with strips of cardboard obtained from packing boxes. Glass is in universal use for windows which range in size from large plates several feet square to mere glass-covered slits a few inches wide. The amount of light admitted varies also with the nature of the glass and the extent to which it is kept clean. The shanties are fairly uniform in size and shape, ranging from 3′ by 6′ to 15′ by 25′; most of them are from 6′ by 9′ to 12′ by 15′. They are usually one-room affairs, sometimes divided into kitchen and bedroom, sheltering one or two men. Ceilings are low, from 6′ to 8′ in height; thus the shacks heat up rapidly and, when stoves or drafts are defective, become smoky little sweatboxes. Except for a few dugout-like dwellings, and the piano-cratelike skyward venture of one Hooverite who has attempted vertical expansion, domestic life goes on at one dead level; there are no cellars nor attics. Building regulations require the homes to be built at least one foot off the ground, but these prescriptions were made when the community was already fairly well settled. Some of the more recent shacks comply in exaggerated fashion, their floors being several feet above the earth; others are set flat upon the ground. According to the "mayor," space under the flooring facilitates the pursuance of rats by cats. In a few cases, lumber and nails involved a cash nexus. One man sticks to his story that his estate "set him back $52.00," f.o.b., unassembled. In general, selling prices have varied with desire for immediate cash, desire to leave town quickly, state of intoxication, and other personal factors. The range is from $4.00 to $30.00; the average, from $10.00 to $15.00. . . .

Though Hooverville will never be classed among the leading health resorts of the Pacific Northwest, it is surprising that morbidity is not more noticeable. Opinions as to the community's healthfulness vary among its inhabitants. The fresh sea breeze, active outdoor life, and the feeling of contentment that comes with having a home where one may do as he pleases, are stressed as advantages or even reasons for health improvement. On the other hand, there are a few who levy negative criticism. It is one man's conviction that "you can never get rid of a cold in this place." Another found the village too damp for his rheumatism and moved out. . . .

Hooverville is first of all a men's town; of its 639 human inhabitants only seven are females. The population age range is 18–73, with the average at 45.4 years. Two-thirds (69.4 percent) are more than 40. Filipino residents, somewhat younger than the others, average 34.9, with only one-third (36.1 percent) past 40. Eliminating this group, 77.6 percent of the population is upward of 40, with the average at 47.9. The respective ages of the seven women are 28, 38, 39, 42, 49, 67, and 73. There are no children because, in the words of the local "mayor," "Hooverville is no place for kids." In 1933 a woman who attempted to bring her small son and daughter of 15 to her shanty was expelled from the community by the police and given living quarters elsewhere by a relief agency.

In its racial composition, Hooverville forms an ethnic rainbow. White, black, red, yellow, and brown brush frayed elbows in shabby camaraderie. Civilizations's torchbearers number 455 or 71.2 percent of the total population. The only significant colored groups are the Filipinos, Negroes, and Mexicans, represented by 120, 29, and 25 individuals, respectively. The smattering of other races and hybrids includes two Japanese, two Eskimos, two American Indians, three Costa Ricans, and one Chilean.

The nationality composition of Hooverville's white stock provides ingredients for an ideal "jungle" mulligan. Out of American broth may be fished English mutton, Irish potatoes, German carrots, Scandinavian turnips, Polish cabbage, Balkan rutabagas, Spanish onions, and Italian garlic. And a stiff "spiking" with Russian vodka may be detected. Of the 455 Whites, 29 percent are native-born, 64.2 percent foreign-born. The birthplace of the remaining 31 was not determined. Nordic stocks are numerically predominant among Hooverville's foreign-born Whites, with Scandinavians constituting 59 percent, other Northwest European countries, 10.5, and Canada, 1.4 percent. Poland and the former Austria-Hungary were outstanding Eastern European contributors (15.1 percent).

The 132 native Whites, 29 Negroes, two Indians, and one Mexican who claim United States as their country of birth trace their origins to 35 states. Nearly one-half of these hail from the Middle West, with Michigan, Wisconsin, and Minnesota making the greatest contributions. Eastern states, led by Pennsylvania and New York, rank second in regional origins with 31 destitute delegates, closely followed by the Western states with 28 representatives. Sixteen of this latter group were Washington local boys who failed to make good. Of the 23 Negroes whose state of birth is known, 13 were born in the South.

With the exception of the Filipinos, the foreign-born Hooverites have been seeking their fortunes in American environs for several decades. Of the 316 non-Filipino foreign-born, 85.1 percent have been in the United States 20 years or longer; only nine individuals have been here less than ten years, and none less than five. The high tide of this immigration came from 1900 to 1914 when 217 or 68.7 percent arrived. The Northwest Europeans of Hooverville boast the highest average length of residence, 29.7 years. South and East Europeans follow with 26.3 years, while the Mexicans come next with 20.7. These figures roughly correlate with the temporal position of the several groups in the history of United States immigration. Sharply distinct in length of residence is the Filipino population which averages only 10.8 years. Of the 434 foreign-born Hooverites whose age at time of entry into the United States is known, 12 came as men of 40 or over, 32 as children under 15. The average at time of immigration was 23 years. . . .

With the exception of a bearded hermit of 50, who has been "jungling up" in Hooverville's dugouts and tin shelters for six years, and five others who claim three years, no Hooverite has resided on the community's terrain for more than 36 months. . . .

The native American and the foreign-born Whites vary considerably in emphasis given to migratory, unskilled types of labor and to skilled or more permanent "city" jobs. Of the 304 native White occupations given, 26.3 percent might be described as skilled, white-collar, "steady" with some urban business concern (such as janitor, night watchman, etc.), or as involving the ownership of capital (independent farmer, wholesale fish business for self, etc.); but only 7.7 percent of the 716 foreign-born White occupations given may be considered in these categories. Nearly all of these "skilled" men also claim experience in hard, rough labor. With few exceptions the White Hooverites may be said to represent a sample of the unskilled labor that cut our forests, built our railroads, highways, and bridges, worked our mines, and harvested our crops in the "boom" decades of the 'teens and 'twenties. One of the exceptions is a 73-year-old physician who, no longer able to pay his room and office rent uptown, cast his lot with the Hooverites; another is a 30-year-old former dairy instructor at a Swiss college who found his services little in demand in the United States.

The 120 Filipinos are predominantly cannery workers, farm laborers, and section hands.

The Mexicans, like the Filipinos, are ex-farm and section hands but, unlike the Filipinos, have not generally been employed in canneries. Only three or 5.97 percent of the 51 occupations named by them may be considered as skilled. The Negroes, on the other hand, boast experience in certain personal and domestic services such as porter and kitchen work, and in certain types of skilled labor; 41.9 percent of their 43 occupations may be classified in the latter category.

More than nine-tenths of the Whites had not been regularly employed in the six months preceding the census interview, nearly half had not obtained even an "odd" job during that period; 84.1 percent had not worked for a year, and nearly 80 percent had not worked for two years. On the other hand, only 4.5 percent had been unemployed longer than five years, a fact which points to the depression as a major factor behind their industrial retirement. Filipinos and Mexicans, somewhat more fortunate than the Whites, showed lower unemployment percentages. . . .

If a Mr. Hooverville were chosen to represent the community in a nationwide contest for the selection of the Unknown Rugged Individual for future depression memorials, the man most qualified on a basis of "average" characteristics might well be described by the Associated Press reporters as follows:

"Mr. Hooverville, Seattle's candidate for all-American oblivion, shuffled lackadaisically upon the platform, and the tin pan quartet struck up an enthusiastic 'Washington, My Washington.' Every inch a Nordic, Mr. Hooverville was born in Northwestern Europe and received a grammar school education there. Now in his late forties, he has been a resident of these United States for 28 years, 19 of which were spent in the State of Washington. Until 1931, intermittent labor in the logging, mining, railroad, and construction camps of Seattle's sylvan hinterland afforded him a livelihood and an occasional 'spree' in the city. Save for odd jobs a few days in duration, he has been unable to find work since 1931. Jobless, propertyless, familyless, and savings spent, he came to Hooverville in the fall of 1932 to make that community his home. . . ."

In the spring of 1933, at the "suggestion" of city officials, a group of 100 Hooverites held an open-air caucus to elect a vigilance committee to attend to the policing of the grounds and the settling of minor disputes between residents. The city ultimatum had been that Hooverville either exert its own social pressure as to sanitation and orderly behavior or undergo the fate of Nero's Rome. Jungletown-by-the-Sea seemed to present possibilities of becoming a first-class nuisance in the matter of health, fire, and petty crime. The observable results of the caucus were six Hooverites bearing the title of "Committeeman"; two of these sudden arrivés were White, two Negro, and two Filipino. The numerically preponderant Whites have come to dominate local politics through the ascendancy of an aggressive representative over his fellow committeemen. This local Bismarck, a Texas cowpuncher, has acquired popular reference as the "mayor." He may obtain police wagon or ambulance service in short order by telephoning from the office of a nearby coal yard. . . .

The police power of the local administration involves not only the preserving of peace and order, but also the maintenance of health and safety from fire. The committee's control over the premises is frequently augmented by visits from officials of the city fire department. . . .

A picture of the Hooverville manner of life is incomplete without the background of social attitudes. One of the most refreshing "pockets" of the Hooverville attitudinal "air" is an almost universal geniality, friendliness, and hospitality. In general, the Hooverites are easy to approach and quick to pick up conversational cues. One who happens along at meal time is invited to partake of food and drink. In this respect the Hooverites have carried over to their stable community the traditions of the "jungle." Here is an urban area where mobility and impersonality of contacts have not choked out the flowers of open, unaffected friendliness and kindness toward fellow men.

This spirit of camaraderie bridges racial barriers. White and colored are tolerant if not actually friendly. Although spatial and social segregation of the Filipinos and Mexicans is a general rule throughout the village, this sorting may be the result of acquaintanceship ties and preferences and difficulty of expression in English. Eleven shanties shelter individuals of different color; two Whites live with Negroes, three with Filipinos, and four with Mexicans; one Filipino lives with a Negro, and another with an Eskimo. The attitudes of the Negroes, particularly, show an utter absence of feelings of resentment or inferiority toward the Whites. And only rarely would a White be heard to express antipathy toward the colored races, although several believe the depression is due to the influx of Filipino labor to the United States.

Another striking aspect of the Hooverville attitudinal pattern is a passivity in regard to the national politico-economic order. Although the group is convinced that some form of socialism is both desirable and inevitable, it is not violently bitter about the present state of affairs, nor does it violently agitate for a new system. Such growling and criticism as may occur is directed mainly against the city administration or the W.E.R.A.

Hooverville lacks hope. With so many of its citizens in upper age groups it is not surprising that the community should not manifest eager anticipation of the future. It is by no means uncommon to hear men of 50 years declare, "I never expect to get work again anyhow. They all say I'm too old." The problem of the aged is not one of finding work, but of avoiding illness and the much dreaded charity medical care. Many fear the "black bottle," the container of a deadly potion supposedly used by county hospitals to rid society of its useless old men.

The Hooverites not only lack hope for betterment of their economic state; they actually fear turns for the worse. Suspicion of governmental forces has become a veritable paranoia with many of them. They read personal repression into every movement and word of national and local authorities. They fear destruction of their homes; they fear concentration camps; they fear deportation. For the inquisitive snooper or government agent they entertain negativism. Theirs is a persecution complex. . . .

The slump in the building trades goes on and on; but Hooverville hammers continue to beat a hollow rat-a-tat all out of tune with the business cycle. And not only does Hooverville expand, but similar eruptions break out elsewhere on the urban countenance. In Seattle there are already several "Hoovervilles," two of which threaten to surpass the original. That the same phenomena have occurred in many other American cities suggests a nation-wide movement similar in scope to our westward expansion in the nineteenth century. In place of one long frontier beyond the river or the mountains and far from the older and larger centers of population, there are now many small frontiers along the railroad tracks, as "close in" as local administrations allow. The American pioneer of the 1830s moved west to hew himself a home in expectation of improving his economic condition; his prototype of the 1930s entertains like motives as he "hotfoots" it to the dump to gather materials for a shack in Hooverville.

The Hooverites may be described as ragged epitomes of rugged individualism in a world of "closed" economic resources—individualists because they have no ties with industry, ragged because of this independence, and rugged because they have to be to survive. Ruthlessly, albeit impersonally, rejected by the industrial chameleon that once wooed their services, these men have no way of obtaining money to pay their way in modern society. Not only has the tap been shut off, but the faucet has been disconnected and the pipes have been taken out. The Hooverites are "up against it." A half-century ago, men in such a predicament could have struck out for the frontier; but since that outlet is no longer available, primary needs must be met in some other way. Customary city haunts, and cheap hotels and lodging houses, don't offer free shelter, nor do restaurants big-heartedly provide the nourishment neccessary for metabolic upkeep. These men cannot move in with the Indians; the reservations are full now. The nut crop is limited and uncertain. Thus the only rational course is to remain in the city, where one can cling parasitically to men still embraced by the "long arm of the job." This parasitism may take either the form of scratching for public relief crumbs or foraging in garbage cans, chicken coops, and at the back doors of sympathetic housewives. Hooverites have combined both methods; but their answer to the problem of lodging has been a spurning of the public transient shelter for more comfortable nests of their own creation. At first loosely thrown-together makeshifts patterned after the rough shelters of the hobo "jungle," these expressions of individuality were later reinforced or remodeled to achieve a greater degree of stability and permanency.

Thus has arisen Hooverville to glorify the hobo "jungle" and carry on to new frontiers the traditional American spirit of rugged individualism. And there remains Hooverville, scrap-heap of cast-off men, junk-yard for human junk, an interesting variation of the grimaces of laissez-faire.

CHAPTER 25

The Case of "The People Versus Mr. Urbano Planner Y Administrador"

MARY RACELIS HOLLNSTEINER

INTRODUCTION

In the search for a better life that is every man's due, Filipinos in droves have been streaming to cities. There they hope to realize at least a part of the dream that has guided so many others like them into urban lives less threatened by economic insecurity and enriched by new opportunities and friendships.

For thousands of migrant families the pursuit of the dream dictates that to conserve their meager savings and augment possible earnings, they must settle in the cheapest and most conveniently located housing available. Usually this means becoming a resident of a squatter community located either on public or private land.

Not that the average migrant harbors an aversion to living as a squatter; in his mind that status carries no more onus than that of the *kaingero* who temporarily preempts for his own private use land not currently being utilized by those in charge of it. Often he pays a minimal rent for the land on which he builds his shack; or he may simply rent a house built by someone else. In paying these amounts, he feels he holds legitimate tenure status. Moreover, the new squatter sees all about him other families like himself merely trying to eke out a better living just as all of them once did in the province. "Can we all be so reprehensible as the newspapers allege?"—he reasons. While his initial makeshift shanty may represent somewhat worse housing conditions than he was used to back home, with time and a higher income level he knows he can improve it. Soon he substitutes heavier construction materials for the current ramshackle ones to fashion a real home. As for the unkempt physical environment, one soon gets used to it and becomes almost immune to its initially jarring sights, sounds, and smells.

Condensed from James Hoyt, ed., *Development in the 70's:* Fifth Annual Seminar for Student Leaders (Manila: USIS, 1974), pp. 84–111, by permission of the publisher and the author. This paper has been edited for this publication.

In short, taking on squatter status does not automatically carry with it a feeling of low self-worth as it would among higher-class urban residents. This is especially true when squatting represents the only means available for achieving that anticipated rise in income and social status that marks the central point of the urban dream.

Mr. Urbano Planner y Administrador does not, however, see it that way. His dream revolves around a beautiful and orderly city with law-abiding people facilitating easy management of public affairs. It is easy to understand, therefore, why the estimated 201,000 squatter families in Metropolitan Manila drive him into a state of anger and frustration. Feeling dutybound to eradicate them from view, he adopts a strategy geared to a form of substitute housing more pleasing to the educated eye.

And there lies the crux of the problem. On the one hand, squatters see their needs as primarily those of earning a living and finding better economic opportunities for the family, including security of residential tenure; housing constitutes only a secondary concern, one that can be handled gradually as the family fortunes rise. Urban planners and administrators, on the other hand, perceive the problem primarily as housing. Accordingly, they devise schemes aimed at housing *per se,* when the reality of squatting arises out of multiple problems linked to the common theme of poverty. Solutions that fail to confront all the dimensions of urban poverty, then, fall short. Unless employment or, more broadly, a means of earning a living, family and community integration, essential physical facilities like water, electricity, drainage, sewage, roads and access to stores, markets, schools, churches and medical facilities are sufficiently provided, the problems of squatting will not be alleviated.

With the framework for analyzing the squatter situation thus established, namely, conflicting definitions of the problem, let us review against the background of actual squatter needs the specific approaches undertaken by Mr. Urbano Planner Y Administrador. This should elicit other alternatives for resolving the issue more satisfactorily for both sides. But first an assessment of the scope of urban squatting, especially in the Metro Manila context, is called for.

THE SCOPE OF URBAN SQUATTING

As early as 1963, estimates of slum and squatter populations in various Philippine cities warned of serious problems ahead. Although squatters were counted in with non-squatter slum dwellers to produce a combined figure, one can guess at the magnitude of squatting by assuming that over half of the percentages constitute squatters. A 1964 United Nations-Philippine Homesite and Housing Corporation (PHHC) survey of slum and squatter dwellers resulted in these estimates (Abesamis, 1972:24, Table 2):

	Percentage of City's Total Population 1964	Estimated No. of Households 1964	Projected No. of Households in 1970
Metro Manila	10	50,000	142,000
Iligan	8	1,120	1,700
Davao	9	4,150	5,900
Cagayan de Oro	11	890	1,250
Ozamis	25	1,050	1,300
Baguio	27	2,795	3,800
Cotabato	30	3,840	4,900
Butuan	44	3,940	6,400
Marawi	45	1,130	1,350
Jolo	50	3,260	3,900

In 1968 a survey of squatter and slum dwellers in Metro Manila conducted by a special presidential committee estimated a total number of 128,000 squatter families alone for the metropolitan area. Apparently, the Committee then multiplied this number by a factor of 6 to represent estimated average family size and arrived at its individual total of 768,000 squatters. . . . Assuming steady in-migration and natural increase through squatter births in the city at a combined rate of 12 percent annually, the total by 1973 would come to some 226,000 squatter families, or 1,356,000 individuals. This represents 30 percent of the current Metro Manila population estimated at 4,500,000. . . .

[D]espite the existence of three relocation sites and a possible fourth . . . , despite government-subsidized apartment units going up . . . and despite a government-anticipated 30 percent of the squatter population willing to accept a transportation subsidy to return to their home province, only a small proportion of squatter families can be accommodated under these three schemes. The rest will have to remain where they are, or if driven out, stay with relatives or friends until they can find some other less vulnerable spot in the city for settling down. . . .

SQUATTER STRATEGIES OF THE GOVERNMENT

The basic policy toward squatters in the last quarter-century has been eviction—immediate, imminent, or eventual. National and city governments have built at least four strategies around this sword-of-Damocles approach, namely: (1) toleration through neglect; (2) subsidized return fare to the provinces; (3) multistorey, low-rent urban housing; and (4) relocation to government urban-fringe sites. Let us examine this fourfold record.

Strategy No. 1: Toleration through Neglect

The appearance of poverty has so long been a part of Metro Manila that only gradually after World War II did her residents come to realize that a new kind of slum was evolving. The squatter community entered into their consciousness not so much because of its run-down housing—Manilans were used to that. Rather it was the unkempt surroundings and densely packed, irregular physical layout that jolted the passerby into this now vivid awareness.

And how did this distinction between the ordinary low-income neighborhood and the squatter settlement become so apparent? Through the reluctance of the authorities to risk validation of squatter claims by providing them public services and improvements. In the absence of any realistic plan of action to grapple with poverty in a new guise, the authorities mostly chose to do nothing. This toleration was broken occasionally by attempts to evict the interlopers, usually with little success. Before long they would be back, protected by a predictable decline in vigilance among the authorities or by local politicians who championed their cause at City Hall and supported court injunctions calling for a stay of eviction in exchange for election votes. While the toleration policy benefited squatters in giving them a chance to entrench themselves in their areas, it also offered them no alternative but to accept a deteriorating physical environment. Ironically, in refusing officially to legitimize the squatter presence by supplying standard urban services to their areas, Mr. Urbano Planner y Administrador helped create and perpetuate the very environment of deterioration he abhorred. . . .

Toleration, therefore, can best be thought of as an ambivalent strategy, ranging as it does from sheer neglect at one extreme to eviction or the threat of it at the other. It is really not a strategy at all because it embodies an unplanned, inertia-fostered set of attitudes which only postpones action and intensifies the problem. Its resolution thereby becomes increasingly difficult.

Strategy No. 2: Return to the Province

"If I had to live like those squatters do," reasons Mr. Urbano Planner y Administrador, "I would much rather go back to the province where life is easier and where one can breathe clean, fresh air." But obviously the average squatter does not share his perspective. Interviews with 37 migrant squatters in Magsaysay Village, Tondo, revealed that only 5 said they would like to return to their hometown; fully 21, or 57 percent, replied that they did not want to go back. The rest (30 percent) said that their decision would depend on a number of factors, among them whether jobs or some means of livelihood, or capital for starting a business would be available in the province. When asked where they felt opportunities were greater—in Tondo or their hometown—an overwhelming majority of 83 percent—

30 out of 36—chose Tondo. Only three opted for their hometowns (Hollnsteiner, 1973:Tables 53 and 51, respectively).

Little wonder then that government offers of free, one-way tickets back home to the province attract only a limited response. For one thing many new urbanites simply prefer metropolitan living to the comparatively dull, circumscribed routine of barrio or town life. The primary problem, however, centers on scarce economic opportunities in the countryside and therefore limited prospects of improvement in family levels of living. This applies especially to rural residents with no access to land or capital. Despite steady out-migration, no significant decline seems to occur in the barrio-town labor force because high fertility rates keep replenishing it. Thus the competition for scarce jobs never really abates. A further, if more minor, deterrent to returning arises from the prospects of facing hometown gossip and teasing about one's assumed failure in the city. The assumption seems to be that no one would voluntarily leave the city if he had been able to achieve any measure of success there.

Seen in a perspective of real opportunities, then, the city holds more promises for the average squatter than does the countryside. Relative to other urbanites, squatters know they generally occupy less desirable places in the metropolitan hierarchy. Yet, relative to their hometown counterparts, they perceive their current or at least future prospects as more favorable. The city then is where they choose to remain. If to do so means becoming a squatter, then squatters they will be.

Strategy No. 3: Multi-storey, Low Rent Urban Housing

Shortly before the 1965 presidential elections, the first government-built multi-storey apartment building at Vitas, Tondo, better known in the neighborhood as *ang tenemen*, was opened for habitation. The 252 units rented out at subsidized rates of ₱5.00–₱15.00 per month, the higher floors renting for less than the lower ones. Squatters ousted from that very site some years earlier were supposed to have first priority in moving in. Many actually did take advantage of the offer; others either chose to squat elsewhere or complained of having been eased out of their claim by more politically powerful applicants, usually of non-squatter origin. Succeeding years saw more units constructed at Punta Sta. Ana and Fort Bonifacio. These three examples of subsidized "high-rise" housing offer interesting insights into the dynamics of low-income residence patterns.

Not long after they moved in, a number of residents surreptitiously began selling the rights to their apartments for sums reportedly ranging from a few hundred pesos to, more recently, two and three thousand pesos. Clearly, there are families attracted by multi-storey apartment living, but they seem to come from the somewhat better off sector of the low or low-middle income population who are already confirmed urbanites. The really poor migrant sector retreats from tenement living in favor of squatter status.

The reasons for this shift as gleaned from interviews with squatters and tenement dwellers prove enlightening. First, government housing requires regular payment of rent; otherwise one is eventually evicted—unless, of course, he has political connections. Yet, squatters by and large have sporadic earning capacities, making regular payments of anything a risky proposition. Second, there is the tremendously attractive prospect of acquiring an immediate sum of money with virtually no capital investment by selling one's apartment rights to more affluent families. Illegal though this transaction is, poverty increases the temptation to seize such immediate advantages. The ex-squatter makes the "sale" and moves back into a squatter community once more, somewhat richer for his temporary sojourn in government-subsidized housing.

A third reason squatters avoid apartment dwelling stems from the higher cost of living there, despite the subsidy. Larger financial outlays are required, not only for furniture but also for "modern" cooking equipment. In a virtually windowless apartment with cement flooring, one can hardly make do with the simple bench and floor mat for sleeping that is possible in the wooden or bamboo-floored shanty. Anticipated illness stemming from having only a thin mat separating the sleeper from the cold floor forces tenement residents to buy simple bamboo beds or wooden bunks, or the material to make them. The prohibition of wood or charcoal-burning stoves as fire hazards likewise requires an additional investment, most commonly in a kerosene stove, but occasionally in a gas or electric one.

The sheer physical layout raises problems, too, a fourth reason for avoiding tenement-dwelling. Living on the fourth to sixth floors in a walk-up apartment hardly encourages one to come and go frequently. Mother cannot check on her children playing far below or along the myriad corridors. Yet the small size of the unexpandable apartment units—27 square meters—is hardly conducive to her forcing the children to remain there. . . .

To recapitulate, squatters find tenement dwelling unattractive for several reasons, namely, decreasing neighborliness and interdependence in time of need, higher costs, regular payment requirements, the lure of windfall money from the sale of rights and unfamiliar, even threatening, physical and social environments. With all its faults the squatter settlement, on the other hand, emphasizes mutual helping through frequent contact in the public spaces interspersed between housing clusters. This form of reciprocity is not unimportant when poverty limits the opportunity to contract and pay for needed services.

Squatter housing costs fluctuate according to how much money the resident has in his possession to invest in gradual improvements. The difference lies in his not feeling compelled to invest at any particular time if he cannot or will not. If he lacks money, he need not make improvements just then. The physical setting may be hazardous and unpleasant in some ways, especially during typhoons, but it has its compensations. Although crime and theft also occur, their incidence is minimized by the informal social controls and group surveillance that close community relations foster.

This comparison of tenement and squatter living is purposely one-sided to counterbalance the already well-known attacks on the housing choices of squatters. We are trying to represent here the views of those squatters who reject tenement living in favor of continued squatter status. Having one's own separate house with at least a small lot still constitutes the Filipino dream. There is evidence, nevertheless, that a substantial number of people see merits in tenement living, especially if they could own their unit in true condominium style. The number of non-squatters willing to pay a few thousand pesos for legally hazardous "rights" reflects the applicability of tenements for slightly-higher-income people faced with a housing shortage and definitely averse to becoming squatters.

Strategy No. 4: Relocation to Government Urban Fringe Sites

To Mr. Urbano Planner y Administrador, what could be more ideal for the urban squatter than his own minimal-cost home lot in hilly and breezy Carmona, Sapang Palay, or San Pedro Tunasan? And indeed many a resettled family agrees. The tragedy is that even when a family opts for a permanent commitment to life in one of these relocation sites, its members often find themselves unable to avail of the opportunity. For virtually no viable means of earning a living are present there.

Thus, the new settlers must make one of several painful choices. Working household members can commute to Metro Manila daily. Or they can remain in Manila during the week and go to the relocation area on weekends to join the rest of the family. A few, mostly women, participate in on-site income-raising projects, like embroidery, dressmaking or woodworking, sponsored by government, religious or civic agencies. Many others cope with the situation as best they can by selling vegetables raised on their lots and soliciting periodic assistance from relatives and service agencies. The rest either abandon their settler lots entirely or leave a *bantay,* or caretaker, in charge so as to retain their claim. Or they return to squat elsewhere in Metro Manila or to live temporarily with relatives there until they find a suitable spot. . . .

Large numbers of settlers, then, frustrated at every turn, find no alternative but to return to Manila. A survey by the Central Institute for the Training and Relocation of Urban Squatters (CITRUS) found that of the 5,975 squatter families sent to Sapang Palay beginning in 1960, only 2,-426, or 41 percent, were left by 1969. This represents an average loss of 394 families per year, or 59 percent over the entire period (Abesamis, 1972:8). The same pattern applies to Carmona and presumably San Pedro

Tunasan, although little is known about the more neglected latter site. Not only is this ebb-and-flow process costly in terms of human sensibilities; it also means substantial loss of precious government funds. The Presidential Assistant on Housing and Resettlement Agency (PAHRA) estimates that the cost of relocating one squatter family including the transfer itself, site preparation and initial subsistence allocations comes to some ₱3,000.

Mr. Urbano Planner y Administrador deplores what he perceives as the ingratitude, irrationality, and wastefulness displayed by the returnees. He cannot understand, or refuses to accept, their apparent preference for living in urban squalor when he has made available to them a place of their own outside the city. He does not seem to realize that in dangling before urban squatters a new and, to many, a desirable option indeed, and then in effect snatching it away by making it virtually impossible for them to utilize, he only raises frustration levels further. Beyond that, he interrupts their already perilously low-earning sequence and wastes that portion of the ₱3,000 relocation investment that cannot be utilized by the next settler. . . .

SITES AND SERVICES: THE UNITED NATIONS VERSUS THE PHILIPPINE SCHEME

Efficiency-oriented, cost-conscious Mr. Urbano Planner y Administrador finds it hard to condone the rejection of "clean-and-airy" Carmona, Sapang Palay, or San Pedro by numerous squatters. He continues to insist that these relocation sites correspond to the most recent United Nations thinking on strategies for squatters, namely, the sites-and-services approach. To a certain extent he is right. This approach focuses on developing a site by installing water supply, sewage, paved road systems and other infrastructure components. Individual lots are allocated to squatters—now settlers—on a long-term lease, outright sale, or donation basis. The housing itself is left to the new settlers to provide through a self-help program. Occasionally, government may prime the pump with basic building materials for free or a minimal charge. This scheme thus simulates the real situation whereby squatters construct their own homes gradually over extended periods of time as earnings allow. At the same time it gives government the chief responsibility for making the general community site acceptable for human habitation, recognizing that individuals, especially poor ones, cannot take on such mammoth engineering tasks.

If Carmona, Sapang Palay and San Pedro Tunasan use the UN model, why then have they not achieved the praise that ordinarily would be their due? For two reasons. First, squatters were moved into those settlements *before* the sites or services were sufficiently established and were therefore subjected to enormous stress and suffering. Public water pumps were either insufficient in number or not working, or both. Moreover, settlers were scattered over wide expanses, having to walk great distances and then line up for an hour or more sometimes to take their turn at drawing water. Although roads had been carved out by bulldozers, with perhaps the main entrance road asphalted, most became virtually impassable in the rainy season. The red clay, spattered on trousers and dresses and caked on the shoes of Sapang Palay worker-commuters, amply testified to that. For education-conscious Filipinos, the lack of schools for their children reinforced the decision to leave. At least, their children had to be given a chance to rise in the social hierarchy even if they themselves could not. The absence of a clinic or doctor nearby to settlers used to these urban conveniences, further oriented them away from their proposed new home.

The second feature of Philippine relocation sites that distinguishes them from the UN model stems from the failure to establish genuine economic opportunities for potential residents *before* they were dumped there—at a recent rate of 20 families per day. Nor was the other alternative of organizing commuter systems that most new settlers could afford, actually realized. . . .

FINDING PEOPLE-BASED SOLUTIONS TO URBAN SQUATTING

Since the four basic strategies are unacceptable as they stand, what, then, can be done? Let us review our thoughts thus far.

We started by considering the scope of squatting in major Philippine cities, Metro Manila in particular, and then reviewing government attempts to cope with the problem. Toleration and corresponding neglect, subsidized return to the province, multi-storey, low-rent urban housing and relocation to urban-fringe government sites have all been found wanting in various degrees. Where, then, do genuine solutions lie?

1. Reconceptualizing the Legal and Social Status of the Urban Squatter

The squatter is a fairly recent addition to the urban scene. Our legal system has not caught up with the reality of city populations, nearly one-third of whom are squatters. The United Nations makes the point this way (1971:-162):

> *In many developing countries, coping with problems arising from squatting and slumdwelling is hampered by the fact that the legal system has not kept pace with the rapid rate of urbanization. Slum and squatter areas are denied basic urban services because they have no status in law. A strict adherence to the concept of private property tends to make the legal system a punitive instrument rather than a means for the ordering of human relationships and behaviour to achieve the developmental potentials inherent in all segments of the urban community. . . .*

> *Traditional legal concepts formulated under conditions of low levels of urbanization may be rendered obsolete by current urban conditions, but they remain in the books. Thus, there is a need to redefine the legal system and to analyze it anew in terms of the prevailing conditions. In this endeavor, the spirit rather than the letter of the law should be upheld.*

Clearly some serious rethinking is needed. For when a large sector of society finds itself automatically characterized as lawbreakers, then one should wonder whether something is wrong, not with the people designated as law violators, but rather with the law itself. One might also compare the low status attributed to urban squatters by Mr. Urbano Planner y Administrador with the way he now looks at the rural Filipino. After many decades of often grudgingly-passed legislation favoring social justice and land reform, the peasant farmer has become the man of the hour. The government cannot do enough for him. Not so for his younger helper-brother with no land to till and insufficient capital with which to finance a rural-based business. When the latter then migrates to the city in search of better opportunity, he finds himself subject to a curious phenomenon. In the transfer process, he metamorphoses from a noble Rousseauan, salt-of-the-earth figure to a despicable law-violating person entitled to no tenure rights over the only living space open to him in the city. For bonafide urban residents apparently fall into only three categories: owners, renters, and boarders. Since squatters do not fit anywhere here, they become *ipso facto* illegal.

This illogical characterization of the squatter clearly stems from a lag between the recognition of real conditions, namely, widespread urban squatting, and the persistence of an antiquated legal framework. The latter in turn derives from categories appropriate to other times and places, but still institutionalized in now-outmoded political, economic and social frameworks. Let us acknowledge that the urban squatter is simply a more modern version of the ordinary *tao* person, trying to improve his family's lot in a developing society. In this he differs only slightly from his rural counterpart, the farmer, except perhaps in the pioneering outlook and willingness to confront new unfamiliar city ways.

2. Urban Land Reform

If rural land reform represents the keystone of current government policy, then surely urban land reform should not fall far behind; for basically the man at the focal point of these programs is the ordinary Filipino. Yet by no means has the urban squatter even come close to approximating his country cousin's public status.

There is no question that high urban real estate values, inappropriate land tax schemes and the existence of idle land for speculation, among other reasons, combine to create an untenable living situation for the city's poor. Those who cannot buy, rent, or board are apparently supposed to hover over the land without ever setting foot on it. In reassessing a system that dismisses 30 percent of its population as second-class citizens worthy only of being driven out, one would have to analyze the entire framework of private property, the application of the stewardship concept and the right of the government to expropriate land for social uses.

3. Developing a National Urban Policy

More and more the Philippine version of the worldwide urbanization crisis impinges on our consciousness. It is not just the proliferation of squatter settlements that raises levels of awareness but everything associated with unpleasant city living. Consider crowded buses, traffic jams, garbage piles, flooded streets, waterless taps and brownouts. Focus, too, on the jarring contrast between Makati villages and plush Quezon City subdivisions, on the one hand, and the many slums of the metropolis, on the other. Clearly, runaway urbanization is upon us. It threatens to make city living, especially in Metro Manila, ever more difficult unless we do some hard thinking, undertake significant research, and initiate judicious action.

These measures should revolve around the development of a national urban policy and the programs arising out of it. . . .

Part of this urban policy review will have to concentrate on mechanisms for making metropolitan governments more responsive to local needs. One flaw stems from the Manila Metropolitan area's being made up of 5 cities and 23 towns in 4 provinces, encompassing 4,404,494 residents as of 1970. . . . Thus, the single continguous population expanse that sociologically can be called "a city" reflects in reality 28 separate and distinct administrative units.*

The effects this has had on squatter populations emerge in two ways: (1) eviction drives mounted in one city or town may successfully oust squatters, only to have them resurface in some adjacent metropolitan city or town; and (2) individual city and town mayors interested in providing sites-and-services areas or tenements for their squatter populations hesitate for fear of attracting even more squatters from adjoining units or from the provinces. Obviously, some form of metropolitan-wide decision-making authority can go far toward coordinating mutual efforts at integrating individual city or town squatters legally and humanely into the local body politic. . . .

*This integration was effected in 1975 when the President of the Philippines officially created Metropolitan Manila as a unified entity under a Governor.

4. Creating Sites-and-Services Communities within the City

Of all the strategies devised for housing the urban poor, the sites-and-services approach offers the greatest potential—but only if it provides a variety of genuine economic opportunities appropriate to largely unskilled workers. In effect, this means living right in the city with its huge client population capable of supporting thousands of vendors, *cargadores*, domestic helpers, craftsmen, and the like.

How will this come about so that the poor, too, share significantly in the benefits of their society? Whatever strategy Mr. Urbano Planner y Administrador adopts, he would do well to stress security of tenure for the settler. For this yields the kind of commitment that enhances the healthy growth and maintenance of the community.

The most drastic alternative proposed for in-city sites-and-services settlements is government expropriation of private land, with or without compensation. More moderate are land-exchange schemes, in which the government trades off some of its suburban or rural holdings for private urban parcels. These parcels need not be single, gigantic expanses the size of the 137-hectare Tondo foreshore area, said to be the largest contiguous squatter settlement in Southeast Asia. Instead, several smaller units can be developed in which the government, perhaps with private sector help, puts in the roads, water system, sewage pipes and other infrastructure and service components. Only when they are installed do the settlers move in and build their own houses bit by bit.

Such a plan may entail several kinds of payment schemes on the part of the recipients. Outright donation of the land to its occupants is one mode, although some form of even token payment is usually desirable to establish the occupant's genuine commitment. Direct sale for a modest sum far below the going market rate and spread out into manageable installments, is another alternative. A long-term lease ranks as a possibility in the event that urban land ownership is made prohibitively expensive to all sectors of the population. For those settlers with little money to spare, an exchange system of work-for-land can be devised. Here each household designates members to work on government-supervised site-development labor gangs. They receive no salary or only a partial payment. The rest is credited in terms of its money value to the balance of the family's indebtedness on the lot. The advantages of this scheme lie in its double thrust of enabling the householder to pay off his land, while equivalently giving him employment right in his home territory.

Variation in tenure strategies in hopes of offsetting the tendency to resell one's allotted site to more affluent families may involve the formation of a residents' cooperative, to which departing tenants would have to resell their lot. To alleviate the pressure from other low-income groups on ex-squatters to resell lots, it is advisable simultaneously to construct four-storey apartment buildings catering to the former as long-term urban dwellers more used to apartment living and having access to low but steady earnings. If the potential apartment dweller can have the option either of renting or actually buying an apartment unit, condominium-style, he may take to this form of urban living much more readily than in the past. Often the resistance to apartment dwelling arises not so much from its physical structure but out of the perpetual-renter status that goes with it.

CONCLUSION

And so we rest the case of the People versus Mr. Urbano Planner y Administrador. We have charged the defendant with not taking *people*, especially the interests and outlooks of the ordinary poor, sufficiently into account in preparing his plans and launching his programs. Rather his master grid descends inexorably upon the city populace and the *tao* below must fit into its framework or be crushed. We further charge Mr. Urbano Planner y Administrador with making little effort to establish two-way communication with the urban masses, much less have them actually participate in the decision-making processes, outside of elections, governing their own lives. . . .

Even when facing the urban mass in community groups, Mr. Urbano Planner y Administrador still takes the traditional authority-figure stance of *telling* the people what he will do for them, instead of *working with* them to evolve mutually acceptable schemes. He listens but does not really hear. We are not faulting him on his goodwill which we know he has in large measure. Nor are we convinced he is unaware of some of the elements of squatter life. . . . Where we do point our accusing finger is to Mr. Urbano Planner y Administrador's reluctance, or outright unwillingness, to implement alternatives for urban squatters that seriously consider *their* needs and viewpoints as *they* express them.

The twentieth century has seen the common man emerge as maker of his own destiny. The egalitarian ideas spawned by the French and the American revolutions have undergone a reformation in the Russian, Chinese and Cuban revolutions and in the peaceful evolution of Scandinavian welfare states. Most Southeast Asian countries are likewise intensely engaged in devising their own models for enabling their common men to share in the good life. Economic development alone is not enough, they realize; for annual increases in gross national product become meaningless in modern thinking if their benefits do not accrue more equitably to all the people. Thus, oligarchs the world over are at bay. Affluence in the midst of widespread poverty no longer finds ready acceptance in developing societies.

If Mr. Urbano Planner y Administrador focuses his concern on the Filipino as *tao* with the right to convey his thoughts to, and bargain with, his more educated peer, then the defendant cannot go far wrong. His high-level skills and training will blossom and mature in the service of the common man in the city and help that man share in the heritage that is his due.

We rest our case. Let the public judge where justice lies.

CHAPTER 26

Development Alternatives For the Peruvian *Barriada*

DIEGO ROBLES RIVAS
ENGLISH TRANSLATION BY
EILEEN WALSH

THE PROBLEM

Existing literature has tended to glorify the process of self-help which is taking place in the *barriadas* of Peru and has created an image of possible self-development of these areas. My rather different evaluation is based on analysis of the series of actions the barriada settlers (*pobladores*) have been able to carry out for the purpose of improving their communities. I believe that, within the context of the capitalist economic system, the process of self-help has been incapable of integrating marginal settlers into national development. Rather, it has served to reinforce, through implantation of populist measures of a paternalistic nature via consumption, the existing system of domination.

The barriada represents one of the settlement forms which typify the process of urban domination and rapid dependent urbanization. The barriada is generated by the economic system. Capitalist forms of production are concentrated in urban areas, which are dependent on foreign power centers. This gives rise to expansion of economic activities in certain coastal towns and cities, without corresponding expansion in the interior of Peru, creating an unstable population equilibrium.

The form of industrialization in Peru also has a decisive influence on its labor force and occupational structure. Industrialization in Peru was not initiated as an internal expansion force. Since the 1950s, it has been oriented chiefly toward import substitution to satisfy select demand for immediate consumption goods. This production structure subordinates and conditions the behavior of less-developed classes by imposing disadvantageous conditions on economic satellite or dependent industries. They are forced to group together or go bankrupt and drive out their labor force. This in turn conditions the behavior of that part of the labor force

Reprinted from *Regional and Urban Development Policies* (Vol. 2, *Latin American Urban Research,* F. F. Rabinovitz and F. M. Trueblood, editors) © 1972, pp. 229–37, by permission of the publisher, Sage Publications, Inc. This paper has been edited for this publication.

which is not absorbed by or expelled from the production structure. For them, there remain only opportunities to take part in independent economic activities or work in unstable salary relationships for extremely low income.

The growth of marginality reinforces one of the basic contradictions of the capitalist production system: it opposes growing production and productivity with the decreasing ability of larger and larger population groups to consume. Nevertheless, by relegating to this marginal labor force those occupational roles of least social significance and lowest income, the system has been able to make this labor force functional for its own development.

The degree of expansion and intensification of the urbanization process in Peru is directly related to the degree of penetration of capitalist forms of production. The cities constitute poles of a network of centralized domination, by their role in the process of social marginalization. In Peruvian cities, precapitalistic and capitalistic forms of production more or less accessible from the capital city coexist. Although the city acts as a core of technological expansion, this same technology initially generates a tendency toward marginalization of noncapitalistic forms of production.

The internal domination suffered by large sectors of the Peruvian population has been reinforced by the intervention of political factors. The capacity of social groups and classes to articulate their demands is directly related to their ability to sell their labor. Those with a greater degree of organization within the rules of the game established by the system have better opportunity to do so. In this way, dominant classes attempt to reproduce the conditions of their social position within the class structure. When the interests of the dominant group are endangered by the insurgence of groups from dominated sectors, such mechanisms as the raising of standards required and discriminatory actions based on deeply rooted prejudices are used to keep these groups out.

The subjective factors created around certain urban areas with a degree of selective industrial expansion have caused strong migratory flows of rural population to those urban centers. The objective factors of expulsion from rural areas characterized by low standards of living have forced peasants (*campesinos*) to move to the cities in proportions much larger than existing employment possibilities. This migratory process in turn is characterized by a high percentage of unskilled labor oriented toward urban areas having the highest rate of industrialization, and where service and commercial activities are concentrated. In addition, in such cities' basic infrastructure, there is heavy public or private investment.

These characteristics complete the complex picture of accelerated urbanization. In urban areas, disequilibrium is shown by nonincorporation and expulsion of the labor force from profitable sectors of the productive apparatus (the marginalization process). These processes have limited the income level of the population, giving rise to a situation in which the population is unable to participate in the urban land and housing market and is forced to form the barriada.

THE CAPITALIST PROGRAM

The barriada is not an isolated phenomenon or separate from the city in which it is located, but rather is dependent on the latter to achieve its development. Yet, the initial collective and insurgent nature of the barriada is transformed by the system into a totality of individual interests conditioned by the participation of external agencies.

Students of urban problems have interpreted the barriada in different ways. One is to consider it an anomalous form of urban development, a position implying application of assistance measures in health, housing, and education, and security measures through repressive action ranging from prohibition of barriada formation to massive eradication programs. Other studies concerned with formation of the barriada emphasize the positive aspects of the settlers' actions—their ingenuity, degree of acculturation, capacity for organization, ability to construct their own houses and necessary services, investment and savings capacity—but at the same time consider the barriada as marginal to the general urban context which conditions or limits the barriada's development possibilities.

Documentation exists showing that the barriada is not marginal geographically, economically, socially, or politically, and that it cannot be considered to be a form of collective development even though in an initial stage, the invasion of the site, collective interests are uppermost. Rather, once the poblador has assured tenure of his land, the initial process breaks up and evolves toward forms constituting a totality of individual interests. This does not differ basically from the achievements of other social groups experiencing similar conditions within the structure of urban domination.

The action of the barriada as a collective project is dysfunctional to the system whenever it involves confrontation with power groups. This is evident in the invasion stage in which private property, defined as one of the system's foundations, is attacked. The barriada is unable to follow a course of permanent rebellion, given that its population is economically, socially, culturally and politically dependent. In order to resolve the barriada's problems within the system, institutional support is chosen as the best strategy.

Once this institutional support has been obtained, a consolidation process is initiated in the barriada. External agencies play an important role, acting as intermediaries between power groups and pobladores. The consolidation process consists of three stages: The first stage is in the barriada's initiation as a collective project located in the "legal" city, with prior action to organize groups participating in the project. The majority resides in deteriorated zones of the city, such as the *tugurios,* internal barriadas which are immensely overcrowded, and, to a lesser extent, other old, peripheral barriadas of high population density. These conditions produce a crisis leading to a position of insurgency. This is capitalized on by leaders who possess knowledge both of the situation of the group and the mechanisms of control of the "legal" city. The slow process of social mobilization begins with organization and identification of groups participating in the project, selection of the invasion site and obtaining of the economic resources necessary for the various negotiations.

The invasion usually takes place on a national or local holiday, preferably during the night or early morning, in order to give time to form an organization. Everyone assumes important roles such as those of defense, communal cooking, surveying and distribution of lots, identification of participants, diffusion of daily news, storage of materials, and construction of temporary shelters using light materials. The opportunity which appears to the insurgent group for invasion of lands on the urban periphery is closely related to the lack of intervention as a consequence of agreement among power groups, or their incapacity to assign new roles to external agencies and to implement action programs consistent with a policy of rapid urban development. Insurgence breaks into the mechanisms of control exercised by power groups, showing their flexibility in the face of apparent and momentary challenges. It arises from periods of internal crisis within the dominant elite and from its capacity for bargaining among power groups themselves or with the population making demands.

In order to achieve their objectives, pobladores take advantage of all possible resources in starting the new settlement, the barriada. In order to analyze costs incurred, it is necessary to distinguish among those concerned with organization and mobilization of the pobladores; investment of their savings in improvement of the area; and the contribution of their labor in housing construction and in installation of public services and other communal facilities.

The directing groups generate educational, consciousness-raising and organizational activities leading to establishment of the "Asociación de Pobladores" and work-groups. It should take advantage of the experience of existing associations, channeling them to the benefit of the new community seeking solution to its immediate problem, that of obtaining a stable residential site within the city itself or nearby. In order to organize, unite, and channel the community's immediate aspirations, and to be able to take advantage of the legal, administrative, commercial, and political mechanisms of the city, a great deal of effort, time, and ability is demanded of pobladores. This initial cost, in advance of formation of the settlement, is entirely borne by the pobladores.

Another contribution made by the pobladores is investment of individual savings, created by their labor and lack of consumption, accumulated long before formation of the barriada. These savings are invested by pobladores in the different stages of consolidation of the settlement. The greater part of these savings is invested in building materials offered by city markets.

Through various commercial mechanisms the pobladores are drained of their capital in favor of other urban power groups. In order to build his house and improve his settlement, the poblador's invested capital must be complemented by his own labor. This obligation in the form of a new investment in labor appears as the result of the poblador's decapitalization in buying building materials and tools. Yet, in this way, he is able to occupy and use his dwelling as it evolves in stages over a relatively long period of time. Since this process of building a house with only the poblador's individual contribution is extremely difficult and slow, pobladores are forced to obtain the help of relatives and to organize themselves in temporary work groups in order to take advantage of mutual aid and to make their own efforts more effective.

A second stage defined as transitional can be identified by the initiation of action by external agencies, public and private, which make contact with the community in order to satisfy the demands of the population in terms of immediate needs.

A third stage follows the point at which the collective nature of the project is disrupted and becomes, rather, the sum of individual actions channeled by certain external agencies. These agencies achieve a certain amount of penetration of and influence within the population and organize it according to their own criteria. These need not coincide with those of the population itself. The action of these external agencies consists mainly in social demobilization activities developed in the face of the possibility that the pobladores might unite, discover their basic interests, and organize for collective insurgence. The latter is linked to the persistence of the role of external agencies within the system and, more basically, to the persistence of the system itself.

External agencies have acted autonomously in the barriada in support of or in coordination with state agencies, according to the type of interest arising at the appearance and development of the barriada. These interests have been linked to the objectives of each power group and include assuring social peace, increasing popular consumption, making available an industrial reserve army in good condition, protecting high-priced urban land in the presence of possibilities for speculation, and attending to the demands of middle and upper social groups. External agencies have interpreted the phenomenon of the barriada in a restricted way—within a technical, social, economic, and political view not in harmony with pobladores' demands for productive means for their authentic mobilization. Instead, social strategies of external agencies are directed at promoting actions whose final objectives are softening the system's internal contradictions, lessening existing tensions, and retarding social change by reinforcing the status quo.

The most general cause of the barriada's appearance as a form of collective insurgence is the deterioration in the exercise of power by the particular ruling political group. This gives rival groups the option of approaching the masses in search of new political loyalties based on promises of solutions to the populations's immediate problems carried out in a climate of expectancy. These promises are confirmed by the new group's exercise of power through the use of populist measures of consumption. These measures take different forms in different settlements and in this way produce a demonstration effect which can be capitalized politically through reinforcement of popular loyalties.

The same thing occurs at moments of economic crisis or boom in Peru, localized in certain areas of the national territory and during which their wealth is either concentrated or consumed. Such a process changes some of the most important activities of the city and in extreme cases changes its incidence in the network of urban interrelations. Collective insurgence also occurs in periods in which external domination is accentuated. This translates into a policy of internal investment, and transfer of national capital abroad, swelling the bank accounts of a few. This situation is related to the impossibility of increasing Peru's internal consumption. Generalized poverty within the population is a direct result.

Three elements intervene in the barriada consolidation process: a power elite whose interests are linked to the production structure, commerce, and land-tenure system; external agencies which fulfill roles assigned to them by the power elite and which organize the population in terms of immediate needs which can be satisfied by consumption; and barriada pobladores who wish to be integrated and to improve their standards of living. Among these elements there is no possibility of agreement or dialogue, either between the power elite and pobladores or between the latter and external agencies. Their various development goals do not coincide, since within the system of domination the pobladores have been considered as not pertaining to the legal city.

EVALUATION

Given the form in which the self-help process operates in the barriada, it is impossible for the poblador to change the system of domination. The productive system does not allow the poblador to participate directly in the market for modern goods and services. As a consequence, the poblador is forced to resort to self-help. But the self-help system employed in welfare tasks directed toward community development decapitalizes the poblador through consumption.

Successive governments have carried out various measures for the purpose of responding to conditions created by the problem of insurgency. The responses have not been oriented toward policies of structural modification but rather have intensified the marginality and poverty of the majority of the Peruvian population. This is reflected in the migration process and in the accelerated urbanization of the nation's urban centers, which have, in a short time, been altered both qualitatively and quantitatively.

Power groups have attempted to manipulate with superficial palliatives the climate of expectancy created by population increase in these cities, by settlements on the urban fringe, and by the process of slum-building, because of their fear of losing social position and status to the dominated groups. As urbanization has become accentuated, the dominant elite, in search of "equilibrium," has imposed assistantial-paternalistic measures on the marginal population based on solution of immediate problems.

Housing and other services or facilities which the marginal dweller "maintains" as permanent necessities have served to condition, within the view of the poblador, technicians, and external agencies, the need to solve the fundamental problems of the majority of the population. The housing-oriented view of the problem has served only to neutralize and demobilize. This distortion has allowed the system and the shifting political elite to incorporate the poblador in mutual-aid programs limited to assistance.

The poblador considers himself author of a "great work" in constructing his own community and in having his land tenure securely legitimized by provisional title. But mutual aid in the barriada is restricted to immediate action and is not oriented toward the poblador's basic interests, such as increase in income-levels, opportunity for stable occupation and active participation in the urban production structure. Self-help restricted to welfare increases the poblador's propensity to consume, distorted by the play of interests of the dominant production groups.

The added value represented by capital created by socially and economically dependent groups only partially benefits these groups. It does not produce a process of accumulation in their favor. Marginal workers, who can generally be classified as underemployed, create added value by their work in labor-intensive productive activities. The added value is transferred to dominant groups, and the depressed are then decapitalized.

This takes place, first, through consumption, pressuring these groups by means of systematic advertising and the opportunity for installment buying to widen the consumer goods market; second, through savings, capturing the savings of this population to the benefit of the banks of the dominant elite; and, lastly, through investment in which pobladores are oriented toward nonproductive activities and housing for the purpose of increasing consumption of products.

These rules of the game respond to the interests of the dominant class and not to the interests of the marginal population. For the dominant elite, the marginal population constitutes a reserve labor force for industry, a large mass of consumers who must be guided, and potential savers who could nourish the economic system. The cumulative effect of these relations has become an ideology of domination in which there is no development alternative for the dominated. While the pobladores make great efforts to increase their consumption, savings, and investment, the transfer of capital in favor of the dominant groups is much greater. In this way, the present system of domination continues to be reinforced.

The great question thus becomes: Is progress for the marginal population possible without altering the terms of the system and ending domination? It is clear that, while the system of domination which has created this unjust social order continues, development of large sectors of the population who do not actively participate in the task of national transformation is impossible. Although a phase of capital accumulation and of economic expansion could exist, a process of national development cannot take place without participation of these sectors in wealth generated and in management and control of the means of production. Only this will give rise to a new social order.

The process of development requires structural changes replacing the system of domination with a new social order allowing full participation of all the population. Within a scheme of domination, only negotiations can exist between the population of the barriadas, which is attempting to satisfy its immediate needs for water, lighting and housing, and external agencies, public and private, which act in accordance with roles assigned by the dominant elites.

In order to overcome the system of domination, it is necessary to establish the possibility of direct negotiations between the population organized functionally by productive activities and the state, and replace the policy of technical assistance oriented toward community development with technical assistance in production. These actions should be incorporated into a national development plan, which coordinates measures taken at the community level with those on a national scale.

Once participation is assured, the population organized in terms of specific functions must have guaranteed access to the decision-making apparatus. This would imply powers of direct negotiation between this population and the state in terms of accelerated and self-supported development, primarily maintained by the internal potential of Peru. Increase in the participation of these populations in decision-making implies total transformation of the educational system, permitting a type of education suited to the structural change of society, to development, and to the workers. The rise in the standard of living of marginal populations will imply rational participation of workers in the management and profits of business and also the development and protection of new cooperative firms.

The strategy for incorporating the poblador into the process of national development depends on six steps. First, it requires definition of a population policy in order to resolve the problem of marked population disequilibria. Next, it requires establishment of orientation and control mechanisms for urban expansion which include programs of land distribution and provision of basic infrastructure for marginal groups, within a short- and medium-term urban development plan which contemplates integration of marginal groups into the socioeconomic development process of Peru. Third, direction and organization of poblador participation in programs of urban living and housing, production and services, within a strategy leading to structural change is needed. Fourth, selection and implementation of technological levels guaranteeing high consumption of labor and allowing the poblador to qualify for integration into the process of socioeconomic development should follow. Fifth, introduction of financial mechanisms attracting the savings of organized pobladores and permitting them to finance programs according to their own interests and true abilities would occur. Finally, the establishment of training programs for leaders, volunteers, and technicians directed toward popular cooperation and social mobilization supported by public and private organizations is needed.

CHAPTER 27

Housing-Settlement Types, Arrangements for Living, Proletarianization, and the Social Structure of the City

ANTHONY LEEDS

As one scans the literature on cities from various disciplines, one observes that they [cities] tend to be conceived, as in architecture and urban planning, as physical entities and apparatus (buildings, open spaces, sewage systems . . .), or, as among the social science disciplines, as stage sets or backgrounds in front of which various categories of behavior . . . (kinship, migration, political behavior . . .) are to be observed. . . .

My own view has increasingly moved in the direction of seeing what we call "a city" . . . as a combination of socio-politico-economic structures and the physical apparatus . . . used in their operation. The physical apparatus reflects the social and ideological order, if always in a laggard manner, because its mere physical concretization lends itself to perpetuation while the social order is changing around it. Consequently, the primary interest in the study of cities . . . is not so much the physical apparatus as the social structure which, as a whole, underlies it.

One theme of this paper, then, is to show, for at least some cases, that the physical apparatus is . . . a reflection or crystallization of the social order and its city subsystem.

A second theme is to suggest that the societal orders of capitalistically constructed societies and their social structural manifestations in cities necessarily involve proletarianization . . . [which is] a dual process in that it is largely a result of the struggle for self-maintenance of elites . . . and the capitalist competition for private property. . . .

I turn, first, to some physical aspects, especially housing, of Rio de Janeiro, Brazil, and other cities, and then to some of the social-structural aspects involved in residence and its differential location in the city.

Reprinted from *Latin American Urban Research,* 4 (W. A. Cornelius and F. M. Trueblood, editors) © 1974, pp. 67–99, by permission of the publisher, Sage Publications, Inc., and the author. This paper has been heavily edited for this publication.

SETTLEMENT SPECIALIZATION IN RIO

One of the striking things in Rio, as in a large number of the cities of Latin America ... is the specialization of housing settlement types. ... [T]he familiar types include "better residential areas," the assumedly "middle and upper-middle class" settlements ... ; and the *favelas* or squatter settlements. ...

Favelas in Brazil, and, more generally, squatter settlements anywhere in the world, are spoken of as a "problem" in a manner analogous to the "problems" of "urban ghettos," of "slums," of "rural migrants," ... (etc.) so often found on the minds and tongues of concerned people. ... Essentially, these all refer to different aspects of the same problem—the proletarianization discussed in this paper.

Favelas are conceived of as a problem—as have been Lima's *barriadas*, San Juan's *arrabales*, Caracas' *ranchos* ... , etc.—because, allegedly, their populations are comprised, at one evil extreme, of assassins, thieves, ... narcotic addicts; at another evil extreme, of communists or other kinds of politically and socially rebellious menaces; at a third, less-evil extreme, of poor, benighted, uneducated, ill-adapted, hick-like immigrants; or, at the best of extremes, of reasonable human beings, but sad and poverty stricken, living in hovels. ...

I have elsewhere shown [Leeds and Leeds, 1970 *inter alia*] that almost all these conceptions are either false or drastic distortions ... but I *do* want to emphasize that favelas, "shanty-towns," squatments," unauthorized urban settlements, call them what you will, are usually the most striking *visible* form of residential settlement to be observed.

Not so visible and often difficult to discover or distinguish is a series of other low-income settlement types or residential neighborhoods which remain—as if they did not exist—almost totally undescribed not only in the literature regarding Rio but that respecting other Latin American cities where equivalent housing exists.

First among these housing-settlement types, comprising perhaps a quarter- to a half-million people, or about five percent of the population of Rio and a much larger percentage in Lima, are the rooming houses ... called, in the former, *cabeça de porco* or *casa de comodo* and, in the latter, *casas subdivididas* ... , a single large building ranging from perhaps twenty to perhaps eighty or one hundred one- and two-room apartments ... occupied by households composed of nuclear, subnuclear, or slightly extended families. ...

A second settlement type in Rio ... is called an *avenida* or *vila* with various qualifying adjectives such as *proletária, de lavadeiras,* etc. In Lima, the equivalent is the *callejon;* in Mexico, the *vecindad,* ... etc. It consists of a series of horizontal one- and two-room renting units, for all of which three or four toilets and a similar number of faucets and washtanks serve. There is what must be a condominium area—the central, elongated courtyard and entryways. . . . In Rio, the older variant of these known as *cortiços* . . . had a very high proportion of single-man occupancy. . . . Average household size was about 3 in contrast to the 4.6–4.7 for favelas and a roughly similar size for the rooming houses. Here, as in the rooming houses, rents are paid.

A third settlement type in Rio is the *parque proletário* or, in Chile, the *vila de emergência*—a government temporary . . . [project] meant to supply roofs and walls to persons without shelter because of urban renewal or drastic accidents to squatter settlements such as flooding or burning. . . . I have seen no estimate of the number of people in *parques proletários.* The housing units, like favelas, are occupied mainly by family groupings, but getting into them is a complex matter involving application to bureaucrats or agencies for assignment to a housing unit—a much more constraining procedure than entering a favela, although no capital is required.

A fourth type is called, in Rio, the *conjunto,* with equivalent terms in other countries. . . . A major feature of the conjuntos populated by low-income people in Rio and, I understand, elsewhere, is their occupational specialization, because each was separately built by an agency, labor union, association, or other corporativist group catering for its membership. . . . Thus, all over Rio and other cities are scattered occupational-residential enclaves. This fact takes on special significance in view of the scaled salaries for the entire Brazilian labor force which tend to set rather clear parameters to family income or family capital resources. . . . By contrast, neither the occupational specialization nor the set range of incomes holds for the favelas or other housing-settlement types. . . .

A fifth type is constituted by the vast "popular"—i.e., proletarian—housing Levittowns (in Rio also sometimes called conjuntos, but more commonly *vilas*). . . . The populations . . . have been removed from other parts of the city by act of man or God—by favela removal, urban renewal, flood, landslide or other disaster. In these vilas, the so-called "embryo" houses are sold to the residents who are selected by virtue of their alleged "capacity to pay" the calculated amortization rates based on the government's cost of building, but who are unable to afford better housing to replace the housing they lost. Thus, family capital inflows of the residents fall within rather narrow ranges, and out of them, they ostensibly pay the amortization payments. I say "ostensibly," because in Rio's vilas about sixty to eighty percent are defaulting at present. . . .

A sixth type is . . . referred to in Rio as the "*subúrbios,*" in general characterized by vast expanses of fairly humble, separate, privately owned houses on official streets which have little or no paving, often no lights, a poor water supply, few or no sewers or other urban services. They are largely in parts of the city farthest from the center, but are not properly suburbs in the North American usage of the term since they fall largely *inside* the juridical boundaries of the central city. . . .

A seventh type consists of the slums proper (*tugurios*)—areas of once-good but now-decaying housing and urban services; rented rooms, apartments and houses; pensions with room and board, mainly for men, cheap travellers' hotels, bordellos, and so on. There are large areas of these in all major Latin American cities, huge expanses in Rio and perhaps even vaster ones in Lima . . . about ten percent of the population of Rio and perhaps considerably more in Lima [lives in them]. . . .

The eighth type is comprised of the squatter settlements. . . . The single uniform criterion that distinguishes squatter settlements from other types of housing settlement in the city is that they constitute "illegal" occupation of the land in that their occupancy is based neither on their owning it nor renting it from legal owners.

All the other criteria often used for distinguishing squatter settlements from other types of housing-settlement in the city only apply some of the time. . . . Aside from occupancy, it is usually the case that they are not planned. . . . Lima, however, is remarkable for the number and size of "illegal" barriadas which were planned before invasion and some that seem to have been regulated after the settling took place.

Because of the developmental pattern of housing improvement to be found in them, it is highly misleading . . . to call squatter settlements "shanty-towns," although many are so. . . . In a number of cases, the squatter settlements, over time, are transformed into regular parts of the city with standard housing by the residents themselves.

It is rarely true that their populations are primarily "true rural" migrants though many people have come from more or less rural areas, although mostly through step-migration. It is not the case that the populations are uniformly made up of marginal labor, lumpenproletariats, or mere proletariats, but, rather, display a range of stratification reaching even up to upper-middle professional, bureaucratic, and business "levels" in some of the larger and more evolved squatter settlements. . . . It is not the fact that they are uniform and close-knit communities even in those squatments which have fairly tight-knit residents' associations.

In sum, though squatter settlements comprise a single housing-settlement type by virtue of their origin and the common characterizing feature of illegal occupancy of the land, and, consequently, their special jural status before the law and the public authority, nevertheless, as a universe, they display a much more varied range for virtually all characteristics of sociological interest than any of the other housing-settlement types and, individually, most of them are quite heterogeneous housing and social areas. Perhaps twenty percent of the jural city of Rio and twenty-two to twenty-five percent of the Rio conurbation . . . live in squatments at present. In Lima, today, upwards of forty percent of the entire population . . . live in barriadas, while nearly fifty percent of Caracas' people are in the barrios. . . .

Thus, approximately seventy percent of the population of the city of Rio de Janeiro and a comparable percentage in Lima and Caracas live in housing, which, with the exception of a few more evolved squatments, is almost exclusively proletarian.

Let me emphasize that each of the settlement types discussed has a characteristic physical apparatus easily recognizable by the trained eye. These apparatuses, plus the specialized physical apparatuses of the labor structure of industry, services, transportation, etc. . . . and of the administrative organization . . . together comprise the physical trappings of the city which . . . , in much of our thinking, tends to be *identified* with and understood *as* the city. Throughout what follows, I shall tease out the social relations which accompany and underlie these apparatuses and their spatial distribution.

[*Editors' note:* We have, with great reluctance, omitted large sections from this lengthy and subtly reasoned article. A section on "Alternative Life Arrangements" discusses how poor households make decisions among the various possible living arrangements open to them, sacrificing certain needs to obtain others, and how the same household over time may move from one type of settlement to another, depending upon the relative salience of needs and fluctuations in their resources. A following section explores the "Consequences of Choices Among Arrangements for Living," demonstrating that the types of proletarian zones breed solidarities and divisions both within and across them. Divisions are often fostered by "the elites who control the communications system and job market in particular," and who, by virtue of their power to relocate zones of proletarian settlement, often disrupt the social relations of the poor.]

CONSTRAINTS ON CHOICE

The key constraint on choice of any given housing-settlement type is financial, whatever the state of the domestic cycle. It is not only the major constraint on specific individual choices but the one that allows no exit from the entire set of choices of proletarian arrangements for living. In Brazil, today, a large proportion of the labor force officially receives only the minimum salary. A minimum income needed by a household to pay for rent, basic food supplies, clothing, transportation, and medical care comes to between 900 and 1200 new *cruzieros* (3–4 minimum salaries). Perhaps a maximum of five to ten percent of proletarian households reach that level of income, the majority taking in between 300 and 500 new cruzieros per month from wages and other sources.

The results are clear. Such a wage system, with the income structure it entails, sets the parameters for the set of choices as a whole ("the choice set"). . . .

The wage-structure is maintained by national policy, national administrative acts, and national agencies, all controlled by national strategic elites. . . . All significant policy- and decision-making offices and administrative posts are held by elite members of the upper class who are linked by blood, affinal, co-parentage, friendship, and other personal networks in bonds which create the class boundaries that exclude all members of the proletariat. . . .

Underlying the highly inequitable wage policy of countries like Brazil and Peru—the capitalist, semi-colonial dependencies of the great metropolitan capitalist countries—are the systems of capitalist-oriented private profits and private property. This is not the place to spell this out in detail. . . . What I wish to emphasize, here, however, is that this kind of dependent capitalist system determines a wage structure which necessarily involves proletarianization, or is indeed identical with it. Proletarianization is built into the capitalist system . . . but it is even more strongly delineated, less alleviated by "affluence," less ameliorated by great masses of better-paid, highly skilled wage earners, less softened by opportunities for upward mobility, less responsive to political protest and electoral expression, and generally more repressive in the "underdeveloped," dependent societies than in the metropoles like Great Britain and the United States.

Insofar as proletarianization is attached to capitalism and to industrialism and insofar as it is intensified by action of the State, all three of which tend institutionally and operationally to be centered in cities, it is a peculiarly *city* phenomenon, even though its extension can be found in the agrarian sectors as well. In its mass housing and settlement aspect, it is intrinsically a part of the city in capitalistically structured societies. Insofar as housing, settlement, work, householding, and carrying out one's daily activities in the city are linked, proletarianization also involves the emergence of a *social* system or subsystem in a given city, complementing its role in the societal structure.

Since proletarianization results from the relationship with the owners and controllers of strategic resources, capital, and the State, it is, in fact, a single process which *necessarily* involves the evolution of complementary roles—a dialectic development. The complementary aspect of the process might be called "elitization," or the continuous construction of upper-class power, of self-identification, of boundary conditions to exclude the proletariat, and of the proletariat needed for its own maintenance and progress.

Put another way, the private-property and private-profit systems necessarily involve a process of complementary role set development—the evolution of the proletariats and the elites; and the city, *as an organization for production,* with all its productive apparatuses, is the primary locus of this process.

The structural underpinning of this process is the economic, social and political exclusiveness of the elites . . . [which] in all these spheres evolve means for their own maintenance of which the wage system already mentioned is perhaps the single most effective one. However, there are a large number of ancillary ones [such as special hiring and firing arrangements, the system of work permits, the high cost of educational materials, patron-client relationships etc.] which operate together with the wage system to maintain the boundaries. In effect, they set the parameters within which the proletarian arrangements for living choices can be made. . . .

[Editors' note: Here there is a lengthy discussion of "Elite Cleavage and Coalitions with Proletarian Groups" which helps set the stage for the final section of the paper, entitled:]

IMPLICATIONS FOR PLANNING

. . . [T]he physical city is, to a great degree, a time-linked crystallization of the *total* social order of the city—of the interactions and interests of elites and proletariats. The physical city, as seen on the ground, not on the planners' drawing board, is unintelligible without understanding the proletarianization process and proletarian action. Generally, and certainly in capitalist societies if not more widely, proletarian process and action are either disregarded altogether, thought of with respect to specific characteristics, such as squatter settlements, as aberrations, or thought of only piecemeal, as with respect to favelas but not other types of housing. . . .

In consequence, planners, who virtually without exception are recruited from the elites, see the physical city in only a partial way. As elite personnel, they see the city as elite process. They see the future city for which planning is to be done in terms of the extrapolated future of the upper class or, more likely, those of its subsegments more closely linked with the professions and with government. Since most of them see the city only partially, they necessarily plan partially.

Planning partially means that the planning is only for some of the social roles of the city. Since the social order of the city *inherently* involves interactions with, and action by, the part(s) *not* planned for, it is necessarily the case that processes, events, and situations immediately linked with the city-as-a-social-process are not accounted for in the plans. At very worst, the accounted for and the nonaccounted for are in direct conflictual contradiction and lead to intensified social struggle and deteriorating city scapes—as in the case of Rio today . . . and possibly major North American cities like Washington and New York. At best, there may be an accidental coincidence of interests for a short term, as perhaps in the case of Lima, where still today, after continuous talk since the 1968 takeover of instituting an urban reform, none yet exists.

In general, then, any urban planning which does not take account of the entire social order of the city is bound to fail. . . . In societies sharply divided along class lines, where official positions such as those of planners are filled only by members of one class, the failures are likely to be of even greater magnitude—i.e., the case of Brasilia (Epstein, 1973). The implication would appear to be that successful urban planning requires the elimination of class. I tend to think that this, too, would be an overly optimistic view. The problem is that a plan foresees only a range of possibilities in the future development and then only for a limited time. Once the plan is put into effect, it *literally* concretizes the state of knowledge, the conception of the possibilities, and the form of social organization that existed at the time of planning. . . . Possibly we shall have to come to the point of view that the social process is *itself* the planning process.

SECTION V

GOALS AND POLICIES

INTRODUCTION

Just as this book began with the theme of inequality, so it ends with the theme of equality. The United Nations, meeting in Vancouver, Canada, in June 1976, at Habitat: United Nations Conference on Human Settlements, deliberated for weeks after a year of preparatory conferences and national studies to set forth the goals to be sought in planning for improved urban *and* rural environments. These goals are incorporated in the *Declaration of Principles* issued at the end of the conference. Beneath the turgid prose of the United Nations is a solemn and important message whose chief thrust is social justice and greater equity. The declaration places the problems of human settlements squarely in the international and national arenas, by noting that "the problems of human settlements are not isolated from the social and economic development of countries and they cannot be set apart from existing unjust international economic relations." It further places the blame for many of the unacceptable conditions of human settlements, found "particularly in developing countries," on such factors as "inequitable economic growth, reflected in the wide disparities in wealth which now exist between countries and between human beings," as well as upon population growth, uncontrolled urbanization, and involuntary migration. While calling upon the international community to implement the new economic order which envisages a more equitable distribution of wealth *among* nations, it also calls upon the individual national governments to adopt policies for the better planning of human settlements and for the improvement of standards of living (including shelter) for its most disadvantaged people.

It is within the realm of the latter that national policies toward urbanization fall. These are the focus of the second paper in this section, prepared by the Secretariat of the U.N. Habitat Conference. A variety of policies and institutions to cope with urban problems in developed and developing nations are described. These national policies have been devised within a wide range of political and economic systems which may be more or less competent to deal with them, but what is striking is that, regardless of system, the problems appear to be quite similar. Some urban problems seem universal and no system can be said to have "solved" them once and for all. One wonders whether it is industrialization *per se* (with its increased scale, complexity, technological interdependence) which generates these problems or whether it is ineradicable social inequities which persist, even under ostensibly different kinds of economic and political systems?

This is what makes the Chinese experiment, with which this volume concludes, so important. Mingione, an Italian sociologist with considerable experience in Communist China, poses the question as follows when, at the beginning of his article, he asks "whether the processes of rapid urbanization, depopulation of the countryside, centralization of production and lack of regional balance—which have characterizd the industrial development of western societies—are characteristics of industrialization *per se* or are the result of the economic, social and political system of capitalism." While it is certainly premature to answer this question here, his account can only intensify our curiosity and our interest in the outcome of this first major radical alternative to the patterns of urbanization we have hitherto known in the developed west and the developing Third World.

CHAPTER 28

Declaration of Principles

HABITAT: UNITED NATIONS CONFERENCE ON HUMAN SETTLEMENTS

Aware that the conference was convened following recommendation of the United Nations Conference on the Human Environment and subsequent resolutions of the General Assembly, particularly resolution 3128 (XXVIII) by which the nations of the world expressed their concern over the extremely serious condition of human settlements, particularly that which prevails in developing countries,

Recognizing that international co-operation, based on the principles of the United Nations Charter, has to be developed and strengthened in order to provide solutions for world problems and to create an international community based on equity, justice and solidarity,

Recalling the decisions of the United Nations Conference on the Human Environment, as well as the recommendations of the World Population Conference, the United Nations World Food Conference, the Second General Conference of the United Nations Industrial Development Organization, the World Conference of the International Women's Year; the Declaration and Programme of Action adopted by the sixth special session of the General Assembly of the United Nations and the Charter of Economic Rights and Duties of States that establish the basis of the New International Economic Order,

Noting that the condition of human settlements largely determines the quality of life, the improvement of which is a prerequisite for the full satisfaction of basic needs, such as employment, housing, health services, education and recreation,

Recognizing that the problems of human settlements are not isolated from the social and economic development of countries and that they cannot be set apart from existing unjust international economic relations,

Being deeply concerned with the increasing difficulties facing the world in satisfying the basic needs and aspirations of peoples consistent with principles of human dignity,

Recognizing that the circumstances of life for vast numbers of people in human settlements are unacceptable, particularly in developing countries, and that, unless positive and concrete action is taken at national and

Reprinted from United Nations, *Declaration of Principles,* Vancouver Declaration on Human Settlements, June 1976, pp. 2–9.

international levels to find and implement solutions, these conditions are likely to be further aggravated, as a result of:

> *Inequitable economic growth,* reflected in the wide disparities in wealth which now exist between countries and between human beings and which condemn millions of people to a life of poverty, without satisfying the basic requirements for food, education, health services, shelter, environmental hygiene, water and energy;
>
> *Social, economic, ecological and environmental deterioration* which are exemplified at the national and international levels by inequalities in living conditions, social segregation, racial discrimination, acute unemployment, illiteracy, disease and poverty, the breakdown of social relationships and traditional cultural values and the increasing degradation of life-supporting resources of air, water and land;
>
> *World population growth* trends which indicate that numbers of mankind in the next 25 years would double, thereby more than doubling the need for food, shelter and all other requirements for life and human dignity which are at the present inadequately met;
>
> *Uncontrolled urbanization* and consequent conditions of overcrowding, pollution, deterioration and psychological tensions in metropolitan regions;
>
> *Rural backwardness* which compels a large majority of mankind to live at the lowest standards of living and contribute to uncontrolled urban growth;
>
> *Rural dispersion* exemplified by small scattered settlements and isolated homesteads which inhibit the provision of infrastructure and services, particularly those relating to water, health and education;
>
> *Involuntary migration,* politically, racially, and economically motivated, relocation and expulsion of people from their national homeland,

Recognizing also that the establishment of a just and equitable world economic order through necessary changes in the areas of international trade, monetary systems, industrialization, transfer of resources, transfer of technology, and the consumption of world resources, is essential for socio-economic development and improvement of human settlement, particularly in developing countries,

Recognizing further that these problems pose a formidable challenge to human understanding, imagination, ingenuity and resolve, and that new priorities to promote the qualitative dimensions to economic development, as well as a new political commitment to find solutions resulting in the practical implementation of the New International Economic Order, become imperative:

I. OPPORTUNITIES AND SOLUTIONS

1. Mankind must not be daunted by the scale of the task ahead. There is need for awareness of and responsibility for increased activity of the national Governments and international community, aimed at mobilization of economic resources, institutional changes and international solidarity by:

(a) Adopting bold, meaningful and effective human settlement policies and spatial planning strategies realistically adapted to local conditions;
(b) Creating more livable, attractive and efficient settlements which recognize human scale, the heritage and culture of people and the special needs of disadvantaged groups especially children, women and the infirm in order to ensure the provision of health, services, education, food and employment within a framework of social justice;
(c) Creating possibilities for effective participation by all people in the planning, building and management of their human settlements;
(d) Developing innovative approaches in formulating and implementing settlement programmes through more appropriate use of science and technology and adequate national and international financing;
(e) Utilizing the most effective means of communications for the exchange of knowledge and experience in the field of human settlements;
(f) Strengthening bonds of international co-operation both regionally and globally;
(g) Creating economic opportunities conducive to full employment where, under healthy, safe conditions, women and men will be fairly compensated for their labour in monetary, health and other personal benefits.

2. In meeting this challenge, human settlements must be seen as an instrument and object of development. The goals of settlement policies are inseparable from the goals of every sector of social and economic life. The solutions to the problems of human settlements must therefore be conceived as an integral part of the development process of individual nations and the world community.
3. With these opportunities and considerations in mind, and being agreed on the necessity of finding common principles that will guide Governments and the world community in solving the problems of human settle-

ments, the Conference proclaims the following general principles and guidelines for action.

II. GENERAL PRINCIPLES

1. The improvement of the quality of life of human beings is the first and most important objective of every human settlement policy. These policies must facilitate the rapid and continuous improvement in the quality of life of all people, beginning with the satisfaction of the basic needs of food, shelter, clean water, employment, health, education, training, social security without any discrimination as to race, colour, sex, language, religion, ideology, national or social origin or other cause, in a frame of freedom, dignity and social justice.
2. In striving to achieve this objective, priority must be given to the needs of the most disadvantaged people.
3. Economic development should lead to the satisfaction of human needs and is a necessary means towards achieving a better quality of life, provided that it contributes to a more equitable distribution of its benefits among people and nations. In this context particular attention should be paid to the accelerated transition in developing countries from primary development to secondary development activities, and particularly to industrial development.
4. Human dignity and the exercise of free choice consistent with over-all public welfare are basic rights which must be assured in every society. It is therefore the duty of all people and Governments to join the struggle against any form of colonialism, foreign aggression and occupation, domination, *apartheid* and all forms of racism and racial discrimination referred to in the resolutions as adopted by the General Assembly of the United Nations.
5. The establishment of settlements in territories occupied by force is illegal. It is condemned by the international community. However, action still remains to be taken against the establishment of such settlements.
6. The right of free movement and the right of each individual to choose the place of settlement within the domain of his own country should be recognized and safeguarded.
7. Every State has the sovereign and inalienable right to choose its economic system, as well as its political, social and cultural system, in accordance with the will of its people, without interference, coercion or external threat of any kind.
8. Every State has the right to exercise full and permanent sovereignty over its wealth, natural resources and economic activities, adopting the necessary measures for the planning and management of its resources, providing for the protection, preservation and enhancement of the environment.

9. Every country should have the right to be a sovereign inheritor of its own cultural values created throughout its history, and has the duty to preserve them as an integral part of the cultural heritage of mankind.
10. Land is one of the fundamental elements in human settlements. Every State has the right to take the necessary steps to maintain under public control the use, possession, disposal and reservation of land. Every State has the right to plan and regulate use of land, which is one of its most important resources, in such a way that the growth of population centres both urban and rural are based on a comprehensive land use plan. Such measures must assure the attainment of basic goals of social and economic reform for every country, in conformity with its national and land tenure system and legislation.
11. The nations must avoid the pollution of the biosphere and the oceans and should join in the effort to end irrational exploitation of all environmental resources, whether non-renewable or renewable in the long term. The environment is the common heritage of mankind and its protection is the responsibility of the whole international community. All acts by nations and people should therefore be inspired by a deep respect for the protection of the environmental resources upon which life itself depends.
12. The waste and misuse of resources in war and armaments should be prevented. All countries should make a firm commitment to promote general and complete disarmament under strict and effective international control, in particular in the field of nuclear disarmament. Part of the resources thus released should be utilized so as to achieve a better quality of life for humanity and particularly the peoples of developing countries.
13. All persons have the right and the duty to participate, individually and collectively in the elaboration and implementation of policies and programmes of their human settlements.
14. To achieve universal progress in the quality of life, a fair and balanced structure of the economic relations between States has to be promoted. It is therefore essential to implement urgently the New International Economic Order, based on the Declaration and Programme of Action approved by the General Assembly in its sixth special session, and on the Charter of Economic Rights and Duties of States.
15. The highest priority should be placed on the rehabilitation of expelled and homeless people who have been displaced by natural or man-made catastrophes, and especially by the act of foreign aggression. In the latter case, all countries have the duty to fully co-operate in order to guarantee that the parties involved allow the return of displaced persons to their homes and to give them the right to possess and enjoy their properties and belongings without interference.
16. Historical settlements, monuments and other items of national heritage, including religious heritage, should be safeguarded against any acts of aggression or abuse by the occupying Power.

17. Every State has the sovereign right to rule and exercise effective control over foreign investments, including the transnational corporations within its national jurisdiction, which affect directly or indirectly the human settlements programmes.
18. All countries, particularly developing countries, must create conditions which make possible the full integration of women and youth in political, economic and social activities, particularly in the planning and implementation of human settlement proposals and in all the associated activities, on the basis of equal rights, in order to achieve an efficient and full utilization of available human resources, bearing in mind that women constitute half of the world population.
19. International co-operation is an objective and a common duty of all States, and necessary efforts must therefore be made to accelerate the social and economic development of developing countries, within the framework of favorable external conditions, which are compatible with their needs and aspirations and which contain the due respect for the sovereign equality of all States.

III. GUIDELINES FOR ACTION

1. It is recommended that Governments and international organizations should make every effort to take urgent action as set out in the following guidelines:
2. It is the responsibility of Governments to prepare spatial strategy plans and adopt human settlement policies to guide the socio-economic development efforts. Such policies must be an essential component of an over-all development strategy, linking and harmonizing them with policies on industrialization, agriculture, social welfare, and environmental and cultural preservation so that each supports the other in a progressive improvement in well-being of all mankind.
3. A human settlement policy must seek harmonious integration or co-ordination of a wide variety of components, including, for example, population growth and distribution, employment, shelter, land use, infrastructure and services. Governments must create mechanisms and institutions to develop and implement such a policy.
4. It is of paramount importance that national and international efforts give priority to improving the rural habitat. In this context, efforts should be made towards the reduction of disparities between rural and urban areas, as needed between regions and within urban areas themselves, for a harmonious development of human settlements.
5. The demographic, natural and economic characteristics of many countries, require policies on growth and distribution of population, land tenure and localization of productive activities to ensure orderly processes of urbanization and arrange for rational occupation of rural space.

6. Human settlement policies and programmes should define and strive for progressive minimum standards for an acceptable quality of life. These standards will vary within and between countries, as well as over periods of time, and therefore must be subject to change in accordance with conditions and possibilities. Some standards are most appropriately defined in quantitative terms, thus providing precisely defined targets at the local and national levels. Others must be qualitative, with their achievement subject to felt need. At the same time, social justice and a fair sharing of resources demand the discouragement of excessive consumption.

7. Attention must also be drawn to the detrimental effects of transposing standards and criteria that can only be adopted by minorities and could heighten inequalities, the misuse of resources and the social, cultural and ecological deterioration of the developing countries.

8. Adequate shelter and services are a basic human right which places an obligation on Governments to ensure their attainment by all people, beginning with direct assistance to the least advantaged through guided programmes of self-help and community action. Governments should endeavour to remove all impediments hindering attainments of these goals. Of special importance is the elimination of social and racial segregation, *inter alia,* through the creation of better balanced communities, which blend different social groups, occupation, housing and amenities.

9. Health is an essential element in the development of the individual and one of the goals of human settlement policies should be to improve environmental health conditions and basic health services.

10. Basic human dignity is the right of people, individually and collectively, to participate directly in shaping the policies and programmes affecting their lives. The process of choosing and carrying out a given course of action for human settlement improvement should be designed expressly to fulfil that right. Effective human settlement policies require a continuous co-operative relationship between a Government and its people at all levels. It is recommended that national Governments promote programmes that will encourage and assist local authorities to participate to a greater extent in national development.

11. Since a genuine human settlement policy requires the effective participation of the entire population, recourse must therefore be made at all times to technical arrangements permitting the use of all human resources, both skilled and unskilled. The equal participation of women must be guaranteed. These goals must be associated with a global training programme to facilitate the introduction and use of technologies that maximize productive employment.

12. International and national institutions should promote and institute education programmes and courses in the subject of "human settlements."

13. Land is an essential element in development of both urban and rural settlements. The use and tenure of land should be subject to public control

because of its limited supply through appropriate measures and legislation including agrarian reform policies—as an essential basis for integrated rural development—that will facilitate the transfer of economic resources to the agricultural sector and the promotion of the agro-industrial effort, so as to improve the integration and organization of human settlements, in accordance with national development plans and programmes. The increase in the value of land as a result of public decision and investment should be recaptured for the benefit of society as a whole. Governments should also ensure that prime agricultural land is destined to its most vital use.

14. Human settlements are characterized by significant disparities in living standards and opportunities. Harmonious development of human settlements requires the reduction of disparities between rural and urban areas, between regions and within regions themselves. Governments should adopt policies which aim at decreasing the differences between living standards and opportunities in urban and non-urban areas. Such policies at the national level should be supplemented by policies designed to reduce disparities between countries within the framework of the New International Economic Order.

15. In achieving the socio-economic and environmental objectives of the development of human settlements, high priority should be given to the actual design and physical planning processes which have as their main tasks the synthesis of various planning approaches and the transformation of broad and general goals into specific design solutions. The sensitive and comprehensive design methodologies related to the particular circumstances of time and space, and based on consideration of the human scale should be pursued and encouraged.

16. The design of human settlements should aim at providing a living environment in which identities of individuals, families and societies are preserved and adequate means for maintaining privacy, the possibility of face-to-face interaction and public participation in the decision-making process are provided.

17. A human settlement is more than a grouping of people, shelter and work places. Diversity in the characteristics of human settlements reflecting cultural and aesthetic values must be respected and encouraged and areas of historical, religious or archaeological importance and nature areas of special interest preserved for posterity. Places of worship, especially in areas of expanding human settlements, should be provided and recognized in order to satisfy the spiritual and religious needs of different groups in accordance with freedom of religious expression.

18. Governments and the international community should facilitate the transfer of relevant technology and experience and should encourage and assist the creation of endogenous technology better suited to the socio-cultural characteristics and patterns of population by means of bilateral or

multilateral agreements having regard to the sovereignty and interest of the participating States. The knowledge and experience accumulated on the subject of human settlements should be available to all countries. Research and academic institutions should contribute more fully to this effort by giving greater attention to human settlements problems.

19. Access should be granted, on more favorable terms, to modern technology, which should be adapted, as necessary, to the specific economic, social and ecological conditions and to the different stages of development of the developing countries. Efforts must be made to ensure that the commercial practices governing the transfer of technology are adapted to the needs of the developing countries and to ensure that buyers' rights are not abused.

20. International, technical and financial co-operation by the developed countries with the developing countries must be conducted on the basis of respect for national sovereignty and national development plans and programmes and designed to solve problems relating to projects, under human settlement programmes, aimed at enhancing the quality of life of the inhabitants.

21. Due attention should be given to implementation of conservation and recycling technologies.

22. In the planning and management of human settlements, Governments should take into consideration all pertinent recommendations on human settlements planning which have emerged from earlier conferences dealing with the quality of life and development problems which affect it, starting with the high global priority represented by the transformation of the economic order at the national and international levels (sixth and seventh special sessions), the environmental impact of human settlements (Stockholm Conference on the Human Environment), the housing and sanitary ramifications of population growth (World Population Conference, Bucharest), rural development and the need to increase food supply (World Food Conference, Rome) and the effect on women of housing and urban development (International Women's Conference, Mexico City).

23. While planning new human settlements or restructuring existing ones, a high priority should be given to the promotion of optimal and creative conditions of human coexistence. This implies the creation of a well-structured urban space on a human scale, the close interconnection of the different urban functions, the relief of urban man from intolerable psychological tensions due to overcrowding and chaos, the creation of chances of human encounters and the elimination of urban concepts leading to human isolation.

24. Guided by the foregoing principles, the international community must exercise its responsibility to support national efforts to meet the human settlements challenges facing them. Since resources of Governments are inadequate to meet all needs, the international community should provide

the necessary financial and technical assistance, evolve appropriate institutional arrangements and seek new effective ways to promote them. In the meantime, assistance to developing countries must at least reach the percentage targets set in the International Development Strategy for the Second United Nations Development Decade.

CHAPTER 29

Policies, Planning and Institutions

HABITAT: UNITED NATIONS CONFERENCE ON HUMAN SETTLEMENTS

URBAN EXPLOSION OR PLANNED GROWTH

... In the affluent parts of the world, there is an often unwanted concentration of activities, things and people in urban agglomerations such as the Eastern Seaboard Megalopolis and the West Coast Megalopolis in the United States of America, or the great urban-industrial complexes of northwestern Europe and Japan. In the developing continents, there is a relentless and chaotic build-up of people and poverty in amorphous metropolitan areas, such as Calcutta, where explosive expansion is due less to the "pull" of industrial and economic growth than to the massive flight of rural people from misery and hunger. In both cases, the agglomeration of production and people continues in certain areas at the expense of the rest. Megalopolis is the ultimate outcome of *laissez-faire* policies in urban affairs.

Yet the urban crisis is also a manifestation of scientific and technological progress. It represents a promise of abundance as well as a challenge to our ability to accept changes in attitudes and values that would enable the world community to use its immense new productivity for truly bettering the human condition throughout the world and closing the gap between the affluence of a few countries and the increasing poverty of the rest. The growth potential of the world society as a whole multiplies with the advances achieved by science and technology, in both the highly industrialized and the developing countries. But the necessary adjustments in the approach to such basic issues as economics, land, and government are very slow to come. Thus, problems multiply everywhere at exceptionally rapid rates. In spite of ever-growing debate and research, the urgency and hugeness of the issues are yet to be grasped by theorists and practitioners, let alone by political leaders or the average man in the street.

Condensed and edited from United Nations, *Policies, Planning and Institutions: Item 10 of the Provisional Agenda,* United Nations Conference on Human Settlements, April 1976, Document A/CONF. 70/A/2, pp.26–53.

The commonest approach in settlement planning was, and still is, to extrapolate from current and past trends. Thus, new designs are proposed for the shape of megalopolis by architects, town planners and futurologists in the spirit of the "space age" and science fiction, but combined with current trends in urbanization. Like their pre-war forerunners, Le Corbusier and Wright, urban planners are fascinated with the new forms of settlement made possible by science and technology. The point they are trying to make is that humans could some day live almost anywhere, in environments of their own making, if they could alter their prejudices and learn to enjoy rather than merely tolerate the effects of science. What these "technological designs" omit are the economic, social, political and cultural instruments that societies must conceive and mobilize in order to build a better human environment. The real challenge is that planning, unlike Utopia, must specify the means of achieving a future state of affairs. . . .

A. SETTLEMENTS IN THE MARKET ECONOMIES

Some of the approaches proposed for dealing with the highly explosive situation in the field of human settlements are explored below. The first examples are taken from market economy industrialized and developing countries. The cases discussed illustrate the phenomenon of the spontaneous agglomeration of activities and people in response primarily to the short-term requirements of economic growth and foreign trade. In the affluent countries, in the absence of any specific national policy for human settlements, action to cope with the resulting social and environmental degradation is piecemeal and ineffective, unable to repair the damage or to arrest the trend toward further degradation. In the less affluent countries, development policies oriented to external markets tend to increase national economic and political dependency, accelerate rural/urban migration, produce disproportionately large metropolitan concentrations and in the process destroy the traditional hierarchy of settlements, the social structure, and national institutions and cultural values. . . .

Deliberate Dispersion of Development [The Case of Japan]

In 1960, against the background of a sustained high rate of economic growth in the post-war period, the Government of Japan formulated a long-range development policy which envisaged doubling the national income in 10 years through intensified industrialization. Subsequently, a regional approach was adopted and nine regions were established, each with its own development target. Twenty-year programs were then prepared for infrastructural development in these regions.

The stated national goals were: continuous and balanced growth of the economy; a rise in levels of living; and conditions in which all citizens throughout the country could enjoy an affluent life and share equally in all the benefits of modern society. In 1960, 1962 and 1964, the United Nations sponsored multidisciplinary discussions on urbanization and regional development in which, in co-operation with Japanese practitioners, administrators, scholars and representatives of the different interest groups, approaches were formulated and actions recommended with regard to a number of urban, metropolitan and regional issues. Some of these recommendations have already been implemented and others are in the process of implementation.

The intended effect of Japan's 10-year plan and the projected further development was to eliminate economic and social disparities among the regions and between town and country; to close the gap between the incomes of agricultural and industrial producers; to overcome the discrepancies in the economic strength of large and small enterprises; and to create full employment. In the process, the number of full-time and part-time agricultural workers was planned to decline from 40 percent of the country's labour force in 1958 to one third that number in 1970. Industry, power and transportation were to be expanded and better co-ordinated nationally.

Japan achieved an amazing post-war recovery, but by the 1960s urban services and utilities in metropolitan areas were at the breaking point. The infrastructures and external economies of the cities would have had to be recreated, at high cost, to allow further uninhibited growth. In order to avoid incurring this enormous outlay, a national policy of deliberate dispersion of investment, particularly investment in physical, economic and social infrastructure, was adopted for the different regions.

Central planning, government guidance, an orientation towards regional development, and local public and private initiatives for the implementation of national objectives are together expected to evolve less-centralized patterns of settlement which will, in time, bring the backward rural areas the benefits of urban life. The basic elements of the policy are:

- The establishment, through sufficient infrastructural investment, within the wider regions of the already overcrowded metropolitan areas, of suitably equipped alternative centres for industry, commerce, culture and residence, so as to reduce the pressure on central cities now caused by rural migration, physical congestion and other social and environmental problems (examples: Tokyo, the National Capital Region, and Osaka/Kobe, the Kinki Region);
- The creation, through even larger infrastructural investment, of adequate economic opportunities and sufficient social and cultural services in the urban and rural areas of as yet underdeveloped regions in order to deflect the flow of rural migration to areas capable of productive growth (example: Nagoya, Chubu Region);

• The strengthening of the economies and infrastructure of smaller towns and cities and rural zones, the expansion of their trade and industry, and the improvement of their living conditions through suitable environmental development, improved social, educational and health services, and enlarged industrial job opportunities, so as to minimize the need for internal migration, and enhance the geographical distribution—the redevelopment of large metropolitan regions and the reinforcement of their economies by judicious investment in capital-intensive industries and services to raise productivity and the quality of life through better environmental design and better social and cultural facilities, so as to obviate or reduce the need for further physical expansion of central cities.

The Japanese policy of geographical dispersion of the means of production and of people, operating within a free enterprise economic system, has been guided by economic incentives and also, necessarily, by economic benefit, often to the detriment of the quality of life and the environment.

Population concentration in urban areas has progressed rapidly in recent years through the transfer of workers from primary industry to secondary and tertiary industry. As a result, the ratio of the population in urban areas to the nation's total population has climbed rapidly from 56.3 percent in 1955 to 72.2 percent in 1970. There are 8 cities with more than one million people and 7 cities with 500,000–1,000,000. There are 21 cities with 300,000 to 500,000 inhabitants and 114 cities with a population of 100,000–300,000. The Tokyo Metropolitan Area (encompassing Tokyo and the three adjacent prefectures) had in 1970 a population of about 22 million within a radius of approximately 50 kilometres, forming a megalopolitan agglomeration of a size and complexity without precedent in the entire history of mankind.

The concentration and expansion of economic activities in the major urban areas has intensified social ills and public hazards. On the other hand, the outflow of population from rural areas has accelerated "depopulation" in the provinces, accompanied by deficiencies in basic facilities and services in the areas of education, medical care, disaster prevention, etc. The local communities are less and less able to perform their normal functions.

In the 1960s, economic efficiency and the growth of the national product were the priorities in Japan. At that time too, 6 to 7 percent of the gross national product went into infrastructural development, a percentage many other countries spend on armaments. This, with national planning, dynamic regional and local development, and the habit of saving, made the income doubling miracle come true. With growing affluence, the challenge looms larger. In the 1960s, familiar models tested for economic efficiency created astonishing productivity. Now, new development concepts, a closer partnership of government and citizens, and fuller integration of all levels in planning and execution are needed in order to identify opportunities and to stimulate and guide further development into socially productive channels.

The nation faces a crucial choice. Development criteria and public investment must shift from economic to social and environmental needs, from the quantity produced to the human qualities of life desired. The tools are there, and so are the capital, talent and know-how. But there must also be social innovation or growing affluence and the misuse of technology may usher in a computerized technocracy, or an era of conformity because citizens are so "programmed" or environmental decay and degradation may reach a point at which there is no hope of repair.

Incipient "Megalopolis" [The Case of Tunisia]

The big cities of the developing countries are experiencing similar concentrations of activities and people. The city of Tunis, a metropolitan agglomeration of about a million people, is a good example of economic development and urban growth within the constraints of external economic dependence.

Tunis is the capital city of a lopsided hierarchy of settlements—a network of smaller towns and villages, some linked to one another through intensive interchanges and others isolated in the predominantly agricultural plains of northern Tunisia. In the nineteenth century, the city's 100,000 inhabitants were one tenth of the country's population. At present, it holds one sixth. The accelerated migration of rural people to the capital was a direct consequence of Tunisia's transition from a pre-industrial economy to a dependent market economy. As a consequence, the old Arab city, the Medina, with its traditional urban pattern, gradually became enveloped by a new city laid out by the colonial authority in rectangular patterns.

The division of the city of Tunis into Arab and European quarters (physically and socially, as well as in terms of amenities and services) reflected the process of outward-oriented development within the limitations imposed by political and economic dependence. In the 1940s, a third element appeared in the form of numerous shanty towns inhabited by landless peasants who had become jobless urban dwellers. This phenomenon was a direct consequence of the concentration of ownership of agricultural land and the intensive mechanization of grain production for the world market. Both of these processes are characteristic of the consolidation of colonial relationships.

By 1956, Tunis had an urban structure consisting of the Medina, whose traditional functions were vanishing, a colonial city which was the centre of political and economic power, and a series of shanty towns containing the rural people who had fled from collapsing rural communities. This urban structure was based on, and an outcome of, social and racial segregation among antagonistic groups: the French and Italian colonists, the Tunisian landlords, the emerging *bourgeoisie* active in commerce and the professions, the working class tied to the colonial economy, the unemployed migrants, and certain ethnic and religious minorities.

Like other newly independent nations, Tunisia faced the problem of transforming its socio-economic structure from that of a dependent colony into a national economy; turning the capital of a colonial society into an integrated national capital; and transforming the country's settlement network that had served the interests of a colonial power into a system designed to promote national development.

Tunisia attempted to do this at first through the planned development of its infrastructural base and through decentralized industrialization. Both elements of this development strategy were intended to reduce and ultimately eliminate regional disparities. A temporary effect of the policy was to ease the pressure of rural migration on the capital city. In the long run, however, the provision of infrastructure and services and industrial decentralization were unable to resolve all the problems of the accelerating rural collapse. Consequently, regionalization and co-operatives were added as supplementary development instruments.

These measures in effect redistributed economic activities to a degree and stabilized the flow of migration for a time. But they were unable to eliminate the grave structural imbalances of an outwardly oriented economy built in the colonial era. The division of Tunis continued, the shanty towns continued to exist and to grow, new prestigious housing estates sprang up at the expense of housing for the lowest income groups, and Tunis obtained some of the structures usually associated with a capital: an Olympic stadium, a congress hall, many office buildings, and so on. However, in spite of these heavy investments, the sorely needed restructuring and integration of Tunis as a national capital did not materialize.

Since 1970, the Government has been implementing a new development policy, relying largely on private initiative aided by new legislation and supportive public investment. The co-existence of the three economic sectors—private, public and co-operative—has in fact resulted in a net advantage to private enterprise. Thus, in the process of reorganizing agricultural co-operatives, a large number of small land-owners have seen their former holdings incorporated in larger units. This, together with agricultural mechanization, has resulted in a new wave of migration. The total effect has been a further transfer of rural people from the interior into Tunisia's coastal areas.

A second strategy instrument of the new economic policy was the 1972 Investment Code enabling foreign-based industries to produce on Tunisian territory goods destined exclusively for the foreign market. The consequence in terms of human settlements was this: in the absence of alternative incentives, investment went to the better-equipped urban areas, particularly to the metropolitan region of Tunis where there exists an abundant supply of labour, a harbour, an airport, automatic telecommunications, office space, hotel facilities, and some higher-standard housing. As a result, a large part of foreign investment under the provisions of this new policy is concentrated in the Tunis metropolitan area, and the excessive growth of the city and its predominance over the rest of the country continues to increase. "Modern" structures are multiplying. At the same time, the shanty towns are extending rapidly in numbers and area. The long-term social cost to the nation of this development model may far outweigh the short-term economic benefit which is its primary motivation.

The policy of creating most favourable conditions for foreign investment has further polarized growth in metropolitan Tunis. The capital city is now becoming more than ever an overgrown centre of production and economic power geared to the requirements of external markets, not to national needs. The former colonial structure has become "externalized" development based on economic dependence. The resulting urbanization patterns cannot but be a continuation of those of the colonial era.

A combination of *laissez-faire* in economic development and orientation to external requirements, and, consequently, economic dependence on external sources, are often determining factors for patterns of settlement in the developing countries. These economic and political factors, together with the consequences of unprecedented population growth and concentration in the metropolitan areas, are giving rise in the developing countries to urban agglomerations that bear all the characteristics of "megalopolitan" development—minus the resources required to cope with the problems it generates.

B. PLANNED GROWTH

Whether uncontrolled growth occurs in a highly industrialized country or in a developing, pre-industrial country, it leads to social and environmental dislocation and human degradation. Other approaches to development have been conceived and tested in a number of countries over the past few decades. Some of them are discussed below, as illustrations of the crucial issues involved and the different instruments that have been applied to implement these approaches under differing conditions, in societies with different economic and political systems, operating at different levels of technology and development, and within the different value-systems of their respective cultures.

Building Socialist Cities [The Case of Eastern Europe]

The planning of urban development in the socialist countries of Eastern Europe is usually an integrated part of planned national development. In the redevelopment of existing large metropolitan cities in the Soviet Union, for example, no encouragement is given to new industries, save for communal services and the building, particularly house-building, industry. New industries are preferably located in new or existing medium-sized or smaller cities, which are then developed to what is regarded as the preferred size of 200,000 to 250,000 inhabitants. Settlements having enterprises that are interrelated in terms of their use of materials, production processes, products, power, transport and utilities, are planned in accordance with regional and sectoral, long-range economic development plans. The distribution of the population increment is guided as far as possible into preferred locations through co-ordinated and selective industrial and agricultural development.

In 1970, the national plan anticipated that over a period of 20 years industrial production would increase about 500 percent and agriculture 250 percent. The plan called for the full electrification, mechanization and automation of industry and agriculture; new production processes, the development of new sources of energy and materials; and a sharp rise in productivity. A balanced pattern of growth, the control of undue concentrations in large cities, and an important reduction in urban/rural differences in incomes and amenities were expected to result from this policy. Regional development within the national plan was to be the general frame of reference for all urban development planning.

An indication of the tremendous volume of construction and amount of resources mobilized for the urban programme in the USSR is the fact that over 800 new towns were built between the early 1920s and the early 1960s, at an annual rate of construction of 3 million dwelling units, or over 14 units per thousand population—the highest rate of housing construction anywhere. Between 1960 and 1980, urban population is expected to rise from 50 percent to 70 percent of the national total. By 1960 already, about half the urban dwellers lived in cities of over 500,000. Since very large cities (as well as very small settlements) are not easy to equip, the *per capita* cost for amenities and services being high in relation to the levels of convenience and comfort provided, the distribution of population and the location of industry and agricultural development are being planned in order to obviate overcrowding and congestion. In agricultural areas also, urban-type settlements are to be established, equipped with the necessary utilities and services, and with educational, cultural, social and health institutions. . . .

Town planning in the socialist countries of Eastern Europe is carried out in advance of, or at least simultaneously with, new economic development. Every plan, whether national, regional or local, global or sectoral, has its physical and social services counterpart. After nearly six decades of effort, a wealth of practical experience exists, and a variety of urban models can be observed at the planning, implementation and real life stages. As a result of this massive experience, as well as growing affluence and rapid scientific and technological progress, the present strong economic and physical bias of settlement planning is likely to be replaced by a more integrated, multi disciplinary approach. The resulting balanced interplay of economic, social, environmental and cultural factors in development planning could then produce viable concepts, theories and models of settlement systems for contemporary industrial society, that would not only be technically efficient but also designed to promote a continuous improvement of the human qualities of life and a considerable reduction in the use of energy and the other scarce resources. . . .

Planning for Change [The Case of Peru]

The current Five-Year Plan of Peru combines a medium-term development model, conceived within a longer-term projection, with relevant legal, financial and political-administrative action, requisite changes in the environmental structure (the urban/rural settlement system) and the institutional and social reforms needed to alter Peru's social and economic structure. The latter is to be achieved by directly involving emerging social groups in the development process (in particular the industrial workers, the urban poor and the rural people). This effort to encourage public participation is intended to bring about a social transformation whereby these formerly passive, inarticulate and oppressed groups will start to participate through their own autonomous organizations in the control of their country's key resources, in the ownership of the means of production, and in the making of strategic decisions that affect Peruvian society as a whole, their own fate, and the fate of their families. To achieve this, the public authorities are also expected to shift from management and control *per se* to a role as productive agents of development and change. The Plan's stated purpose is to eliminate existing inequalities and economic and social disparities, rather than to create an industrial society as such.

The Peruvian Plan is firmly based upon the concept of specialized development in particular geographical areas that are at the same time closely linked with the rest of the nation and neighbouring countries. Present investment resources are insufficient for the total development and total integration of the country. The Plan therefore introduces selective regionalization. Its main objectives are to reduce demographic pressure on metropolitan Lima through the selective development of adequate economic, social and physical infrastructures elsewhere in Peru, and to create alternative growth centres and offer new locations for major projects through the development of additional land and sites equipped to accommodate them. These and similar measures are expected to guide the future pattern of settlements and the use of urban and agricultural land.

The experience of rural communities and of the spontaneously emerging new urban communities provides examples of many promising programmes, such as land reclamation and desalination, water control and irrigation, the building of secondary and rural roads, resettlement schemes, urban development programmes in new communities and in existing squatter towns and slums, and so on.

The main immediate objective of the regionalization programme is decentralization through planned dispersion. Its operational strategies include diverting new investment from the metropolitan region of Lima to other regions, at the same time strengthening existing urban centres in those regions or creating new ones; strengthening the economic and administrative linkages between neighbouring zones, thus taking full advantage of their complementary nature; and institutionalizing the different forms of self-help and mutual aid, thus transforming the abundant but rarely used potential of the lower-income groups for development into concrete and substantial investment inputs.

Infrastructural works in particular, even those on a relatively small scale, have helped to make more agricultural and urban land available, and brought sorely needed communal facilities and services to many areas. As a result, the urban and rural underemployed will be able to find employment in pursuits that are neither typically urban nor typically rural, but instead intended to develop basic utilities, transport and communications networks, hydraulic projects and harbours, and similar programmes that can use large inputs of relatively unskilled labour. In order to raise productivity, the Plan envisages supporting these labour-intensive methods by suitably mechanizing the production process at strategic points. These projects and programmes will provide basic training for large numbers of people of rural origin on the way to becoming skilled industrial workers.

The principal features of the Plan are as follows:

- Public investment in basic industries and infrastructure will grow over the five-year period to a point at which it can ensure self-sustained economic growth and social development.
- Social development is conceived as a process of societal transformation achieved through institutional reform and through education geared to productivity and to the acceptance of change as the vehicle of progress.
- Public participation in local and regional development within the larger national Plan will be stimulated and co-ordinated, both territorially and sectorally, in line with vital local needs and available total resources.
- Public resources will be reserved for strategically important projects, such as the development of natural resources, infrastructure to support agriculture and industry, and strategically located settlements.
- Full employment will be achieved by combining a variety of innova-

tive approaches and technologies (for infrastructural development, environmental improvement, etc.) with labour-intensive techniques.
- Imported technology will decline in importance as a result of the intensive development of human capacities, through self-management and participation in urban and rural settlement activities, and the merger and conversion of industrial plants together with the joint provision of essential services and facilities.

These and similar measures will barely begin to cope with the longstanding problems of regional inequality that are inherent in the existence of an over-extended and overcrowded metropolis in the midst of underdevelopment. The situation is one of environmental decay and social dislocation, growing diseconomies and rapidly increasing economic, social and cultural disparities between the centre and the periphery. The regional development strategy embodied in the Plan seeks to restructure Peru's economy and society by creating political and administrative institutions and a system of settlement designed to reverse the current trends toward centralization, environmental degradation and depressed living conditions in town and country alike. A more equitable distribution of income is expected to result, particularly in areas that are rich in natural resources and human capacities, where existing urban centres can be strengthened and improved or new ones created.

C. NEW DEPARTURES

The gap in incomes and living conditions between the affluent and the developing parts of the world is widening. Currently, seven-eighths of the world's wealth is produced, and largely consumed, by one-third of its population. By the end of the century, according to United Nations estimates, the other two-thirds will have to provide as much urban employment in tolerable environments as the industrialized countries have done in the whole course of their history. And they will have to do this with only one-eighth of the world's gross product. To tackle this task, innovative approaches are needed.

In some countries, for example, "squatting" is becoming an accepted form of rapid urbanization. The new social organization, which has emerged on some squatter settlements, co-operative in nature, relying strongly on self-help, self-management and direct action, is beginning to be used in a genuine partnership between governments and the people in a number of countries. In other cases, public participation is becoming institutionalized within the general development process, and in connection with the development and improvement of the quality of life in settlements more particularly. Experience to date shows, however, that such efforts at deliberate urbanization and at mobilizing the people's resources and skills for development purposes require very careful planning. As instruments for implementing national settlement policies, programmes of this kind need to be fully integrated with the other elements of settlement strategies at the local, regional and national level.

Rural Development [The Case of China and Mexico]

Development planning in China is motivated by social criteria and a concern for the human qualities of life. Rural development and a reduction of the differences in levels of living between town and country are primary goals. Thus, industry is being dispersed as much as possible throughout the countryside. In some cases, entire large enterprises with their equipment and staffs have been transferred out of the very big cities, to the benefit both of their original community and of their adopted one. This deliberate strategy of development dispersion combined with the integration of town and country, industry and agriculture, is consistently pursued. Together with programmes for the social and cultural development of the individual, the family and the community, it has already begun to create highly innovative patterns of settlement and new forms of social relations. These, in conjunction with the new institutional structure resulting from them, are conducive to the attainment of higher levels of living and an improved quality of life.

China's rural development policies may be illustrated by this description of a hypothetical village, "Peipan," in hilly eastern Hunan. "Peipan" is a village of about 1,000 people (200 households) which has a cultivated area of 200 hectares. Peipan is part of a commune of several villages, the basic administrative unit for the implementation of national agricultural policy. The land in the commune is owned collectively by its members, but a small amount is privately owned. The annual and long-term production goals for the commune are set by the central Government, but day-to-day operations are locally controlled.

Apathy towards increasing production, which was a feature of Chinese village life before the revolution, has largely been overcome. One of the instruments of this change has been the grant of small private plots to each family for raising pigs and cultivating vegetables and bamboo or fruit trees in their spare time. Families may barter or sell this produce at local markets.

The produce of the commune proper is either distributed among its members or exported to the cities. The share of the workers is determined according to their work and productivity (the "work point" system). Most of the rest of the produce is exported to the cities and towns. These exports provide the capital out of which most agricultural development has been financed. This economic self-reliance has been a major feature of Chinese agricultural development policy.

About 70 percent of Peipan's 200 hectares is irrigated land. Since the land is hilly, much of it is terraced and therefore fragmented into many small plots. Traditionally, animal labour was used to drive irrigation mechanisms. In years of drought, it was common for the cattle population to decline substantially from overwork and undernourishment. Today, most of the irrigation and drainage is mechanized. Irrigation from ponds and tube wells generally provides an ample supply of water for rice, which is the main crop. Farm machinery increasingly supplements human and animal labour in the fields. A great deal of work is devoted to maintaining the terraces, ensuring proper drainage, and collecting and spreading manure.

Sixty percent of the nitrogen fertilizer and about 80 percent of the phosphorous and potassium fertilizers come from organic sources, principally animal dung and night soil. The rest is provided by chemical fertilizers. The total nitrogen used is 12 metric tons per year, which averages about 60 kilogrammes per hectare of cultivated land. Rice, being the favoured crop, gets larger than average applications of manure and fertilizer. This rate of fertilizer use is quite high, comparable to that in many industrialized countries. That most of this fertilizer comes from organic sources reflects the remarkable and careful husbanding of organic manure that has characterized Chinese agriculture for centuries.

Though irrigation and field operations in Peipan are becoming progressively mechanized, human and animal labour still play a large role in the village economy. About 75 horsepower is available from the 150 draft animals, which are utilized 100 to 200 workdays a year, yielding a useful energy output of about 250 million Btu. Draft animals are used primarily for transportation, ploughing and other fieldwork. Irrigation is mechanically powered, partly by electricity and partly by coal and oil-driven engines. Electricity use in rural China is 10 kwh *per capita* per year and is growing at about 10 percent a year. To accelerate rural electrification, decentralized stations (500 kw) are used to complement the large-scale transmission and distribution systems.

Rural development has been pursued in other countries in somewhat different ways. "Arango," a hypothetical village in northern Mexico, may serve as prototype to illustrate what has been called the green revolution.

Arango was first established as a collective *ejido* (village); the land is now held or leased by individual *ejidatarios* (heads of households). The *ejidos* as collectives withered away in the 1940s and 1950s because of government neglect, mismanagement, lack of effective extension services, the desire of the peasants to have their land, and the preferential lending policies of the Ejido Bank to individual *ejidatarios.* The 80 *ejidatarios* of Arango (supporting 420 people) have 380 hectares of irrigated land, with an average holding of 4.8 hectares. The smallest farm is four hectares and the largest seven—a remarkably small spread compared to other parts of Latin American and the Third World in general. There are also 700 hectares of unirrigated land. By the standards of most of the rural population of the Third World, Arango is a rich community. It illustrates the dramatic difference that high-yielding seeds (particularly wheat), irrigation water, fertilizers, and fuel have made in the lives of millions of farmers in Mexico and elsewhere.

The Laguna region in which Arango is situated has an arid subtropical climate with hot summers and relatively cool winters. The annual rainfall is only about 30 centimetres. Unirrigated land produces a meagre harvest because of the scanty rainfall and is therefore not cultivated, particularly as surpluses of wheat are produced by the use of high-yielding varieties on the irrigated land. The soil of Arango is shallow and stony, typical of over 50 percent of Mexico. There is sparse shrub and grass cover, some suitable for grazing. The soil is fertile if irrigated, and dry farming is possible in some areas.

Most of Arango's cultivated land is irrigated by an intricate system of canals fed by the Nazas river, which never runs dry; 30 percent is irrigated from the six wells in the *ejido* and the remaining 10 percent from the Aguanaval river which is dry in December and January. Irrigation has freed the *ejidatarios* from the vagaries of the highly variable and scant rainfall in the region and enabled them to use high-yielding seed varieties and fertilizers in a farming cycle that is largely under human control. However, the level of the water table in the region is dropping and there is a danger that it may cease to be usable in the foreseeable future. Farm machinery is widely employed, as are fertilizers and pesticides. Draft animals mingle with machines on providing labour for farms, but most of the 400 horsepower used for farming comes from machines. In addition, there are mechanical irrigation pumps with a total of about 100 horsepower. Animal dung is not used either for fertilizer or fuel.

The people of Arango use substantially more commercial energy than their counterparts in most Third World villages. Each home has a kerosene lamp, half have potable running water, one-fifth have electricity, and one-third have gas stoves. Many families spend $50 or more a year on gas, kerosene, and electricity. Two-thirds of the people still heat their houses and cook by burning agricultural wastes such as cotton stubble. The largest indirect energy input imported into the *ejido* comes in the form of chemical fertilizer. One hundred and twenty kilogrammes per hectare are commonly used for the high-yielding wheat varieties which are in almost universal use in this region.

Most of the agricultural production is exported to the cities either through private traders or through the Ejido Bank. The considerable surplus production is responsible for the relative prosperity of farmers in this region of northern Mexico. In fact, most of the food for Mexican cities and towns is provided by the surpluses of relatively few farmers who, by and large, live in northern Mexico. . . .

Self-reliance and the Ujamaa Concept [The Case of Tanzania]

The *ujamaa* Village Programme in Tanzania is perhaps the world's best example of the use of a human settlement strategy to accomplish comprehensive developmental objectives. More than 90 percent of Tanzania's population is rural, and the programme is one of the very few attempts to produce rural-based self-reliant development. At the same time, the objective of this strategy is to reduce the comparatively high growth rate of Dar es Salaam and to achieve adequate growth in all regional centres, balancing the population growth among them by distributing employment opportunities to regional centres through decentralization. Also, a new capital city, Dodoma, is being planned to replace the colonial capital, Dar es Salaam. Dodoma is planned as the metropolitan centre of a region in which all settlements, rural and urban, including the capital city, are based on the *ujamaa* concept at the neighbourhood and local community level.

In 1976, President Nyerere, outlining the structural reorganization to be introduced in the rural areas of Tanzania, called upon the peasants to organize themselves into viable socio-economic and political communities. These are the "*ujamaa* villages," intended to transform now private and scattered production activities into communal planned production.

Ideally, the *ujamaa* villages are socialist organizations, created by people who decide of their own free will to live and work together for their common good, and governed by those who live and work in them. The peasants form these villages either in existing traditional communities or more typically by moving onto unused land or regrouping scattered homesteads to establish new ones. The villagers are required to own, control and run at least some of the economic activities in their villages communally, and to organize productive activity on that basis.

The *ujamaa* villages are governed by their members, who make their own decisions jointly on all issues of exclusive concern to the village. They are expected to own and run their communal farms and other projects such as shops, flour mills, pre-primary and primary schools, dispensaries, and cultural and recreational activities. The size of an *ujamaa* village depends upon the land available and the number of people in each village ranges from 50 to 4,000, although 500 to 2,500 is normally viewed as the desirable village size. There is no standard pattern of organization which must be followed by all *ujamaa* villages. This is a deliberate attempt to ensure that each village is organized in a way compatible with its environment and to encourage creative local planning.

The main objectives of most *ujamaa* villages are:

- To engender ideological awareness among members, to organize total opposition to exploitation in any form, and to create instead a sense of communal spirit by working together for the benefit of all members;

- To give employment opportunities to every member of the village, thus enabling everyone to earn a fair income;
- To expand the socialist economic undertakings of the village by establishing communal farms, shops, industries, and commercial and service activities;
- To market all the products of the village, including those from private plots;
- To buy or construct the buildings, offices, machines and other equipment needed for the development of the village;
- To co-operate with other *ujamaa* villages or para-statal institutions in commercial undertakings, provided that such co-operation is not contrary to the common good;
- To give the villagers an opportunity to receive adult and primary education, medical treatment, pure water, improved housing and other essential services necessary for an adequate material standard of living and fuller human development;
- To be an example to Tanzanians who are not members, so that they can see the benefit of the *ujamaa* way of living.

Over-all responsibility for running the affairs of the village is vested in a general meeting consisting of all the villagers. The meeting is normally held four times a year. The day-to-day management of the village rests with the village executive committee elected by the general meeting.

Several short- and long-term benefits are expected to be derived from successful implementation of the *ujamaa* village programme. Among these are:

- The creation of self-reliant and self-determining communities. National self-reliance is impossible without deep-rooted self-reliant communities at the local level.
- Avoidance of exploitation and excessive differentiation in wealth, income and power.
- Better utilization of rural labour to realize potential productivity through groups of peasants working together. Realization of this potential requires the specialization of functions, division of labour, work discipline and strong leadership to guide the enthusiasm of group activity into productive channels.
- Openness to technical innovations, through increases in scale, and readier access to farmer education.
- Facilitating national planning, both as to the formulation of over-all goals and as to decentralized implementation.
- Creating communities which can relate effectively to government officials and councils, following the 1972 decentralization of most governmental functions that directly affect individuals.

Movement to villages is now compulsory, but the transformation of a village into an *ujamaa* village is still, and likely to continue to be, voluntary. It is hoped that many of the benefits expected to accrue from living in an *ujamaa* village will also prevail in the ordinary planned villages.

What can be said of the results of the *ujamaa* programme to date? First, the human settlement pattern has been transformed from one dominated by scattered homesteads and hamlets to one of more compact communities. Access to basic health, adult education, primary education and communication facilities has greatly improved. Access to skills within the village community (not simply to outside skilled agents) is also increasing. Access to directly productive knowledge and inputs has improved somewhat but is hampered by inadequate central and village resources and by inadequate communication and control patterns between the villages and "experts."

The food production effects to date are marginal—though probably positive—because weather and relocation difficulties have imposed costs, and reorganization has taken time. Access to food, however, has improved; the change in settlement pattern, for example, was useful in identifying and meeting deficits during the 1973–75 drought.

Mistakes have been made at the village, regional and national levels, both through poor technical planning and through the partial substitution of exhortation and coercion for education and participation. However, these have usually been identified and rapid corrective measures have been taken.

Communal action has risen rapidly with regard to infrastructure, new economic activities (e.g., shops, dairy herds, small workshops) and new crops, but less in respect of staple foods and traditional cash crops, which are largely grown on individual plots. Egalitarianism has also progressed within villages—private plots are unequal in size but not radically so, especially when compared to those in some non-*ujamaa* rural enclaves. Participation within villages has broadened and the degree of rural elite control has declined. The villages have often been able to exert far more influence on the post–1972 decentralized government structure than peasants could apply to the elite-dominated cooperative unions or to the agents of central-government bodies during the pre-*ujamaa* period.

Village self-reliance has been unequal both in terms of goals and in terms of achievements, but in many cases it has risen. Peasant ideological development has begun, especially the realization that basic needs can be met and that they must mobilize pressures and support for party and government agents and institutions in order to safeguard and build on the results to date. Adult education and improved communication have led to broader and deeper individual and community consciousness in a significant number of villages, even though it may be fair to question the present degree of clarity and elaboration.

Each of these assessments is a qualified one—individual cases of total failure or regression, authoritarianism or clientage can be found. Each represents a state of transition, not arrival. To attempt more than an interim assessment would be to distort. Tanzanians do not claim to have achieved participatory, self-reliant, socialist development but to have begun the long transition to it. They do not claim that the *ujamaa* village core of rural development is complete, but that it has begun to emerge. The participatory nature of the transition demands the evolution of sequences and programmes within the strategic framework rather than the laying down of detailed patterns for the year 2000. . . .

STRATEGY COMPONENTS OF NATIONAL SETTLEMENT POLICIES

Once the need for a national policy for human settlements is established, the question arises of its objectives and the instruments by which they are to be realized. It goes without saying that every country has the sovereign right to determine its own lifestyle. The factors that determine the base, the priorities and the direction of development vary from nation to nation and region to region. It follows that every country must evolve its own settlement policies and strategies. However, each country's policies and strategies will depend on its political and socio-economic system, its physical resources and human capacities, and on the technologies and capital resources accessible to it. To be effective, every nation's policies and strategies must be backed by a national commitment to put human settlements among its highest development priorities.

A fundamental element of a coherent national policy is the deliberate development of settlement systems. Regionalization and socio-economic infrastructure development are two further essential components of an integrated policy. The rural-urban dichotomy should be rejected in favor of a continuum of settlements operating as interacting centres of human activity. Settlement planning should be supported by research and training facilities and the international exchange of information. Lastly, there should be a complete mobilization of resources, including land and water, new settlement technologies, capital resources, and public participation. . . .

CHAPTER 30

Territorial Social Problems in Socialist China

ENZO MINGIONE

INTRODUCTION

One of the most important theoretical problems for social scientists dealing with land use is the relationship between the socio-economic development of modern industrial societies and the territorial and regional organization which develops in them. One wonders whether the processes of rapid urbanization, depopulation of the countryside, centralization of production and lack of regional balance—which have characterized the industrial development of western societies—are characteristics of industrialization *per se* or are the result of the economic, social and political system of capitalism. This problem is extremely complex and one cannot hope to resolve it, even by considering an alternative model of industrial development such as the Chinese one which has a territorial organization markedly different from that typically found in western societies. Not only can one not make theoretical generalizations from concrete historical experiences but certain aspects of Chinese socialist development are unique. China is a vast country with ample natural resources; it is still relatively underdeveloped; and it is still predominantly agricultural. Rural communities have traditionally been of primary importance in Chinese history. All the same, an analysis of the territorial aspects of Chinese socialist development is important as a clarifying element in the theoretical debate concerning the compatibility between industrialization and alternative systems of territorial organization. We ask: Is it possible to make rapid and consistent industrial progress *without* paying the very high social costs usually connected with rapid urbanization, namely, increasing inequalities and the creation of relative regional and rural underdevelopment?

Before discussing the basic policies of land development in socialist China, one must quickly review the Marxist position concerning territorial

This paper was delivered at the 30th International Congress of Human Sciences in Asia and North Africa, Mexico City, August 1976, and is printed with the permission of the author. The complete Italian version is available in *Cittá e Campagna in Cina,* edited by CRMP and Enzo Mingione (Milano: Mazzota, forthcoming).

organization under socialism. While in utopian socialism there is a pervasive anti-urban bias, in classical Marxist theory the process of territorial concentration is seen as a necessary stage of social organization and as a fundamental contradiction of capitalist society. The development of the contradiction between town and country, the exploitation of agricultural resources and of the peasants, together with territorial concentration, are, according to Marx, aspects of capitalist accumulation and, at the same time, an important aspect of the struggle between social classes. Implicit in Marx's works is the idea that socialism should tend to eliminate the contradictions and the social division of labor between town and countryside, but there is little reference to the way in which this is to happen. Very practical controversies have ensued.

For example, when an important controversy broke out in the 1930s in the USSR over the process of urbanization, Marxist theoreticians opposed to concentration based their views on a single explicit passage from Engels which said, "Civilization has undoubtedly with its cities left us a heritage which will take much time and effort to eliminate; but it must and will be eliminated even if this process of elimination will be a laborious one." But Stalin and those responsible for Soviet territorial planning held opposite views. They argued that the development of socialism in the USSR should be based on the development and growth of socialist towns, as is demonstrated by the following remarks of Stalin:

> The problem of the elimination of the conflict between town and country, between industry and agriculture, is a well-known problem which Marx and Engels have both posed. The economic basis of this conflict is the exploitation of the countryside by the towns, the expropriation of the peasants and ruining of the majority of the rural population following the course of the development of industry, commerce and the capitalist system of credit. The conflict between town and country in capitalism must, therefore, be considered as a conflict of interests. On these grounds arose a hostile attitude of the countryside toward town dwellers. . . . This does not of course mean that the elimination of the conflict between town and country necessarily leads to the downfall of large towns and cities. Cities will not be destroyed, but new cities will arise and these cities will be centers of greater cultural development, centers not only of big industry but also of the processing of agricultural products, of great development of all branches of the food industry. This will encourage the cultural growth of the country and will determine a levelling out of living conditions in town and country.
>
> [Stalin, *Economic Problems of Socialism in the USSR*]

Even today the process of territorial development in the USSR is based on the growth of towns, on the unequal development of the various regions, on productive specialization of regions, and insofar as it does this, it does not differ radically from the land organization typical of capitalist countries.

The Chinese, on the other hand, have a decidedly different opinion on this question. The Chinese reject urban centralization, denounce big cities like Shanghai which were inherited from the previous colonized regime, and base their process of industrialization more on dispersion, on polyvalence, and on local economic autonomy, rather than on specialization, territorial inequality and large industrial cities. This particular tendency has prevailed, despite numerous contradictions and conflict with an opposing view (represented by the ex-president, Liu Shao Chi) which favored a model of development based on the town, heavy industry, indiscriminate increases in industrial productivity and regional specialization.

Chinese development has gone through periods in which the Maoist policy of decentralization was prevalent, others when Liu Shao Chi's policy of centralized development was prevalent, and also periods during which there has been bitter conflict between these two policies. Chronologically we can distinguish five different periods:

1949–52 the post-revolutionary period of reconstruction
1952–57 the period of the first Five-Year Plan
1957–60 the Great Leap Forward
1960–66 the crisis and its aftermath and
1966— the Cultural Revolution and its subsequent developments.

It is difficult to say which policy prevailed during the period of reconstruction since the country was exhausted by fifty years of external and civil war, deeply divided, and was characterized by an underdeveloped, extremely unproductive agriculture and by huge parasitic cities where millions of refugees and ruined peasants had found refuge. Most probably, the Maoist component supported agricultural reconstruction while the Liu Shao Chi component was more interested in the construction of heavy industry and large cities. Since these two tendencies were complementary at this early stage, it is probable that conflict was minimal and that both collaborated toward the reconstruction of the country.

In 1952, however, began a phase in which the Liu Shao Chi line was prominent and by the end of that period the two positions clashed. The first Five-Year Plan followed the Soviet model closely: namely, rapid accumulation and draining resources from the countryside in order to build up heavy industry; the parallel decapitalization of the countryside to increase industrialization and a centralized stratification between rich and poor peasants; and considerable expansion of the already-existing large industrial cities and mining areas. This policy caused widespread discontent—among the poorer peasants who saw their incomes reduced by massive withdrawals on the part of the state to encourage industrialization and by fierce competition from the richer peasants who were making considerable profits, and among industrial workers who were being subjected to an increasingly oppressive internal hierarchy in labor organization and to a rate of productivity which they did not and could not control.

The controversy in the countryside arose over the question of agricultural development. Mao maintained that one should try to speed up the transformation of agricultural cooperatives into popular communes, the elimination of private ownership and management in agriculture and the process of making the large peasant masses responsible; Liu Shao Chi, on the other hand, wanted to maintain agricultural stratification and private ownership as an incentive toward agricultural production.

During the Great Leap Forward the Maoist line assumed predominance, and brought about the formation of peoples' communes, a considerable return movement to the countryside and increased efforts to carry out a true agricultural revolution, to experiment with new forms of production in all sectors, and to grant greater autonomy to peripheral regions, to small towns of the interior, and to ethnic minorities. However, several consecutive years of severe drought, causing fairly serious famine, and the withdrawal of Soviet technicians and aid subsequent to the break in relations between the two countries, put the Maoist line favoring decentralization into a critical situation and gave the initiative back to those who favored more centralized technocratic development.

The period between the crisis of the early 1960s and the Cultural Revolution saw the spread of alternative experiments in development. In particular, the new-style industrial-agricultural settlement of Taching was begun during this period, as well as the experiment of the Tachai agricultural brigade. However, these experiments remained of secondary importance until the Cultural Revolution when they became a fundamental part of the Maoist line and spread rapidly. They began a long consolidated phase of socialist development which is completely different from any other experience of development—whether capitalist, Third World, or other socialist.

The fundamental points of this Chinese alternative territorial development, to be analyzed briefly in this paper, are as follows. First, the development of the countryside has been directed toward optimal utilization of all productive resources—mainly peasant labor together with modern technology—and has depended on the political activation of the peasants and on the industrialization of rural areas rather than on migration and urbanization. Widespread industrial growth throughout the whole country in various spheres instead of specialization by region has favored the development of small and medium-sized towns and peripheral areas rather than a concentration in the traditional large Chinese cities of the coast. Through a civil and political struggle, millions of people living in urban slums and shantytowns and millions of peasants living in huts or shacks in the countryside have been rehoused. China seems to have succeeded in shifting the axis of industrial development from the coastal strip, where the majority of the population and practically all of the industrial productive capacity of the country were previously concentrated, and has spread this development more evenly throughout the country.

In 1949, the urban population of China was approximately 58 million; by 1957 it had passed 100 million, representing 15.4 percent of the total. After several fluctuations the urban population stabilized in the 1970s at about 18 percent of the total, i.e., 140–150 million. Big cities such as Shanghai and Peking have grown relatively little in terms of population and have been radically redimensioned in terms of land use. Shanghai now has a population of ten million (as compared to five million in 1949), but it is no longer a city in the strict sense of the term because a third of the active population is engaged in agriculture and the increased area covered by the city includes a considerable number of popular communes which make the inhabitants of urban districts self-sufficient as far as food is concerned.

Both our concept of town and metropolis and our concept of country and village have been virtually overturned by the Chinese development and yet the economy of the country appears to remain competitive, not only in comparison to other Third World economies but also *vis à vis* western capitalist and socialist countries. One wonders how long a territorial organization which is becoming progressively more decentralized, like that of the Chinese, can remain compatible with the contemporary technological revolution. But many new technologies have been developed which assist this decentralization. The manufacture of basic electronic parts can be carried out in small local and rural units and these techniques can be transferred to other industries as well; scientific and industrial research can be decentralized and communications and cooperation between economic units spread over large territory can be improved. I am of the opinion that the main obstacles standing in the way of decentralized development and the main solution of the dialectic between town and country and of the social division of labor in a new territorial organization are mainly political rather than technical in nature. The rest of the paper will explore the goals of the Chinese reorganization and some of the means that have been employed to reach them.

AGRICULTURE AS THE BASIS FOR DEVELOPMENT

It is not difficult from an economic point of view to explain the significance of the phrase "make agriculture the basic factor." Difficulties begin when one attempts to explain the political implications of this slogan. After the revolution, China was mainly an agricultural country (over 90 percent of the population was engaged in agriculture) and a process of economic development could be carried out only through an increase in agricultural production and the consequent accumulation of an agricultural surplus that could be used to build up modern industry.

The Chinese aimed, therefore, at consolidating the production structures of the countryside, making sure that this did not at the same time lead to too great an expulsion of the labor force towards towns and industry, since Chinese industry was too weak at this stage to absorb the surplus labor. A policy of labor-intensive agricultural development was therefore stressed. The increases in productivity were to be achieved through greater and better utilization of the vast existing agricultural population. This type of policy is not in itself new nor typical only of Chinese socialist development. Both Japan for a long time and India today followed or are following economic policies based on labor-intensive agricultural development. The crucial difference is in the destination of the agricultural surplus realized in this way and its subsequent utilization. If the agricultural surplus is systematically taken away from the agricultural sector in order to support the indiscriminate growth of capitalist industrial production, as was true in Japan, it is inevitable that things become difficult for the poorer peasants and eventually the process of abandonment of the countryside and the impoverishment of agriculture, typical of capitalist development, gets underway on a large scale. If, on the other hand, the surplus is taken over systematically by a class of rich peasants and by the state in order to maintain a strong bureaucracy and substantial urban overpopulation, as is happening in India, without even starting a process of industrialization, the weaker peasants are ruined and urban and rural poverty are constantly reinforced. These two examples show that an agricultural policy based on high labor intensity is not, in itself, sufficient to overcome social, political, and territorial inequality between town and country.

However, in China most of the agricultural surplus is managed, for the most part, by the peasants themselves and is used to advance agricultural development, to improve production and the standard of living of the peasants and, therefore, ultimately to strengthen peasant demand for industrial products. The remainder of the agricultural surplus is utilized for industrial accumulation. Industry and agriculture can then expand in a coordinated manner and make mutual use of each other. The matter is not quite as simple as we have described it, but it seems clear that it is the political relationships of appropriation of the surplus which define the difference between capitalist and socialist development.

This type of development is not without contradictions in China, even from a strictly economic point of view. The increase in agricultural production and the simultaneous increase in the demand among the peasants for industrial goods can lead (and certainly does in capitalist development) to a scissors-like relationship of agricultural prices compared with industrial prices. If this is not controlled and limited, it rapidly leads to impoverishment of the countryside. If agricultural prices tend to drop while industrial prices rise, the peasants pay more and more while receiving less and less; this process ruins weaker agriculture which has a lower productivity rate. In China, had there not been a centralized policy of maintaining agricultural prices, the peasants of the poorer areas would have been forced to abandon their work and emigrate to the towns, thus triggering again the mechanism of territorial inequality.

Secondly, at a certain stage of development, the mechanism of industrial accumulation, even in socialism, enters into competition with agriculture for the utilization of resources. Any further expansion of industrial activity means the concentration of a new labor force and of raw materials in factories located in towns, and therefore a process of urbanization. Neither can one refuse to satisfy the growing demand originating in the countryside for industrial products. The answer given to this problem by the Chinese is the industrialization of the countryside. An increasingly large part of industrial demand generated in the countryside is satisfied by the small industries of the communes which are springing up all over China. The mechanical workshop of the commune is no longer content just to mend broken machinery or carry out routine maintenance; it now supplies spare parts, produces new agricultural equipment, and has its own foundry. Small chemical plants produce fertilizers needed for local agriculture. Building materials are generally produced locally. Although the rate of productivity of these small decentralized industries is lower than that of the large urban manufacturing plants, real production costs are more or less similar if not lower, since the small units utilize residual labor that otherwise might be wasted, new methods are discovered to use raw materials that otherwise would have been overlooked, and a saving is made on the economic and social costs of transport. All this means a tendency to overcome, in the long run, the division of labor between industry and agriculture and the territorial inequality between town and country, thus sparing China a rapid process of concentration of the population in big cities.

Another aspect of the solution is the administrative and political decentralization of the country and the tendency to lessen the administrative dependence of the peasants on centralized bureaucracy located in cities. An increase in employment in public services and in the number of bureaucrats who can control and oppress is a typical tendency of development, both in developed countries and in most underdeveloped ones. In China there has been an attempt to control this tendency through the creation of popular communes. The peasants administer themselves directly at the levels of the popular communes and brigades, deciding how much to produce and how to coordinate their plans with the demands of regional and national planning. The leaders of the communes and brigades are peasants who carry on with their jobs. In China local autonomy is real; this means that socialist development does not necessarily involve a constant increase in bureaucracy nor does it produce huge administrative centers which grow like parasites.

We do not claim that Chinese development in the last twenty years has cancelled out territorial inequalities entirely, but only that there is a tendency to move in this direction. Contradictions still exist in the difference between poor and rich agricultural land. It is clear that the communes and brigades of elevated stony areas or of plains that cannot be irrigated have a much lower rate of productivity than those of the irrigated and fertile plains of Canton and Peking. On the average, peasants in the former areas are still much poorer than those in better endowed zones, even though they may work harder. On the other hand, however, only development that is not subordinate to the accumulation of profits can afford to devote its attention and resources to peripheral farming and can promote a struggle that, while carrying a high economic cost in the short run, brings enormous advantages over a longer period of time.

SOCIALIST URBANIZATION AND INDUSTRIALIZATION

We have already alluded to the fact that Chinese development has been characterized by a complex dialectic in which in certain periods there prevailed an emphasis on concentrated industrial accumulation, on the superiority of technology and profit, whereas in other periods the Maoist position that favored decentralization prevailed. To sum up the history of this alternation, we may distinguish five periods over the last quarter of a century:

a. the period of economic reconstruction from 1949 to the mid-50s, characterized by the effort to industrialize the big cities inherited from capitalism, using the concentration of resources that the imperialist colonial regime had created;

b. the first period in which the revisionist line was predominant (1952 to the Great Leap Forward of 1958), when industrialization was mainly copied from the Soviet model and dependent upon Soviet aid. During this period the preeminence of town over country was strengthened but some new towns in peripheral areas were also created;

c. the era of the Great Leap Forward (1958–62) based on a rapid and radical process of communalization of the countryside and on a parallel process of decentralization of light industry, with the first examples of courtyard and local industries and the first commune industries; in the latter part of this period, the economic crisis and the partial failure of this policy led to a process of abandonment;

d. the period following the crisis of the early sixties up to the Cultural Revolution (1962–65), in which the revisionist line stressing profit and industrial concentration prevailed. During this period territorial inequalities became more deep-rooted as poor peasants ruined by the rationalization of the countryside began to emigrate toward the large industrial cities where the more productive and expanding industries were located;

e. the era of the Cultural Revolution, during which there was a deep clash between the two lines until the Maoist one eventually prevailed. This led to a policy which favored widespread industrialization of the countryside and of less developed regions and in which politics and mass participation took priority over technology and profit. Centralization of capitalistic production was abandoned.

These alternating phases of the urban and industrial development of China have two essential lessons: first, that class contradictions persist even in socialism; and second, that the socialist process of industrialization can and must work to overcome the serious territorial inequalities that capitalist industrialization, focused on profit rather than the social needs of the masses, brings about. The principal elements of the process of Chinese industrialization are, in my opinion, the following:

1. the attempt to make each region self-sufficient with respect to basic industry;
2. the attempt to overcome the division of labor and specialization between industry, agriculture and public services in the new territorial units;
3. the method of planning from the bottom up rather than from the top down, together with all that this involves, such as the industrialization of the countryside and the creation of a vast network of industries based on demand and local needs, and utilizing the surplus productive capacities that otherwise would have been wasted; and
4. maximum utilization of territorial infrastructures already in existence (towns, transportation, large factories, etc.) and the further development of these to meet immediate needs of the population.

These four policies, working together, have helped the Chinese to overcome the inherent tendencies toward large town and territorial concentration and inequality, while at the same time achieving considerable economic growth. I should therefore like to look more closely at each of these policies.

1. Self-sufficiency

The autonomous development of each region inverts the tendency toward specialization typical of capitalist industrial development, a tendency which eventually causes the underdevelopment of large areas and transforms the interdependence of modern economies into dependence upon a center. The Chinese aim to make every region autonomous, within the limits of natural resources, at least as far as basic iron and steel, mechanical and chemical industries are concerned. This involves stressing the development of peripheral and depopulated areas by decentralizing capital investments and manpower and encouraging local socialist accumulation through the modernization of agriculture and the creation of basic industries connected with growing agricultural demand and the availability of local raw materials. The old capitals of outlying provinces and the parasitic administrative towns—whose old purpose had been to guarantee capitalist-feudal control of the region—are thereby transformed into industrial centers linked to the local economy. New settlements integrating farming, industry, and public services have been established which permit exploitation of local resources without recreating the division of labor and a contradiction within the surrounding countryside. Regions such as Sinkiang and Tibet are not subject to centralized development and are progressively overcoming their relative poverty without paying the social costs of massive emigration, without suffering from imposed "poles of development" which uproot local activities, and without the growth of parasitic public employment designed to compensate for the lack of dynamic industrial growth.

2. Less Division of Labor

The example of Taching, which serves as a model for other new settlements, illustrates the policies designed to overcome the division of labor and thereby reduce the contradiction between town and country. While this territorial experiment, in which light and heavy industry, agriculture and public services are integrated and where the concentration of population creates neither town nor countryside (in our sense of the terms), is extremely complex, it does indicate that the first steps have been taken towards a completely new and alternative system of territorial organization.

Taching originated in 1960 when China, isolated internationally, was stressing self-sufficiency. Geologists maintained that below the frozen surface of the plain of Taching were considerable oil deposits. Numerous teams of workers with their families went to this desolate area to begin drilling. Within a few years, the settlers had established a completely new kind of community. The houses were built directly by groups of industrial workers who followed the solid and inexpensive methods of local tradition. The land was gradually ploughed and the workers take turns at farming, especially during the sowing and harvesting seasons when demand for labor goes up. Women work in farming, workshops and small factories engaged in light industry which were set up. Families of the workers established social services, schools, hospitals, a cinema and a theater, and are now running them. Through job rotation the division of labor has been reduced to a minimum. This is reflected in the spatial organization; the community is a non-centralized but cohesive settlement in which factories, houses, public buildings and farm lands are mixed in an orderly manner. Meanwhile, the production of oil has been increasing on the average of 31 percent annually and a modern refinery and pipeline to link the area to the industrial coastal zone several hundred miles away has been built.

During the Cultural Revolution the Taching experience was taken as a model by the workers in many large factories, following the express instructions of Chairman Mao in his famous directive of 6 May 1966.

3. Decentralized Planning

The fact that the Chinese process of planning works upwards, rather than being imposed by central power, is extremely important. Only through a decentralized process can one put to maximum use local manpower, raw materials, and traditional knowledge and skills. In China it is the peasants of a brigade or the local housewives who set up a new industry in order to produce what they need and in order to utilize manpower that might otherwise be wasted. It is the workers of the individual factories who, within certain limits of course, decide to expand production or to modify the product manufactured, especially when they are in direct contact with the consumer. State and central planning intervene to coordinate or to press for changes which the central administration believes important, but the local unit has the final say in the matter. This allows the productive apparatus to adapt to changing social relations and to particular needs of the masses in a manner that is not possible in a highly centralized system of planning. As agricultural production increases and the amount of labor required diminishes, the peasants can devote more time to industrial and social service activities without needing to emigrate or to wait for the creation of a new large factory in the neighborhood. Industry permeates the countryside without upsetting agriculture and without ruining masses of peasants. Modern technology and higher rates of productivity do not cause the total disappearance of traditional industry and craftsmanship nor does improvement bring about the terrible inequities typical of the countryside surrounding the European or South American "poles of development," because modern industry created by the masses establishes a complementary and functional relationship with local traditions and needs. To sum up, we argue that industrialization planned by the masses is a process that reduces to a minimum the inevitable contradictions connected with growth.

4. Utilization of Existing Resources

An economic policy that aims to avoid waste influences both the process of industrialization and the urban structure and the housing question in China. (We discuss housing later.) The policy of avoiding waste has two very different territorial effects. First of all, it has required the preservation of large territorial concentrations and large-scale productive structures created earlier under capitalism, although it has sought to transform them. The parasitic cities that had become bloated with excess population under the capitalist-feudal system were transformed into centers of modern heavy industrial production. Similarly, the few already existing heavy industries, which had formerly been almost entirely dependent on foreign capital, together with the mining centers, were further developed and not destroyed by the socialist revolution, even if this did strengthen the tendency toward concentration and the division of labor. But this conservation and use of existing structures has not started irreversible and progressive urban concentration because it has been balanced by parallel processes of decentralization. As we have already seen, Chinese development favors the birth of local industries, especially in less densely populated rural and peripheral areas and in underdeveloped regions. This second policy results logically from the principle to keep waste to a minimum, just as much as the policy to integrate existing resources into the system follows from it.

The combined result of these two processes is that in China, unlike other cases of rapid industrial development, the large cities retain their population and their dimensions but, at the same time, rural and peripheral areas, small towns, and economically and demographically underdeveloped regions have, in general, much higher rates of growth than the large urban concentrations. On the whole, therefore, a process of decentralized industrialization is produced which spares China the problems of urban overdevelopment and metropolitanization typical of western countries and of many underdeveloped countries, especially in Latin America.

THE HOUSING QUESTION

Chinese socialism inherited a very serious housing problem. The extensive shantytown of Shanghai was notorious. More than two million people lived in makeshift dwellings there, with no sanitation, enjoying none of the advantages but suffering all the hardships of life in a big city. In all the towns of presocialist China, the housing problem had become increasingly acute in the preceding decades because the civil war and the war against the Japanese had accelerated the urbanization of ruined peasants and had, at the same time, made city administrations and economies even more precarious and inefficient than usual. The great and urgent need for housing in China was thus not a problem of socialist development but rather the heritage from the previous regime.

It was clear that, in order to give accommodations to all who needed them, it was not enough to build new housing estates rapidly and at low cost but it was also necessary to improve the old slums. The extensive shantytowns which provided precarious accommodations for a large part of the urban proletariat were not, therefore, demolished but were improved with massive investments which resulted in great savings per dwelling in comparison to the costs of new construction.

This scheme of restoration and improvement has three main aspects. The first is the progressive reduction of overcrowding through transfer of a considerable number of families from the shantytowns into the large bourgeois residences and then into new housing areas. The second aspect is the construction of urban infrastructures in the shantytown areas by the local or central government. Roads were asphalted, paved or at least rolled flat so that rain would no longer create the muddy lakes that periodically flooded the houses, sewerage and water systems were constructed with if not private at least frequent communal facilities. Still later, electricity was introduced. The third aspect was the supply at a very low price of housing and building materials, so that the dwellers could transform their huts into permanent houses.

The results are evident today in the large Chinese cities and especially in Shanghai. Where once there were overcrowded shanties there are now rows of small houses of two storeys, each a little different from the others but all appearing sufficiently comfortable and permanent. The great advantage of this policy is that it gave all urban proletarian families a home without too high an expenditure of capital which China did not then have at its disposal. The central element of this transformation was the direct mobilization of the masses and the revitalization of popular creativity without which it would have been impossible to house the millions of homeless and shantytown dwellers in so short a time.

The second aspect of the housing policy is the new residential areas. In constructing these housing estates, the chief aims were savings, efficiency, adherence to the social needs of the masses, and mass mobilization. The new housing estates are built not too far from the work places of the residents and it is the rule that a worker who changes his job can find new accommodations in an area close to his factory. These housing estates are far enough away from the factories to avoid their noise and odors and are large enough to include populations employed in a variety of adjacent locations. A second aspect of the new housing estates in China is that, in order to conserve space, savings are made within the individual dwellings but not in the communal facilities. A family dwelling comprises one bedroom for a childless couple or more if the family has children, plus a kitchen and bathroom which are shared with one or two other families. This limited private space is amply compensated for in the communal areas which contain a meeting room for each block, large courtyards full of trees where children can play, a recreation ground, schools, nurseries and sport facilities for each estate.

The estate is not, in fact, intended to be a dormitory where the individual, exhausted by the too intense rhythms of the factory, retires to regain his strength in the isolation of the home but rather an important aspect of the worker's social life in a community with other workers. The home and the housing estate are communal organizations. Residents have the opportunity to meet and discuss, to practice sports together, bring up their children together, and build the new society. Old age pensioners play an active part in the communal life of the estate. They contribute to bringing up the children and to keeping the estate clean and the gardens of the courtyards and avenues well cared for. They are not isolated and when they are ill they are aided and cared for.

The construction of houses takes into account the contribution and experience of the workers and uses cheap materials. The same policy is followed in rural building which is adapted to meet specific needs and takes into account the traditions and the particular skills of the various regions. This maximizes the use of cheap insulating materials and practical heating systems. For example, the peasants of North China still use, even in their new homes, the beds made partly of brickwork, with a source of heat underneath, that have been used for many, many years. Rather than impose urban-type standards on rural housing, the Chinese have sought to incorporate into their new housing the older traditions which have been adapted to the environment.

CONCLUSIONS

When a foreign visitor arrives in China, sooner or later one of the Chinese officials will tell him, "Keep in mind that you have arrived in a poor underdeveloped country." I may have given the reader an opposite impression that China has solved most of her development problems, but this could hardly be the case. The Chinese "experiment" is on its way but is very far from its end. China is still an underdeveloped country and her average income remains considerably below that in the developed countries. From the territorial point of view a full set of contradictions will continue to characterize Chinese development for a long time to come. Coastal cities, and especially Shanghai, remain enormous concentrations of human and material resources and constitute major obstacles to a more equal distribution of economic and social activities. Even though the chemical and petrochemical industries are pushing westward toward the interior of China and new steel and heavy mechanical industries have been located chiefly in the central part of the country, and despite the proliferation of rural manufacture throughout the country, the regional imbalance remains. From 30 to 40 percent of the industrial capacity of China is concentrated in four or five eastern conurbations (Canton, Shanghai-Hangchow, Tientsin-Peking, the northeast Anchan-Fouchoun). Differences in average productivity and income between the irrigated wealthy communes and the poor stony-desert-highland communes may prove intractable to further attempts to equalize regional differences. Housing standards, although far better than those in other parts of the less developed Third World, remain substantially below those of average working class housing in the west.

Personally, I am optimistic and appreciate the enormous effort the Chinese are making to initiate a completely different pattern of industrialization and economic development which has important consequences for the territorial division of labor. But I agree with Engels that the process of creating this alternative territorial organization is extremely complex and will take a long time. The Chinese experience is at its beginning, having started for all practical purposes on a large scale only since the 1966 outburst of the Cultural Revolution. It has a long way to go.

INDEX TO BIBLIOGRAPHY BY KEY NUMBER OF SOURCE

Basic Works on Social Change, Economic Development, Modernization

Works on the Theory of Political Economy-Dependence and/or Applications

Basic Statistics

Bibliographic References

Basic Works on Urbanization Theory and Cross-Regional Studies of Urbanization

Specific Case Studies of Cities

Urban Structure

Labor Force

Housing—Squatter Settlements, Slums

Migration

Asia

Africa

Latin America

BIBLIOGRAPHY

1. Abesamis, Felix. 1972. "Relocation and Employment." *Papers and Proceedings of the Workshop on Manpower and Human Resources, Part II.* University of the Philippines and the National Economic and Development Authority.
2. Abrams, Charles. 1964. *Man's Struggle for Shelter in an Urbanizing World.* Cambridge, Mass.: M.I.T. Press.
3. ______. 1965. *Squatter Settlements: The Problem and the Opportunity.* Special Report to the U.S. Agency for International Development. New York: Department of Housing and Urban Development.
4. Abu-Lughod, Janet L. 1961. "Migrant Adjustment to City Life: The Egyptian Case." *American Journal of Sociology* 67: 22–32.
5. ______. 1969. "Testing the Theory of Social Area Analysis: The Ecology of Cairo, Egypt." *American Sociological Review* 34: 198–212.
6. ______. 1971. *Cairo: 1001 Years of the City Victorious.* Princeton: Princeton University Press.
7. ______. 1975. "Moroccan Cities: Apartheid and the Serendipity of Conservation." In *African Themes,* edited by Ibrahim Abu-Lughod. Ann Arbor: Edwards Brothers.
8. ______. 1976. "Developments in North African Urbanism: The Process of Decolonization." Pp. 191–211 in *Urbanization and Counter-Urbanization,* Vol. 11, Urban Affairs Annuals edited by Brian Berry. Beverly Hills, Cal.: Sage.
9. Adams, Robert McC. 1966. *The Evolution of Urban Society.* Chicago: Aldine.
10. African Studies Center. (Serial publication.) *African Urban Notes.* East Lansing: Michigan State University.
11. Ajaegbu, Hyacinth I. 1972. *African Urbanization: A Bibliography.* London: International African Institute.
12. Alexander, Robert J. 1962. *Labor Relations in Argentina, Brazil and Chile.* New York: McGraw-Hill.
13. Alonso, S., and E. Meyer. 1972. "Pôles d'influence et espaces dépendants," *Revue Tiers-Monde* XIII, 50: 329–78.
14. Amin, Samir. 1967. *Le développement du capitalisme en Côte d'Ivoire.* Paris: Editions de Minuit.
15. ______. 1971. "Le modèle théorique de l'accumulation dans le monde contemporain, centre et périphérie." Dakar: I.D.E.P.
16. ______. 1972. "Underdevelopment and Dependence in Black Africa—Origins and Contemporary Forms." *The Journal of Modern African Studies* X, 4: 503–24.
17. ______. 1973. *Neo-Colonialism in West Africa.* Harmondsworth, England: Penguin.
18. ______. 1974. *Accumulation on a World Scale: A Critique of the Theory of Underdevelopment.* 2 vols. New York: Monthly Review Press.
19. Apter, David. 1971. *Choice and the Politics of Allocation.* New Haven, Conn.: Yale University Press.
20. Armstrong, W. R., and T. G. McGee. 1968. "Revolutionary Change and the Third World City: A Theory of Urban Involution." *Civilisations* XVIII, 3: 353–78.
21. Arrighi, Giovanni. 1967. *The Political Economy of Rhodesia.* The Hague: Mouton.
22. Arrighi, Giovanni, and John Saul, eds. 1973. *Essays on the Political Economy of Africa.* New York: Monthly Review Press.
23. Artle, Roland. 1970. *Urbanization and Economic Growth in Latin America.* Berkeley: University of California, Institute of Urban and Regional Development.
24. Balán, Jorge, Harley L. Browning, and Elizabeth Jelin. 1973. *Men in a Developing Society: Geographic and Social Mobility in Monterrey, Mexico.* Austin: University of Texas Press.
25. Baran, Paul. 1957. *The Political Economy of Growth.* New York: Monthly Review Press.
26. Baran, Paul, and Paul Sweezy. 1966. *Monopoly Capital: An Essay on the American Economic and Social Order.* New York: Monthly Review Press.
27. Barnet, Richard J., and Ronald E. Müller. 1974. *Global Reach.* New York: Simon & Schuster.
28. Barratt Brown, M. 1974. *The Economics of Imperialism.* Harmondsworth, England: Penguin.
29. Beckford, George L. 1972. *Persistent Poverty: Underdevelopment in Plantation Economies of the Third World.* New York: Oxford University Press.
30. Bendix, Reinhard. 1967. "Tradition and Modernity Reconsidered." *Comparative Studies in Society and History* 9 (April): 292–346.
31. Berry, Brian. 1973. *The Human Consequences of Urbanization.* New York: St. Martin's Press.
32. ______. ed. 1976. *Urbanization and Counter-Urbanization.* Vol. 11, Urban Affairs Annuals. Beverly Hills, Cal.: Sage.
33. Beyer, Glenn H. 1967. *The Urban Explosion in Latin America.* Ithaca: Cornell University Press.
34. Binet, Jacques. 1976. "Urbanism and its Expression in the African City." *Diogenes* 93: 81–104.
35. Bodenheimer, Susanne. 1970. "Dependency and Imperialism: The Roots of Latin American Underdevelopment." *NACLA Newsletter* IV (May-June): 18–27.
36. Bonilla, Frank. 1961. "Rio's *Favelas:* The Rural Slum Within the City." *American Universities Field Staff Reports,* East Coast South America Series 8, 3 (August).
37. ______. 1962. "The *Favelas* of Rio: The Rundown Rural *Barrio* in the City." *Dissent* IX: 383–86.
38. Bose, Ashish. 1970. *Urbanization in India: An Inventory of Source Materials.* Bombay: Academic Books Ltd.
39. ______. 1971. "The Urbanization Process in South and Southeast Asia." Pp. 81–109 in *Urbanization and National Development,* edited by Leo Jakobson and Ved Prakash. Beverly Hills, Cal.: Sage.
40. ______. 1973. *Studies in India's Urbanization 1901–1971.* Bombay: Tata McGraw-Hill.
41. Bovill, E. W. 1958. *The Golden Trade of the Moors.* London: Oxford University Press.
42. Brand, Richard R. 1972. "The Spatial Organization of Residential Areas in Accra, Ghana, with Particular Reference to Aspects of Modernization." *Economic Geography* 48 (July): 284–98.
43. Braudel, Fernand. 1973. *The Mediterranean and the Mediterranean World in the Age of Philip II.* 2 vols. New York: Harper & Row. Translated from French. Original edition 1949.
44. Breese, Gerald. 1966. *Urbanization in Newly Developing Countries.* Englewood Cliffs, N.J.: Prentice-Hall.
45. ______. ed. 1969. *The City in Newly Developing Countries: Readings on Urbanism and Urbanization.* Englewood Cliffs, N.J.: Prentice-Hall.
46. Brett, E. A. 1971. "Dependency and Development: Some Problems Involved in the Analysis of Change in Colonial Africa." *Cahiers d'Études Africaines* XI, 44: 538–64.
47. ______. 1973. *Colonialism and Underdevelopment in East Africa: The Politics of Economic Change, 1919–1939.* New York: NOK Pub. Ltd.
48. Brookfield, Harold C. 1975. *Interdependent Development.* Pittsburgh: University of Pittsburgh Press.
49. Browning, Harley L. 1958. "Recent Trends in Latin American Urbanization." *Annals of the American Academy of Political and Social Sciences* 316 (March): 111–20.

50. Burawoy, Michael. 1976. "The Functions and Reproduction of Migrant Labor." *American Journal of Sociology* 81 (March): 1050–87.
51. Caldwell, J. C. 1969. *African Rural-Urban Migration: The Movement to Ghana's Towns.* New York: Columbia University Press.
52. Caplow, Theodore, Sheldon Stryker, and S. E. Wallace. 1964. *The Urban Ambiance: A Study of San Juan, Puerto Rico.* Totowa, N.J.: Bedminster Press.
53. Cardoso, Fernando Henrique. 1965. "Análises sociológicas del desarrollo económico." *Revista Latinoamericana de Sociología* I (July): 178–98.
54. ______. 1972. "Dependency and Development in Latin America." *New Left Review* 74 (July-August): 83–95.
55. ______. 1973a. "Associated Dependent Development: Theoretical and Practical Implications." Pp. 142–76 in *Authoritarian Brazil: Origins, Policies and Future,* edited by Alfred Stepan. New Haven, Conn.: Yale University Press.
56. ______. 1973b. "Imperialism and Dependency in Latin America." Pp. 7–16 in *Structures of Dependency,* edited by Frank Bonilla and Robert Girling. Nairobi, (E. Palo Alto) California.
57. ______. 1973c. "Notas sobre estado e dependencia." Sao Paulo: Centro Brasileiro de Análise e Planejamento (Caderno 11).
58. Cardoso, Fernando Henrique, and Enzo Faletto. 1969. *Dependencia y desarrollo en América Latina,* Mexico City: Siglo Veintiuno Editores.
59. Cardoso, Fernando Henrique, and Jose Luis Reyna. 1968. "Industrialization, Occupational Structure, and Social Stratification in Latin America." Pp. 19–56 in *Constructuve Change in Latin America,* edited by Stewart Cole Blasier. Pittsburgh: University of Pittsburgh Press.
60. Carlos, Manuel, and Lois Sellers. 1972. "Family, Kinship Structure and Modernization in Latin America." *Latin American Research Review* 7: 95–124.
61. Casanova, Pablo González. 1970. *Sociología de la exploitación.* 2nd ed., Mexico City: Siglo Veintiuno Editores.
62. Castells, Manuel. 1972. *La question urbaine.* Paris: Maspero. English edition 1976, London: Edward Arnold.
63. ______. 1976. "Urban Sociology and Urban Politics: From a Critique to New Trends of Research." Pp. 291–300 in *The City in Comparative Perspective,* edited by John Walton and Louis Masotti. New York: Halsted Press, John Wiley.
64. Chandler, Tertius, and Gerald Fox. 1974. *3000 Years of Urban Growth.* New York: Academic Press.
65. Chase-Dunn, Christopher. 1975. "The Effects of International Economic Dependence on Development and Inequality: A Cross-National Study." *American Sociological Review* 40 (December): 720–38.
66. Chayanov, A. V. 1966. *The Theory of Peasant Economy.* Homewood, Ill.: Dorsey Press.
67. Chilcote, Ronald H., and Steve Gorman, Cis Le Roy, and Sara Sheehan. 1975. "Internal and External Issues of Dependency: Approach, Pedagogical Method, and Critiques of Two Courses on Latin America." *Review of Radical Political Economics* VI: 80–94.
68. Chilcote, Ronald H., and Joel C. Edelstein, eds. 1974. *Latin America: The Struggle with Dependency and Beyond,* Ch. 1, "Alternative Perspectives of Development and Underdevelopment in Latin America." Cambridge, Mass.: Schenkman Publishing.
69. Chodak, Szymon. 1973. *Societal Development.* New York: Oxford University Press.
70. Clarke, J. I., and W. B. Fisher, eds. 1972. *Populations of the Middle East and North Africa: A Geographical Approach.* New York: Africana Publishing Corporation.
71. Cockcroft, James D., André Gunder Frank, and Dale L. Johnson. 1972. *Dependence and Underdevelopment: Latin America's Political Economy.* Garden City, N.Y.: Doubleday.
72. Cohen, Abner. 1969. *Custom and Politics in Urban Africa: A Study of Hausa Migrants in Yoruba Towns.* Berkeley: University of California Press.
73. Cohen, Benjamin J. 1973. *The Question of Imperialism: The Political Economy of Dominance and Dependence.* New York: Basic Books.
74. Cooley, Charles H. 1962. *Social Organization.* Glencoe, Ill.: Free Press.
75. Cornelius, Wayne A. 1969. "Urbanization as an Agent in Latin American Political Instability: The Case of Mexico." *American Political Science Review* 63, 3 (September): 833–57.
76. ______. 1974. "Urbanization and Political Demand Making: Political Participation Among Migrant Poor in Latin American Cities." *American Political Science Review* 68, 3 (September): 1125–46.
77. ______. 1975. *Politics and the Migrant Poor in Mexico City.* Stanford, Cal.: Stanford University Press.
78. Daland, Robert T. 1969, "Urbanization Policy and Political Development in Latin America." *American Behavioral Scientist* XII, 5: 22–33.
79. Davidson, Basil. 1959. *The Lost Cities of Africa.* Boston: Little, Brown.
80. Davis, Kingsley. 1960. "Colonial Expansion and Urban Diffusion in the Americas." *International Journal of Comparative Sociology* 1: 43–66.
81. ______. 1969. *World Urbanization 1950–1970.* Volume I. *Basic Data for Cities, Countries and Regions.* Population Monograph Series 4. Berkeley: University of California, Institute of International Studies.
82. ______. 1972. *World Urbanization 1950–1970.* Volume II. *Analysis of Trends, Relationships, and Development.* Population Monograph Series 9. Berkeley: University of California, Institute of International Studies.
83. De Blij, H. J. 1963. *Dar es Salaam.* Evanston, Ill.: Northwestern University Press.
84. Dethier, Jean. 1973. "Evolution of Concepts of Housing, Urbanism and Country Planning in a Developing Country: Morocco, 1900–1972." Pp. 197–243 in *From Madina to Metropolis,* edited by L. Carl Brown. Princeton: The Darwin Press.
85. Dietz, Henry. 1969. "Urban Squatter Settlements in Peru: A Case History and Analysis." *Journal of Inter-American Studies* XI: 353–70.
86. Dos Santos, Theotonio. 1968. *El nuevo carácter de la dependencia.* Santiago: Cuadernos de Estudios Socio-Economicos (10), Centro de Estudios Socio-Economicos, Universidad de Chile.
87. ______. 1970. "The Structure of Dependence." *American Economic Review* LX (May): 231–36.
88. Durkheim, Emile. 1933. *The Division of Labor in Society.* New York: Macmillan. English translation by G. Simpson.
89. Du Toit, Brian M. 1976. "Language and Ethnicity as Factors in African Urban Migration in Southern Africa." In *Migration, Ethnicity and Change: Migratory Models and Adaptive Strategies,* edited by Brian M. Du Toit. The Hague: Mouton.
90. Du Toit, Brian M., and Helen I. Safa, eds. 1975. *Migration and Urbanization: Models and Adaptive Strategies.* The Hague: Mouton.
91. Dwyer, D. J., ed. 1974. *The City in the Third World.* New York: Barnes & Noble.

92. Eddy, Elizabeth M., ed. 1968. *Urban Anthropology: Research Perspectives and Strategies.* Athens, Ga.: University of Georgia Press.
93. Eisenstadt, S. N. 1964. "Social Change, Differentiation, and Evolution." *American Sociological Review* 29 (June): 375–86.
94. Elkan, Walter. 1960. *Migrants and Proletarians: Urban Labour in the Economic Development of Uganda.* London: Oxford University Press.
95. El-Shakhs, Salah, and Robert Obudho, eds. 1974. *Urbanization, National Development and Regional Planning in Africa.* New York: Praeger.
96. Emmanuel, Arghiri. 1972. *Unequal Exchange: A Study of the Imperialism of Trade.* New York: Monthly Review Press.
97. Epstein, A. L. 1958. *Politics in an Urban African Community.* Manchester: Manchester University Press
98. ______. 1967. "Urbanization and Social Change in Africa." *Current Anthropology* 8: 275–96.
99. Epstein, David G. 1973. *Brasilia, Plan and Reality: A Study of Planned and Spontaneous Urban Settlement.* Berkeley: University of California Press.
100. Esseks, J. 1971. "Economic Dependence and Political Development in New States of Africa." *Journal of Politics* XXXIII (November): 1052–75.
101. Estaban, Juan Carlos. 1961. *Imperialismo y desarrollo económico: la Argentina frente a nuevas relaciones de dependencia.* Buenos Aires: Editorial Palestra.
102. Fanon, Frantz. 1963. *The Wretched of the Earth.* New York: Grove Press. Originally *Les damnés de la terre.*
103. ______. 1965. *A Dying Colonialism.* New York: Monthly Review Press.
104. ______. 1967. *Black Skin, White Masks.* New York: Grove Press.
105. Fava, Sylvia Fleis. 1968. *Urbanism in World Perspective: A Reader.* New York: Thomas Y. Crowell.
106. Feit, Edward. 1970. "Urban Revolt in South Africa: A Case Study." *The Journal of Modern African Studies* 8: 55–72.
107. Field, Arthur J., ed. 1970. *City and Country in the Third World: Issues in the Modernization of Latin America.* Cambridge, Mass.: Schenkman Publishing.
108. Fieldhouse, D. K. 1961. " 'Imperialism': An Historiographical Revision." *Economic History Review* 14, (2nd): 187–209.
109. Frank, Andre Gunder (Andrew Gunder). 1966. "The Development of Underdevelopment." *Monthly Review* XVIII (September): 17–31.
110. ______. 1967. *Capitalism and Underdevelopment in Latin America: Historical Studies of Chile and Brazil.* New York: Monthly Review Press. (Hardcover.)
111. ______. 1969a. *Capitalism and Underdevelopment.* . . . (Paper.)
112. ______. 1969b. *Latin America: Underdevelopment or Revolution. Essays on the Development of Underdevelopment and the Immediate Enemy.* New York: Modern Reader/Monthly Review Press.
113. ______. 1970. "Urban Poverty in Latin America." Pp. 215–34 in *Masses in Latin America,* edited by Irving Horowitz. New York: Oxford University Press.
114. ______. 1972. *Lumpenbourgeoisie: Lumpendevit; Dependence, Class, and Politics in Latin America.* New York: Monthly Review Press.
115. Franklin, S. H. 1965. "Systems of Production: Systems of Appropriation." *Pacific Viewpoint* VI, 2: 145–66.
116. ______. 1969. *The European Peasantry. The Final Phase.* London: Methuen.
117. Frazer, J. M. 1952. "Housing and Planning in Singapore." *Town Planning Review* 23 (April): 5–25.
118. Freyssinet, Jacques 1966. *Le concept de sous-développement.* Paris: Mouton.
119. Friedl, John, and Noel Chrisman, eds. 1975. *City Ways: A Selective Reader in Urban Anthropology.* New York: Thomas Y. Crowell.
120. Friedmann, John. 1961. "Cities in Social Transformation." *Comparative Studies in Society and History,* IV (November): 86–103.
121. ______. 1968. "The Strategy of Deliberate Urbanization." *Journal of the American Institute of Planners* 34, 6: 364–73.
122. ______. 1969. "The Role of Cities in National Development." *American Behavioral Scientist* XII, 5: 13–21.
123. ______. 1972. "A General Theory of Polarized Development." Pp. 82–107 in *Growth Centers in Regional Economic Development,* edited by Niles Hansen. New York: Free Press.
124. ______. 1973. "Urbanization and National Development: A Comparative Analysis." Pp. 65–90 in his *Urbanization, Planning and National Development.* Beverly Hills, Cal.: Sage.
125. Friedmann, John, and Tomás Lackington. 1967. "Hyperurbanization and National Development in Chile." *Urban Affairs Quarterly* II, 4: 3–29.
126. Friedmann, John, and Flora Sullivan. 1972. "The Absorption of Labor in the Urban Economy. The Case of Developing Countries." Los Angeles: University of California, School of Architecture and Urban Planning. Mimeo.
127. Friedmann, John, and Robert Wulff. 1976. *The Urban Transition.* London: Edward Arnold.
128. Frisbie, W. Parker. 1977. "The Scale and Growth of World Urbanization." In *Cities in Change: Studies on the Urban Condition,* edited by John Walton and Donald E. Carns. 2nd ed. Boston: Allyn and Bacon.
129. Furtado, Celso. 1963. *The Economic Growth of Brazil: A Survey from Colonial To Modern Times.* Berkeley: University of California Press.
130. ______. 1964. *Development and Underdevelopment.* Berkeley: University of California Press.
131. ______. 1973a. "The Concept of External Dependence in the Study of Underdevelopment." Pp. 118–23 in *The Political Economy of Development and Underdevelopment,* edited by Charles K. Wilber. New York: Random House.
132. ______. 1973b. "The Brazilian 'Model' of Development." Pp. 297–306 in *The Political Economy of Development and Underdevelopment,* edited by Charles K. Wilber. New York: Random House.
133. Gakenheimer, Ralph. 1967. "The Peruvian City of the Sixteenth Century." Pp. 33–56 in *The Urban Explosion in Latin America,* edited by Glenn H. Beyer. Ithaca: Cornell University Press.
134. Garlick, Peter C. 1971. *African Traders and Economic Development in Ghana.* Oxford: Clarendon Press.
135. Geertz, Clifford. 1963a. *Peddlers and Princes: Social Development and Economic Change in Two Indonesian Towns.* Chicago: University of Chicago Press.
136. ______. 1963b. *Agricultural Involution: The Processes of Ecological Change in Indonesia.* Berkeley: University of California Press.
137. Germani, Gino. 1967. "The City as an Integrating Mechanism: The Concept of Social Integration." Pp. 175–89 in *The Urban Explosion in Latin America,* edited by Glenn H. Beyer. Ithaca: Cornell University Press.
138. ______. ed. 1973. *Modernization, Urbanization, and the Urban Crisis.* Boston: Little, Brown.
139. Gibson, Jeffry R. 1970. *A Demographic Analysis of Urbanization: Evolution of a System of Cities in Honduras, El Salvador and Costa Rica.* Ithaca: Cornell University Dissertation Series No. 20.

140. Golay, Frank, Ralph Anspach, M. Ruth Pfanner, and Eliezer Ayal. 1969. *Underdevelopment and Economic Nationalism in Southeast Asia.* Ithaca: Cornell University Press.
141. Goldrich, Daniel M. 1964. "Peasants' Sons in City Schools: An Inquiry into the Politics of Urbanization in Panama and Costa Rica." *Human Organization* 23, 4: 328–33.
142. Goldrich, Daniel M., et al. 1967–68. "The Political Integration of Lower-class Urban Settlements in Chile and Peru." *Studies in Comparative International Development* 3, 1: 3–22.
143. Gugler, Josef. 1970. *Urban Growth in Sub-Saharan Africa.* Kampala: Makerere Institute of Social Research.
144. Gutkind, Peter C. W. 1963. *The Royal Capital of Buganda: A Study of Internal Conflict and External Ambiguity.* The Hague: Mouton.
145. ______. 1972. "The Socio-Political and Economic Foundations of Social Problems in African Urban Areas: An Exploratory Conceptual Overview." *Civilisations* 22: 18–34.
146. ______. 1974. *Urban Anthropology: Perspectives on 'Third World' Urbanization and Urbanism.* New York: Harper & Row, Barnes & Noble Import Division.
147. Hagen, Everett E. 1962. *On the Theory of Social Change.* Homewood, Ill.: Dorsey Press.
148. Hall, Peter. 1966. *The World Cities.* New York: McGraw-Hill.
149. Hallaron, Shirley. 1976. *Urbanization in the Developing Nations—A Bibliography Compiled for the 1960's and 1970's.* Champaign-Urbana: University of Illinois.
150. Hamdan, Gamal. 1964. "Capitals of the New Africa." *Economic Geography* 40: 239–53.
151. Hance, William A. 1970. *Population, Migration and Urbanization in Africa.* New York: Columbia University Press.
152. Hanna, William J., and Judith L. Hanna. 1971. *Urban Dynamics in Black Africa.* Chicago: Aldine.
153. Hardoy, Jorge Enrique. 1970. *Urban Land Policies and Land Use Control Measures in Cuba.* New York: United Nations Centre for Housing, Building and Planning.
154. ______. ed. 1975. *Urbanization in Latin America: Approaches and Issues.* Garden City, N.Y.: Anchor Books.
155. Hardoy, Jorge E., and Richard Schaedel, eds. 1969. *The Urbanization Process in America from Its Origins to the Present Day.* Buenos Aires, Argentina: Editorial del Instituto Torcuato Di Tella.
156. Harris, Marvin. 1956. *Town and Country in Brazil.* New York: Columbia University Press.
157. Hart, Keith. 1970. "Small-scale Entrepreneurs in Ghana and Development Planning." *Journal of Development Studies* VI, 4: 104–20.
158. ______. 1973. "Informal Income Opportunities and Urban Employment in Ghana." *The Journal of Modern African Studies* XI, 1: 61–89.
159. Harvey, David. 1973. *Social Justice and the City.* London: Edward Arnold. Baltimore: The Johns Hopkins University Press.
160. ______. 1975. "The Geography of Capitalist Accumulation: A Reconstruction of the Marxian Theory." *Antipode* 7, 2 (September): 9–21.
161. Hauser, Philip M., ed. 1958. *Urbanization in Asia and the Far East.* Calcutta. UNESCO.
162. ______. 1971. "A New Approach to the Measurement of the Work Force in Developing Areas." Cyclostyled paper for limited circulation.
163. ______. 1972. "Population Change and Developments in Manpower, Labour Force, Employment and Income." Cyclostyled paper prepared for UN ECAFE Seminar on Population Aspects of Social Development, Bangkok.
164. Hauser, Philip M., and Leo F. Schnore, eds. 1965. *The Study of Urbanization.* New York: John Wiley.
165. Herrick, Bruce H. 1965. *Urban Migration and Economic Development in Chile.* Cambridge, Mass.: M.I.T. Press.
166. Hinkelammert, Franz. 1970. "La teoría clásica del imperialismo, el subdesarrollo y la acumulación socialista." *Cuadernos de la Realidad Nacional* 4 (June): 137–60.
167. Hla, Myint U. 1972. *Southeast Asia's Economy: Development Policies in the 1970's.* New York: Praeger.
168. Hobson, John A. 1965. *Imperialism: A Study.* Ann Arbor: University of Michigan Press.
169. Hollnsteiner, Mary R. 1973. "City? Province? or Relocation Site? Options for Manila's Squatters." *Philippine Sociological Review* 21, 3–4.
170. Hopkins, Nicholas. 1972. *Popular Government in an African Town: Kita, Mali.* Chicago: University of Chicago Press.
171. Horowitz, Irving L. 1966. *Three Worlds of Development.* New York: Oxford University Press.
172. ______. 1970. "Personality and Structural Dimensions in Comparative International Development." *Social Science Quarterly* 51 (December): 494–513.
173. Hoselitz, Berthold F. 1960. *Sociological Aspects in Economic Growth.* Glencoe, Ill.: Free Press.
174. Hutton, John, ed. 1972. *Urban Challenge in East Africa.* Nairobi: East African Publishing House.
175. Iliffe, J. 1969. "The Age of Improvement and Differentiation (1907–45)." Pp. 123–60 in *A History of Tanzania,* edited by I. N. Kimambo and A. J. Temu. Nairobi: East African Publishing House. Evanston, Ill.: Northwestern University Press.
176. Inkeles, Alex. 1966. "The Modernization of Man." Pp. 138–50 in *Modernization: The Dynamics of Growth,* edited by Myron Weiner. New York: Basic Books.
177. ______. 1969. "Making Men Modern: On the Causes and Consequences of Individual Change in Six Countries." *American Journal of Sociology* 75 (September): 208–25.
178. International African Institute. 1956. *Social Implications of Industrialization and Urbanization in Africa South of the Sahara.* Paris: UNESCO.
179. International Labor Organization. 1972. *Employment, Incomes and Equality: A Strategy for Increasing Productive Employment in Kenya.* Geneva: I.L.O.
180. Jaguaribe, Helio, Aldo Ferrer, Miguel S. Wionczek, and Theotónio Dos Santos. 1970. *La dependencia politico-económica de América Latina.* 2nd ed. Mexico City: Siglo Veintiuno Editores.
181. Jakobson, Leo, and Ved Prakash, eds. 1971. *Urbanization and National Development.* Beverly Hills, Cal.: Sage.
182. de Jesus, Carolina Maria. 1962. *Child of the Dark: The Diary of Carolina Maria de Jesus.* Translation by David St. Clair. New York: E. P. Dutton.
183. Jones, Rodney W. 1974. *Urban Politics in India: Area, Power, and Policy in a Penetrated System.* Berkeley: University of California Press.
184. Jones, Ronald, ed. 1975. *Essays on World Urbanization.* London: G. Philip.
185. Kahl, Joseph A. 1968. *The Measurement of Modernism.* Austin: University of Texas Press.
186. Kamerschen, David R. 1969. "Further Analysis of Overurbanization." *Economic Development and Cultural Change* 17, 2: 235–53.
187. Karpat, Kemal H. 1976. *The Gecekondu: Rural Migration and Urbanization.* Cambridge, London, New York, Melbourne: Cambridge University Press.

188. Kerr, Clark, John T. Dunlop, Frederick Harbison, and Charles A. Myers. 1960. *Industrialism and Industrial Man: The Problems of Labor and Management in Economic Growth.* New York: Oxford University Press.
189. Khalaf, Samir, and Per Kongstad. 1973. *Hamra of Beirut: A Case of Rapid Urbanization.* Leiden: Brill.
190. Kilby, Peter. 1971. *Entrepreneurship and Economic Development.* New York: Free Press.
191. King, Anthony D. 1976. *Colonial Urban Development: Culture, Social Power and Environment.* London: Routledge & Kegan Paul.
192. Kohl, James, and John Litt, eds. 1974. *Urban Guerrilla Warfare in Latin America.* Cambridge, Mass.: M.I.T. Press.
193. Kornhauser, David. 1976. *Urban Japan: Its Foundations and Growth.* London and New York: Longman.
194. Kornhauser, William. 1959. *The Politics of Mass Society.* New York: Free Press.
195. Krapf-Askari, Eva. 1969. *Yoruba Towns and Cities.* Oxford: Clarendon Press.
196. Kuper, Hilda, ed. 1965. *Urbanization and Migration in West Africa.* Berkeley: University of California Press.
197. Kuznets, Simon. 1966. *Modern Economic Growth.* New Haven, Conn.: Yale University Press.
198. Laclau, Ernesto. 1971. "Feudalism and Capitalism in Latin America." *New Left Review* 67 (May-June): 19–38.
199. Lagos Matus, Gustavo. 1963. *International Stratification and Underdeveloped Countries.* Chapel Hill, N.C.: University of North Carolina Press.
200. Lapidus, Ira, ed. 1969. *Middle Eastern Cities.* Berkeley: University of California Press.
201. Latif, A. H. 1974. "Factor Structure and Change Analysis of Alexandria, Egypt, 1947 and 1960." Pp. 338–49 in *Comparative Urban Structure,* edited by Kent P. Schwirian. Lexington, Mass.: D. C. Heath.
202. Leeds, Anthony. 1965. "Brazilian Careers and Social Structure: A Case History and Model." Pp. 379–404 in *Contemporary Cultures and Societies in Latin America,* edited by D. B. Heath and R. N. Adams. New York: Random House.
203. ______. 1968. "The Anthropology of Cities: Some Methodological Issues." Pp. 31–47 in *Urban Anthropology,* edited by Elizabeth M. Eddy. Athens, Ga.: University of Georgia Press.
204. ______. 1971. "The Concept of the 'Culture of Poverty.'" In *The Culture of Poverty: A Critique,* edited by Eleanor B. Leacock. New York: Simon & Schuster.
205. Leeds, Anthony, and Elizabeth Leeds. 1970. "Brazil and the Myth of Urban Rurality: Urban Experience, Work, and Values in 'Squatments' of Rio de Janeiro and Lima." Pp. 229–85 in *City and Country in the Third World,* edited by Arthur J. Field. Cambridge, Mass.: Schenkman Publishing.
206. Lenin, Vladimir. *The Development of Capitalism in Russia,* first published 1899.
207. ______. 1939. *Imperialism: The Highest Stage of Capitalism.* New York: International Publishers.
208. ______. 1967. *Selected Works in Three Volumes.* Moscow: Progress Publishers.
209. Lerner, Daniel. 1965. *The Passing of Traditional Society: Modernizing the Middle East.* New York: Free Press.
210. Levy, Marion J. 1966. *Modernization and the Structure of Societies.* 2 vols. Princeton, N.J.: Princeton University Press.
211. Lewis, Oscar. 1961. *The Children of Sanchez: Autobiography of a Mexican Family.* New York: Random House.
212. ______. 1965. "Further Observations on the Folk-Urban Continuum and Urbanization with Special Reference to Mexico City." Pp. 491–503 in *The Study of Urbanization,* edited by Philip M. Hauser and Leo F. Schnore. New York: John Wiley.
213. ______. 1966. *La Vida: A Puerto Rican Family in the Culture of Poverty.* New York: Random House.
214. Leys, Colin. 1974. *Underdevelopment in Kenya: The Political Economy of Neo-colonialism, 1964–1971.* Berkeley: University of California Press.
215. Linz, Juan. 1973. "Early State-Building and Late Peripheral Nationalism against the State: The Case of Spain." Pp. 32–116 in *Building States and Nations,* edited by S. N. Eisenstadt and Stein Rokkan. Beverly Hills, Cal.: Sage.
216. Little, Kenneth. 1965. *West African Urbanization: A Study of Voluntary Associations in Social Change.* London: Cambridge University Press.
217. ______. 1973. *African Women in Towns.* London: Cambridge University Press.
218. ______. 1974a. *Urbanization, Migration, and the African Family.* Reading, Mass.: Addison-Wesley.
219. ______. 1974b. *Urbanization as a Social Process: An Essay on Movement and Change in Contemporary Africa.* London: Routledge & Kegan Paul.
220. Lloyd, Peter C. 1973. "The Yoruba: An Urban People?" Pp. 107–23 in *Urban Anthropology,* edited by Aidan Southall. New York: Oxford University Press.
221. Lloyd, Peter C., A. L. Mabogunje and B. Awe, eds. 1967. *The City of Ibadan: A Symposium on its Structure and Development.* London: Cambridge Universiy Press.
222. López Segrera, Francisco. 1972. *Cuba: capitalismo dependiente y subdesarrollo.* Havana: Casa de las Américas.
223. Lubeck, Paul. 1975. "Unions, Workers and Consciousness in Kano, Nigeria: A View From Below." In *The Development of an African Working Class,* edited by R. Sandbrook and R. Cohen. London: Longman.
224. Luttrell, W. L. 1972. "Location Planning and Regional Development in Tanzania." Pp. 119–49 in *Towards Socialist Planning* (Tanzanian Studies No. 1), edited by J. F. Rweyemamu. Dar es Salaam: Tanzania Publishing House.
225. Luxemburg, Rosa. 1951. *The Accumulation of Capital.* London: Routledge & Kegan Paul.
226. Lynch, Kevin. 1972. *What Time is This Place?* Cambridge, Mass.: M.I.T. Press.
227. Lynch, Owen M. 1969. *The Politics of Untouchability: Social Mobility and Social Change in a City in India.* New York: Columbia University Press.
228. Mabogunje, Akin L. 1968. *Urbanization in Nigeria.* London: University of London Press.
229. McClelland, David C. 1963. "Motivational Patterns in Southeast Asia with Special Reference to the Chinese Case." *Journal of Social Issues* 19 (January): 6–19.
230. ______. 1967. *The Achieving Society.* New York: Free Press.
231. McGee, T. G. 1967. *The Southeast Asian City: A Social Geography of the Primate Cities of Southeast Asia.* London: G. Bell and Sons, Ltd.
232. ______. 1971a. *The Urbanization Process in the Third World: Explorations in Search of a Theory.* London: G. Bell and Sons, Ltd.
233. ______. 1971b. "Catalysts or Cancers? The Role of Cities in Asian Society." Pp. 157–81 in *Urbanization and National Development,* edited by Leo Jakobson and Ved Prakash. Beverly Hills, Cal.: Sage.
234. ______. 1973a. "Peasants in the Cities: A Paradox, A Paradox, A Most Ingenious Paradox." *Human Organization* XXXII, 2: 135–42.
235. ______. 1973b. *Hawkers in Hong Kong.* Hong Kong: University of Hong Kong, Centre of Asian Studies.
236. McNulty, Michael L. 1969. "Urban Structure and Development: The Urban System of Ghana." *The Journal of Developing Areas* 3: 159–76.
237. Magdoff, Harry. 1969. *The Age of Imperialism: The Economics of U.S. Foreign Policy.* New York: Monthly Review Press.

238. Maine, Henry. 1907. *Ancient Law.* London.
239. Malloy, James M. 1971. "Generation of Political Support and Allocation of Costs." Pp. 23–42 in *Revolutionary Change in Cuba,* edited by C. Mesa Lago. Pittsburgh: University of Pittsburgh Press.
240. Mangin, William. 1967. "Latin American Squatter Settlements: A Problem and a Solution." *Latin American Research Review* II, 3: 65–98.
241. ______. ed. 1970. *Peasants In Cities: Readings in the Anthropology of Urbanization.* Boston: Houghton Mifflin.
242. ______. 1973. "Sociological, Cultural and Political Characteristics of Some Urban Migrants in Peru." Pp. 315–50 in *Urban Anthropology,* edited by Aidan Southall. Chicago: Aldine.
243. Marris, Peter. 1962. *Family and Social Change in an African City.* Evanston, Ill.: Northwestern University Press.
244. Marx, Karl. 1904. *A Contribution to the Critique of Political Economy.* Translated from the second German edition by N. I. Stone. Chicago: Charles H. Kerr.
245. Mascarenhas, Adolfo, and Claes Claeson. 1972. "Factors Influencing Tanzania's Urban Policy." *African Urban Notes* 6: 24–42.
246. Mayer, Philip. 1961. *Townsmen or Tribesmen: Conservatism and the Process of Urbanization in a South African City.* Cape Town: Oxford University Press.
247. Meadows, Paul, and Ephraim H. Mizruchi, eds. 1976. *Urbanism, Urbanization and Change: Comparative Perspectives.* 2nd ed. Reading, Mass.: Addison-Wesley.
248. Merrington, John. 1975. "Town and Country in the Transition to Capitalism." *New Left Review* 95 (September-October): 71–92.
249. Meyer, John W., John Boli-Bennett, and Christopher Chase-Dunn. 1975. "Convergence and Divergence in Development." In *Annual Review of Sociology, 1975,* edited by Alex Inkeles, James Coleman, and Neil Smelser. Palo Alto, Cal.: Annual Reviews.
250. Miller, Frank C. 1973. *Industrialization in Mexico: Old Villages in a New Town.* Menlo Park, Cal.: Cummings Publishing.
251. Mills, C. Wright. 1956. *The Power Elite.* New York: Oxford University Press.
252. Miner, Horace, ed. 1967. *The City in Modern Africa.* New York: Praeger.
253. Mitchell, J. Clyde, ed. 1969. *Social Networks in Urban Situations: Analyses of Personal Relationships in Central African Towns.* Manchester: Manchester University Press.
254. ______. 1970. "The Causes of Labor Migration." Pp. 23–37 in *Black Africa,* edited by John Middleton. Toronto, New York: Macmillan.
255. Mitchell, J. Clyde, and A. L. Epstein. 1959. "Occupational Prestige and Social Status Among Urban Africans in Northern Rhodesia." *Africa* 29: 22–40.
256. Moore, Barrington, Jr. 1966. *Social Origins of Dictatorship and Democracy.* Boston: Beacon.
257. Moore, Russell Martin. 1973. "Imperialism and Dependency in Latin America: A View of the New Reality of Multinational Investment." *Journal of Inter-American Studies and World Affairs* XV (February): 21–35.
258. Morrison, Donald G., et al. 1977. *Black Africa: A Comparative Handbook.* 2nd ed. New York: Free Press.
259. Morse, Richard M. 1958. *From Community to Metropolis: A Biography of Sao Paulo.* Gainesville: University of Florida Press.
260. ______. 1962. "Latin American Cities: Aspects of Function and Structure." *Comparative Studies in Society and History* 4 (July): 473–93.
261. ______. 1965. "Recent Research on Latin American Urbanization: A Selective Survey with Commentary." *Latin American Research Review* 1: 35–75.
262. ______. 1971. "Trends and Issues in Latin American Urban Research, 1965–1970." *Latin American Research Review* 6, Part I (Spring): 3–52; Part II (Summer): 19–75.
263. ______. 1972. "A Prolegomenon to Latin American Urban History." *Hispanic American Historical Review* 52: 359–94.
264. ______. 1974. "The Development of Urban Systems in the Americas in the Nineteenth Century." *Journal of Interamerican Studies and World Affairs* 17: 4–26.
265. Myrdal, Gunnar. 1957. *Rich Lands and Poor.* New York: Harper & Row.
266. ______. 1968. *Asian Drama.* 3 vols. New York: Random House.
267. Nash, Manning. 1967. *Machine Age Maya: The Industrialization of a Guatemalan Community.* Chicago: University of Chicago Press.
268. Navarrete, Ifigenia M. de. 1967. "Income Distribution in Mexico." Pp. 133–72 in *Mexico's Recent Economic Growth,* edited by Enrique Pérez López et al. Austin: University of Texas Press.
269. Nelson, Joan M. 1969. *Migrants, Urban Poverty and Instability in Developing Nations.* Cambridge, Mass.: Harvard University Center for International Affairs.
270. Nutini, Hugo G. 1972. "The Latin American City: A Cultural-Historical Approach." Pp. 89–96 in *The Anthropology of Urban Environments,* edited by Thomas Weaver and Douglas White. Washington, D.C.: The Society for Applied Anthropology.
271. O'Brien, Philip J. 1975. "A Critique of Latin American Theories of Dependency." Pp. 7–27 in *Beyond the Sociology of Development,* edited by Ivar Oxaal, Tony Barnett, and David Booth. London: Routledge & Kegan Paul.
272. Odell, Peter R., and David A. Preston. 1973. *Economies and Societies in Latin America: A Geographical Interpretation.* London: John Wiley Ltd.
273. O'Donnell, Guillermo. 1972. "Modernizacion y golpes militares: Teoría, comparacion y el caso Argentino." *Desarrollo economico* 47 (October-December): 519–66.
274. Oliver, Paul, ed. 1971. *Shelter in Africa.* New York: Praeger.
275. Onibokun, G. A. 1971. "Housing Finance in Nigeria." *Town Planning Review* 42: 277–92.
276. Orellana, C. L. 1973. "Mixtec Migrants in Mexico City: A Case Study of Urbanization." *Human Organization* 32: 273–83.
277. Oxaal, Ivar, Tony Barnett, and David Booth, eds. 1975. *Beyond the Sociology of Development: Economy and Society in Latin American and Africa.* London: Routledge & Kegan Paul.
278. Parsons, Talcott. 1964a. *The Social System.* New York: Free Press.
279. ______. 1964b. "Evolutionary Universals in Society." *American Sociological Review* 29 (June): 339–57.
280. Peattie, Lisa R. 1968. *The View from the Barrio.* Ann Arbor: University of Michigan Press.
281. ______. 1969. "Social Mobility and Economic Development." Pp. 400–410 in *Planning Urban Growth and Regional Development: The Experience of the Guayana Program of Venezuela,* by Lloyd Rodwin et al. Cambridge, Mass.: M.I.T. Press.
282. Pedersen, Poul O. 1975. *Urban-Regional Development in South America: A Process of Diffusion and Integration.* The Hague: Mouton.

283. Petras, James, ed. 1973. *Latin America: From Dependence to Revolution.* New York: John Wiley.
284. Petras, James, and Maurice Zeitlin, eds. 1968. *Latin America: Reform or Revolution? A Reader.* Greenwich, Conn.: Fawcett.
285. Pickvance, Christopher G., ed. 1976. *Urban Sociology: Critical Essays.* London: Methuen. New York: St. Martin's Press.
286. Plotnicov, Leonard. 1967. *Strangers to the City: Urban Man in Jos, Nigeria.* Pittsburgh: University of Pittsburgh Press.
287. Poncet, J. 1973. "La régionalisation en Tunisie." *Revue Tiers-Monde* XIV (July-September): 597–614.
288. Portes, Alejandro. 1971. "The Urban Slum in Chile: Types and Correlates." *Land Economics* 47: 235–48.
289. ______. 1972. "Rationality in the Slum: An Essay of Interpretive Sociology." *Comparative Studies in Society and History* 14: 268–86.
290. ______. 1973. "Modernity and Development: A Critique." *Studies in Comparative International Development* 8 (Fall): 247–79.
291. Portes, Alejandro, and Harley Browning, eds. 1976. *Current Perspectives in Latin American Urban Research.* Austin: University of Texas Press.
292. Portes, Alejandro, and John Walton. 1976. *Urban Latin America: The Political Condition from Above and Below.* Austin: University of Texas Press.
293. Porzecanski, Leopold. 1973. *A Selected Bibliography on Urban Housing in Latin America.* Exchange Bibliography No. 412. Monticello, Ill.: Council of Planning Librarians.
294. Powdermaker, Hortense. 1962. *Copper Town: Changing Africa.* New York: Harper & Row.
295. Pratt, Raymond B. 1971. "Community Political Organization and Lower Class Politization in Two Latin American Cities." *The Journal of Developing Areas* 5: 523–42.
296. Prebisch, Raul. 1950. *The Economic Development of Latin America and Its Principal Problems.* New York: United Nations.
297. ______. 1959. "Commercial Policy in the Underdeveloped Countries." *American Economic Review Papers and Proceedings* 49 (May): 251–73.
298. Quijano, Aníbal. 1967. "La urbanizacion de la sociedad en Latinoamerica." *Revista Mexicana de Sociología* 29 (October-December): 669–703.
299. ______. 1968. "Tendencies in Peruvian Development and in the Class Structure." Pp. 289–328 in *Latin America,* edited by James Petras and Maurice Zeitlin. Greenwich, Conn.: Fawcett.
300. ______. 1971. *Nationalism and Colonialism in Peru: A Study in Neo-Imperialism.* New York: Monthly Review Press.
301. ______. 1972. "Imperialism and International Relations in Latin America." Paper delivered in Lima, Peru. Pp. 67–91 in *Latin America and the United States: The Changing Political Realities,* edited by Julio Cotler and Richard R. Fagen. Stanford, Cal.: Stanford University Press, 1974.
302. ______. 1974. "Redefinición de la teoría de dependencia." Unpublished paper, Lima, Peru.
303. Ray, Talton F. 1969. *The Politics of the Barrios of Venezuela.* Berkeley: University of California Press.
304. Redfield, Robert. 1965. *Peasant Society and Culture.* Chicago: University of Chicago Press.
305. Rhodes, Robert I., ed. 1970. *Imperialism and Underdevelopment: A Reader.* New York: Monthly Review Press.
306. Riddell, J. B., and M. E. Harvey. 1972. "The Urban System in the Migration Process: An Evaluation of Step-Wise Migration in Sierra Leone." *Economic Geography* 48: 270–83.
307. Roberts, Bryan R. 1968. "Politics in a Neighbourhood of Guatemala City." *Sociology* II (London), 2: 185–204.
308. ______. 1970. "The Social Organization of Low-Income Families." Pp. 345–82 in *Masses in Latin America,* edited by I. Horowitz. New York: Oxford University Press.
309. ______. 1973a. "Migracion urbana y cambie en la organización provincial en la Sierra Central de Perú." *Ethnica* VI (Barcelona, Spain).
310. ______. 1973b. *Organizing Strangers: Poor Families in Guatemala City.* Austin: University of Texas Press.
311. ______. 1974a. "The Social History of a Provincial Town: Huancayo, 1890–1972." In *Social and Economic Change in Peru,* edited by R. Miller, C. Smith, and J. Fisher. Monograph 6, University of Liverpool.
312. ______. 1974b. "Interrelationships of City and Provinces in Peru and Guatemala." Pp. 207–36 in *Latin American Urban Research* IV, edited by Wayne A. Cornelius and Felicity M. Trueblood. Beverly Hills, Cal.: Sage.
313. ______. 1976. "The Provincial Urban System and the Process of Dependency." In *Current Perspectives in Latin American Urban Research,* edited by Alejandro Portes and Harley Browning. Austin: University of Texas Press.
314. Robson, William, and D. E. Regan. 1972. *Great Cities of the World: Their Government, Politics and Planning.* London: Allen and Unwin.
315. Rodney, Walter. 1972. *How Europe Underdeveloped Africa.* London and Dar es Salaam: Bogle-L'Ouverture and Tanzania Publishing House.
316. Rogler, Lloyd H. 1967. "Slum Neighborhoods in Latin America." *Journal of Inter-American Studies* IX, 4: 507–28.
317. Rosenstein-Rodan, P. N. 1972. "The Have's and the Have-Not's Around the Year 2000." Pp. 29–38 in *Economics and World Order: From the 1970's to the 1990's,* edited by Jagdish N. Bhagwati. New York: Macmillan.
318. Ross, Marc H. 1970. "Urbanization and Political Participation: The Effect of Increasing Scale in Nairobi." Boston: African Studies Association paper.
319. ______. 1973. "Community Formation in an Urban Squatter Settlement." *Comparative Political Studies* 6: 296–328.
320. ______. 1975. *Grass Roots in an African City: Political Behavior in Nairobi.* Cambridge, Mass.: M.I.T. Press.
321. Rosser, Colin. 1973. *Urbanization in Tropical Africa: A Demographic Introduction.* New York: International Urbanization Survey of the Ford Foundation.
322. Roth, Guenther. 1975. "Socio-Historical Model and Developmental Theory." *American Sociological Review* 40 (April): 148–57.
323. Russell, Charles A., et al. 1974. "The Urban Guerrilla in Latin America: A Select Bibliography." *Latin American Research Review* 9: 37–79.
324. Rweyemamu, Justinian F. 1971a. "The Causes of Poverty in the Periphery." *The Journal of Modern African Studies* IX (October): 453–55.
325. ______. 1971b. *The Structure of Tanzanian Industry.* Dar es Salaam: University of Dar es Salaam, Economic Research Bureau.
326. ______. 1973. *Underdevelopment and Industrialization in Tanzania: A Study of Perverse Capitalist Industrial Development.* London: Oxford University Press.
327. Sable, Martin H. 1971. *Latin American Urbanization: A Guide to the Literature, Organization and Personnel.* Metuchen, N.J.: Scare Crow Press.
328. Safa, Helen Icken. 1974. *The Urban Poor of Puerto Rico: A Study in Development and Inequality.* New York: Holt, Rinehart

and Winston.

329. Sánchez-Albornoz, Nicolás. 1974. *The Population of Latin America: A History.* Berkeley: University of California Press.
330. Santos, Milton. 1971. *Les villes du tiers monde.* Paris: M.–Th. Génin.
331. ______. 1972. "Economic Development and Urbanization in Underdeveloped Countries. The Two Flow Systems of the Urban Economy and their Spatial Implications." University of Toronto unpublished paper.
332. Savit Bacha, Claire. 1971. "A dependencia nas relações internacionais: uma introducão a experiencia brasileira." Rio de Janeiro: Master's Thesis, Instituto Universitário de Pesquisas do Rio de Janeiro.
333. Schmitter, Phillipe C. 1971. "Desarrollo retrasado dependencia externa y cambio politico en América Latina." *Foro Internacional* XII (October-December): 135–74.
334. ______. 1972. "Paths to Political Development in Latin America." Pp. 83–150 in *Changing Latin America: New Interpretations of its Politics and Society,* edited by Douglas A. Chalmers. New York: Proceedings of the Academy of Political Science (XXX, 4), Columbia University.
335. Schnaiberg, Allan. 1970. "Measuring Modernism: Theoretical and Empirical Explorations." *American Journal of Sociology* 76 (December): 399–425.
336. Schnore, Leo F. 1965. "On The Spatial Structure of Cities in the Two Americas." Pp. 347–98 in *The Study of Urbanization,* edited by Philip M. Hauser and Leo F. Schnore. New York: John Wiley.
337. Schulze-Gaevernitz. 1906. *Britischer Imperialismus und englischer Freihandel zu Beginn des 20–ten Jahrhunderts.* Leipzig.
338. Schwirian, Kent P., ed. 1974. *Comparative Urban Structure: Studies in Ecology of Cities.* Lexington, Mass.: D. C. Health.
339. Scientific American, Inc. 1973. *Cities: Their Origin, Growth and Human Impact.* San Francisco: W. H. Freeman.
340. Scobie, James. 1974. *Buenos Aires: Plaza to Suburb 1870–1910.* New York: Oxford University Press.
341. Scott, Robert E. 1973. "National Integration Problems and Military Regimes in Latin America." Pp. 285–356 in *Latin American Modernization Problems,* edited by Robert E. Scott. Champaign-Urbana: University of Illinois Press.
342. Seers, Dudley. 1969. "The Meaning of Development." *International Development Review* 11 (December): 2–6.
343. Sheriff, A. M. H. 1972. "Trade and Development—The Role of International Trade in the Economic History of the East African Coast Before the Sixteenth Century." Mimeo.
344. Sjoberg, Gideon. 1960. *The Preindustrial City: Past and Present.* New York: Free Press.
345. Skinner, Elliott P. 1974. *African Urban Life: The Transformation of Ouagadougou.* Princeton: Princeton University Press.
346. Smelser, Neil J. 1968. *Essays in Sociological Explanation.* Englewood Cliffs, N. J.: Prentice–Hall.
347. Soja, Edward. 1976. "Spatial Inequality in Africa." *Comparative Urbanization Studies.* Los Angeles: University of California, School of Architecture and Urban Planning.
348. Solow, Anatole A. 1967. "Housing in Latin America: The Problem of the Urban Low-Income Families." *Town Planning Review* 38: 83–102.
349. Southall, Aidan, ed. 1973. *Urban Anthropology: Cross-Cultural Studies of Urbanization.* New York: Oxford University Press.
350. Southall, Aidan, and Peter Gutkind. 1957. *Townsmen in the Making: Kampala and its Suburbs.* Kampala, Uganda: East African Institute of Social Research.
351. Spoehr, Alexander, ed. 1963. *Pacific Port Towns and Cities: A Symposium.* Honolulu: Bishop Museum Press.
352. Stallings, Barbara. 1972. "Economic Dependency in Africa and Latin America." *Comparative Politics Series* III. Beverly Hills, Cal.: Sage.
353. Stavenhagen, Rodolfo. 1965. "Classes, Colonialism and Acculturation." *Studies in Comparative International Development* 1, 6: 53–77.
354. ______.1968. "Seven Fallacies About Latin America." Pp. 13–31 in *Latin America,* edited by James Petras and Maurice Zeitlin. Greenwich, Conn.: Fawcett.
355. Stein, Stanley J., and Barbara H. Stein. 1970. *The Colonial Heritage of Latin America: Essays on Economic Dependence in Perspective.* New York: Oxford University Press.
356. Stren, Richard. 1972. "Urban Policy in Africa: A Political Analysis." *African Studies Review* 15: 489–516.
357. Sunkel, Osvaldo. 1969. "National Development Policy and External Dependence in Latin America." *Journal of Development Studies* VI (October): 23–48.
358. ______.1972. *Capitalismo transnacional y desintegración nacional en América Latina.* Buenos Aires: Nueva Visión.
359. ______.1974. "External Economic Relationship and the Process of Development: Suggestions for an Alternative Analytical Framework." Pp. 27–39 in *Latin American-U.S. Economic Interactions,* edited by R. B. Williamson, W. P. Glade, and K. M. Schmitt. Washington, D. C.: American Enterprise Institute for Public Policy Research.
360. Sunkel, Osvaldo, and Pedro Paz. 1970. *El Subdesarrollo Latinoamericano y la teoría del desarrollo.* Mexico City: Siglo Veintiuno Editores.
361. Supan, Alexander G. 1906. *Die territoriale Entwicklung der europäischen Kolonién.* (The Territorial Development of the European Colonies.) Gotha.
362. Sween, Joyce, and John Walton. 1971. "Urbanization, Industrialization and Voting in Mexico: A Longitudinal Analysis of Official and Opposition Party Support." *Social Science Quarterly* 52: 721–45.
363. Tax, Sol. 1953, *Penny Capitalism: A Guatemalan Indian Economy.* Smithsonian Institute, Institute of Social Anthropology, Publication No. 16. Washington, D. C.: United States Printing Office.
364. Terrill, Ross. 1975. *Flowers on an Iron Tree: Five Cities of China.* Boston: Atlantic-Little, Brown.
365. Thompson, E. P. 1963. *The Making of the English Working Class.* New York: Random House.
366. Todaro, M. P. 1969. "A Model of Labor Migration and Urban Unemployment in Less Developed Countries." *American Economic Review* 59: 138–48.
367. ______.1971. "Income Expectations, Rural–Urban Migration and Employment in Africa." *International Labour Review* 104, 5: 387–414.
368. Toennies, Ferdinand. 1957. *Community and Society.* East Lansing: Michigan State University Press.
369. Torres Rivas, Edelberto. 1974. "Poder nacional y sociedad dependiente: Notas sobre las clases y el estado en Centroamerica." *Revista Paraguaya de Sociologia* 29 (January-April): 179–210.
370. Turner, John. 1965. "Lima's *Barriadas* and *Corrolones:* Suburbs vs. Slums." *Ekistics* 19 (Greece), 112: 152–56.
371. ______.1967. "Four Autonomous Settlements in Lima, Peru." Paper presented at the Latin American Colloquium, Department of Sociology, Brandeis University, May.
372. ______.1968. "Uncontrolled Urban Settlement: Problems and Policies." *International Social Development Review* (United Nations) 1: 107–30. Also pp. 507–534 in *The City in Newly Developing Countries,* edited by Gerald Breese. Englewood

Cliffs, N. J.: Prentice-Hall, 1969.

373. ______.1976. *Housing By People: Towards Autonomy in Building Environments.* London: Marion Boyars.
374. Turner, John, and Robert Fichter, eds. 1972. *Freedom to Build.* New York: Macmillan.
375. Turner, Roy, ed. 1962. *India's Urban Future.* Berkeley: University of California Press.
376. Turnham, David, and Ingelies Jaeger. 1971. *The Employment Problem in Less Developed Countries.* Paris: OECD.
377. Tyler, William G., and J. Peter Wogart. 1973. "Economic Dependence and Marginalization: Some Empirical Evidence." *Journal of Interamerican Studies and World Affairs* XV (February): 36–45.
378. Ullman, Edward. 1958. "Regional Development and the Geography of Concentration." *Papers and Proceedings of the Regional Science Association* IV: 179–98.
379. United Nations, Conference on Human Settlements. 1976. *Global Review of Human Settlements, Item 10 of the Provisional Agenda.* HABITAT: United Nations Conference on Human Settlements, A/CONF.70/A/1.
380. United Nations, Council on Trade and Development. 1976a. *New Directions and New Structures for Trade and Development, Report of the Secretary-General of UNCTAD to the Conference, UNCTAD 4.* Nairobi: UNCTAD 4.
381. ______. 1976b. *Handbook of International Trade and Development Statistics 1976.* UNTD/STAT. 6. New York: United Nations.
382. United Nations, Department of Economic and Social Affairs. 1969. *Growth of the World's Urban and Rural Population, 1920–2000.* Population Studies, No. 44. New York: United Nations.
383. ______. 1971. *Improvement of Slums and Controlled Settlements: Report of the inter-regional seminar,* Medillin, Columbia. New York: United Nations.
384. ______. 1973. *Multinational Corporations in World Development.* ST/ECA/190. New York: United Nations.
385. ______. 1976. *Global Review of Human Settlements: Statistical Annex.* Statistical Office, Dept. of Economic and Social Affairs, United Nations. A/CONF.70/A/1/1 ADD1.
386. United Nations, Economic Commission for Africa. 1970. *Statistical Yearbook.* Economic Commission for Africa, E/CN.14.
387. ______. 1972. *Urbanization in Africa: Levels, Trends and Prospects.*
388. United Nations, Economic Commission for Latin America. 1950. *The Economic Development of Latin America and its Principal Problems.* New York: United Nations. Same as Prebisch, No. 296.
389. U.S. Department of State. 1975. *World Military Expenditures and Arms Trade, 1963–73.* Washington, D.C.: Government Printing Office.
390. van den Berghe, P. C. 1964. *Caneville: The Social Structure of a South African Town.* Middletown, Conn.: Wesleyan University Press.
391. Vatuk, Sylvia. 1972. *Kinship and Urbanization: White Collar Migrants in North India.* Berkeley: University of California Press.
392. Wagley, Charles, and Marvin Harris. 1968. "A Typology of Latin American Subcultures." Pp. 81–117 in *The Latin American Tradition: Essays on the Unity and the Diversity of Latin American Culture,* edited by Charles Wagley. New York: Columbia University Press.
393. Wallerstein, Immanuel. 1974. *The Modern World-System—Capitalist Agriculture and the Origins of the European World-Economy in the Sixteenth Century.* New York: Academic Press.
394. Walton, John. 1972. "Political Development and Economic Development: A Regional Assessment of Contemporary Theories." *Studies in Comparative International Development* 7: 39–63.
395. ______. 1977. *Elites and Economic Development: Comparative Studies on the Political Economy of Latin American Cities.* Austin: University of Texas Press.
396. Walton, John, and Louis Masotti, eds. 1976. *The City in Comparative Perspective: Cross-National Research and New Directions in Theory.* New York: Halsted Press, John Wiley.
397. Weaver, Thomas, and Douglas White, eds. 1972. *The Anthropology of Urban Environments.* The Society for Applied Anthropology Monograph Series, Monograph Number 11.
398. Webb, R. 1974. "Government Policy and the Distribution of Income in Peru, 1963–1973." Ph.D. Dissertation, Harvard University.
399. Weber, Adna Ferrin. 1965. *The Growth of Cities in the Nineteenth Century.* (Originally published in 1899.) Ithaca: Cornell University Press.
400. Weber, Max. 1951. *The Religion of China.* New York: Free Press.
401. ______. 1958. *The City.* New York: Free Press.
402. Weeks, J. 1973. "Uneven Sectoral Development and the Role of the State." *Bulletin of the Institute of Development Studies* 5 (University of Sussex), 233: 76–82.
403. Weiner, Myron. 1966. "Introduction." Pp. 1–14 in *Modernization: The Dynamics of Growth,* edited by Myron Weiner. New York: Basic Books.
404. Werlin, H. H. 1966. "The Nairobi City Council: A Study in Comparative Local Government." *Comparative Studies in Society and History* VIII (January): 181–98.
405. Wheatley, Paul. 1971. *The Pivot of the Four Quarters: A Preliminary Inquiry into the Origins and Character of the Ancient Chinese City.* Chicago: Aldine.
406. Whiteford, Andrew H. 1961. *Two Cities in Latin America: A Comparative Description of Social Classes.* Garden City, N.Y.: Doubleday.
407. Wingo, Lowdon, Jr. 1967. "Recent Patterns of Urbanization among Latin American Countries." *Urban Affairs Quarterly* 2 (March): 81–109.
408. Wolpe, Howard, 1974. *Urban Politics in Nigeria: A Study of Port Harcourt.* Berkeley: University of California Press.
409. The World Bank. 1975. *World Bank Atlas: Population, Per Capita Product and Growth Rates.* Washington, D.C.: International Bank for Reconstruction and Development.
410. Worsley, Peter. 1972. "Frantz Fanon and the 'Lumpenproletariat.'" Pp. 193–230 in *The Socialist Register 1972,* edited by Ralph Miliband and John Savile. London: Merlin.
411. Youssef, Nadia Haggag. 1975. *Women and Work in Developing Societies.* Berkeley: University of California, Institute of International Studies.
412. Zelinsky, Wilbur. 1971. "The Hypothesis of the Mobility Transition." *The Geographical Review* LXI: 219–49.